# Regenerative Stochastic Simulation

This is a volume in
STATISTICAL MODELING AND DECISION SCIENCE

Gerald J. Lieberman and Ingram Olkin, editors
*Stanford University, Stanford, California*

A list of titles in this series appears at the end of this volume.

# Regenerative Stochastic Simulation

**Gerald S. Shedler**

*IBM Research Division*
*Almaden Research Center*
*San Jose, California*

ACADEMIC PRESS, INC.
*Harcourt Brace Jovanovich, Publishers*
Boston San Diego New York
London Sydney Tokyo Toronto

THIS BOOK IS PRINTED ON ACID-FREE PAPER. ♾



ACADEMIC PRESS, INC.
1250 Sixth Avenue, San Diego, CA 92101-4311

*United Kingdom Edition published by*
ACADEMIC PRESS LIMITED
24–28 Oval Road, London NW1 7DX

**Library of Congress Cataloging-in-Publication Data:**

Shedler. G.S. (Gerald S.), date.
Regenerative stochastic simulation/Gerald S. Shedler.
p. cm. — (Statistical modeling and decision science)
Includes bibliographical references and index.
ISBN 0-12-639360-5 (alk. paper)
1. Simulation methods. 2. Stochastic processes. 3. Decision -making—Statistical methods. I. Title. II. Series
T57.62.S475 1992
620'.001'13—dc20

92-23205
CIP

PRINTED IN THE UNITED STATES OF AMERICA

92 93 94 95 **EB** 9 8 7 6 5 4 3 2 1

# Contents

# Preface

It is frequently useful to view a complex stochastic system as making state transitions when events associated with the occupied state occur. Often the system is a discrete-event system in that the stochastic state transitions occur only at an increasing sequence of random times. This book develops probabilistic methods for simulation of discrete-event stochastic systems. The discussion emphasizes stochastic modeling, estimation procedures that rest on limit theorems for regenerative stochastic processes, and engineering applications to computer, communication, manufacturing, and transportation systems. The focus is on simulations with an underlying stochastic process that can be specified as a generalized semi-Markov process. Representation as a generalized semi-Markov process leads directly to an algorithm for generating sample paths of the underlying stochastic process of a simulation and is the basis for a rich theory of analysis of simulation output. This representation is particularly advantageous in connection with nonMarkovian systems.

The most obvious methodological advantage of discrete-event simulation is that in principle it is applicable to stochastic systems of arbitrary complexity. However, it is often a decidedly nontrivial matter in practice to obtain, at reasonable cost, information that is both useful and accurate. The difficulties arise primarily from the inherent variability in a stochastic system, and it is necessary to seek theoretically sound and computationally efficient methods for carrying out the simulation. Apart from issues related to implementation, important concerns for simulation relate to efficient methods for generating sample paths of the underlying stochastic process, the design of simulation experiments, and the analysis of simulation output. Since the results of a simulation are based on observation of a stochastic system, it is essential to provide an assessment of the precision of the estimates that are obtained. Assessment of the statistical precision of an estimator requires careful analysis of the simulation output. In general, the desired precision takes the form of a confidence interval.

A regenerative stochastic process has the characteristic property that there exists a sequence of random times at which the process probabilistically restarts. Under mild regularity conditions, the regenerative property guarantees the existence of a limiting distribution for the process provided that the expected time between regeneration points is finite. Moreover, the behavior of a regenerative process in a regenerative cycle determines the limiting distribution of the process as a ratio of expected values. These results have important implications for the analysis of simulation output. When applicable, the "regenerative method" for simulation analysis is attractive because it provides point and interval estimates with desirable properties and avoids some of the difficulties associated with alternative methods.

The presentation is self contained. Some knowledge of elementary probability theory, statistics, and stochastic models is necessary for an understanding of the theory and the examples. Many of the arguments use results often contained in a first year graduate course on stochastic processes. A brief review of the necessary material is in Appendix A.

I am grateful to the IBM Corporation for support of my work and for the resources of the Almaden Research Center. This book is based on lecture notes for a graduate course on simulation that I have taught for a number of years in the Department of Operations Research at Stanford University. I have benefitted from the interest and encouragement of Peter Glynn, Peter Haas, Donald Iglehart, and Gerald Lieberman and from the comments and suggestions of my students.

Los Gatos, California
May, 1992 Gerald S. Shedler

## Chapter 1

# Discrete-Event Simulations

Simulation is a controlled statistical sampling technique that can be used to study complex stochastic systems when analytic and/or numerical techniques do not suffice. The focus of this book is on simulations of discrete-event stochastic systems; namely, simulations in which stochastic state transitions occur only at an increasing sequence of random times. The discussion emphasizes simulations on a finite or countably infinite state space.

Simulation methods are adapted to experimentation with and study of the behavior of a specified discrete-event stochastic system by observing its evolution over time. In the course of a simulation relevant quantities are measured and, using statistical techniques, inferences are drawn about characteristics of well defined random variables associated with the system. The most obvious methodological advantage of discrete-event simulation is that in principle it is applicable to systems of arbitrary complexity. It is often difficult merely to debug a complex simulation program, and it is a decidedly nontrivial matter in practice to obtain, at reasonable cost, information that is both useful and accurate. The difficulties arise primarily from the inherent variability in a stochastic system, and it is necessary to seek theoretically sound and computationally efficient methods for carrying out the simulation. Apart from implementation considerations, important concerns relate to methods for generating sample paths of the underlying stochastic process, the design of simulation experiments, and the analysis of simulation output. Since results are based on observation of a stochastic system, it is essential to provide an assessment of the precision of the estimates that are obtained.

## 1.1. Methodological Considerations

The definition of a suitable state of the system is implicit in the specification of a simulation. Heuristically, the *state* of the system maintains sufficient information so that state transitions that occur over time along with the random times at which these state transitions occur determine the characteristics of the system that are of interest. For a given system and specified characteristics of the system, there can be more than one suitable state definition. The *underlying stochastic process* of the simulation records the state of the system as it evolves over continuous or discrete time, and we require a procedure for generating sample paths of the process. Most of the stochastic processes encountered in the context of discrete-event simulation have piecewise constant sample paths. For such processes it is sufficient to have a means for generating successive state transitions and corresponding state transition times. The underlying stochastic process of a discrete-event simulation makes state transitions in accordance with the occurrence of *events* associated with the occupied state. This point of view facilitates development of an algorithm for generating sample paths of the process.

To generate sample paths of the underlying stochastic process of a simulation, it is necessary to have some means of obtaining "random numbers" that can be treated as samples of random variables. A *random number generator* (or *pseudo-random number generator*) is an algorithm that produces, in a serial manner, sequences of numbers that follow a specified probability distribution and possess the appearance of randomness. Although only relatively few numbers may be needed, the algorithm should be capable of producing many numbers. "Probability distribution" means that a probability statement can be associated with the occurrence of each number that the algorithm produces. The probability distribution usually is taken to be the uniform distribution on the interval (0,1). In principle, uniform random numbers can be transformed by means of the inverse distribution function into random numbers with any specified nonuniform distribution. However, many methods for direct generation of nonuniform random numbers are also available.

With respect to "appearance of randomness," it may be somewhat surprising that the basis of most commonly used algorithms for generating uniform random numbers is a recurrence relation in which each number produced is a deterministic function of the preceding number. True randomness requires independence of successive numbers; however, the algorithm produces a deterministic, dependent sequence. If the parameters of the recurrence relation are chosen properly, such algorithms for generating uniform random numbers do produce sequences that appear statistically to be random. This appearance of randomness is the origin of the term "pseudorandom numbers." Since results obtained from a simulation depend critically on an acceptable appearance of randomness, it is important to subject a proposed uniform random number generator to thorough statistical testing. Although it is not necessary to dwell on the details of the rather specialized techniques for statistical testing of uniform random number generators, the analyst should be certain that a proposed random number generator has been tested successfully.

In addition to an algorithm for generating sample paths of the underlying stochastic process, statistical techniques are needed in order to obtain meaningful estimates for pertinent characteristics of the system. An important consideration is the question of what constitutes an adequate experiment;

namely, how much and what kind of sampling needs to be done. Point estimates based on small samples can be very misleading. It is necessary to decide what measurements (observations) to make and how to combine them to obtain meaningful point estimates for the pertinent characteristics of the system.

The usual purpose of a simulation is to provide quantitative information in order to make a choice among alternatives. By observing the behavior of the system under a number of different circumstances, inferences are to be drawn concerning differences in system behavior. It is important to be able to distinguish real differences from apparent differences that are due to random fluctuation. It is essential to provide some indication of the variability or statistical precision of an estimate obtained from a simulation. Estimates of greater precision obtainable with equal computational work (cost) are clearly preferable. The assessment of the statistical precision of an estimator requires careful design of the simulation experiments and analysis of the simulation output. The basic technique for assessment of the precision of a statistical experiment is replication. In the context of simulation, replication means observation of a finite portion of many sample paths (realizations) of the underlying stochastic process. However, in the presence of special stochastic structure, meaningful point and interval estimates can be obtained from a finite portion of a single sample path.

A great deal of effort has gone into the development of programming languages for simulation. The methods implemented in program packages for analysis of simulation output should rest on sound theoretical foundations. However, program packages for simulation typically compute by rote and report a large variety of point estimates. Seldom do these reports indicate the variability or statistical precision of the estimates. An experienced simulation analyst knows that it is assuredly easier to produce reams of simulation output than it is to make any sense out of it. In general, the desired statistical precision takes the form of a confidence interval. Issues the analyst must consider include the initial conditions for the system being simulated, the length of the simulation run, the number of replications of the experiments, and the length of the confidence interval. The observations made in the course of a simulation are generally far from being independent and identically distributed (i.i.d.) as in classical statistics. These issues have received much attention over the last fifteen years, leading to a theory of simulation analysis. Based on limit theorems for stochastic processes, this theory is central to the present discussion.

## 1.2. State Definitions

A key step in carrying out a simulation is the definition of a suitable state of the system. The state must maintain sufficient information so that pertinent quantities can be measured and combined to yield meaningful estimates for the characteristics of system that are of interest. There is of course the additional requirement that it must be possible to construct an algorithm for generating sample paths of the underlying stochastic process of the simulation.

**Example 2.1.** (Machine availability) Consider a machine with one component that is subject to failure. A major overhaul of the machine is required to effect repair of a failed component. In an attempt to achieve high availability of the machine, the operator maintains a pool of spare components and a repair facility that can repair one component at a time. A spare immediately replaces a component that fails, and the failed component moves to the repair facility. Relative to the time between failures and the time for repair, the time to install a spare component is small. The operator contemplates the purchase of additional spare components in order to achieve higher machine availability.

Now consider the following stochastic model. A fixed number $N$ ($\geq 2$) of customers (jobs) circulate in a closed network of queues with two single-server centers; see Figure 1.1. Service at center 1 represents the lifetime of a component, and service at center 2 represents repair time. The number of spare components is $N - 1$. After completion of service at center 1, a customer moves instantaneously from center 1 to the tail of the queue at center 2, and after completion of service at center 2, the customer moves back to the tail of the queue at center 1. The queueing discipline at each center is first-come, first-served (FCFS). To complete the specification of the model, we assume that (i) the lifetimes and repair times of each component are distributed as mutually independent positive random variables; and (ii) the successive lifetimes and repair times, respectively, of each component are identically distributed.

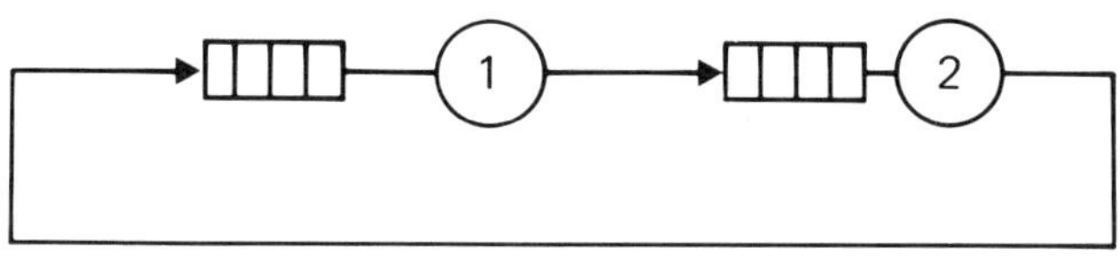


**Figure 1.1.** Machine availability.

The machine is available if and only if there are less than $N$ components at the repair facility. Accordingly, set

$$A(t) = \begin{cases} 1 & \text{if the server at center 1 is busy at time } t \\ 0 & \text{otherwise} \end{cases}$$

and define machine availability $A$ as

$$A \stackrel{\text{def}}{=} \lim_{t \to \infty} E\left[\frac{1}{t} \int_0^t A(u)du\right].$$

In addition to machine availability, characteristics (such as the expected value) of the workload at the repair facility may be of interest. This workload is measured by the number of customers waiting or in service at center 2. Observe that the value of $A(t)$ alone does not determine the number of customers at center 2 at time $t$. Moreover, it is not clear how to generate sample paths of the process $\{A(t)\colon\ t \geq 0\}$ directly. It is therefore necessary to incorporate other information into the state if we wish to estimate characteristics of the workload.

One possibility is to keep track of the number of customers waiting or in service at center 1 along with the number of customers waiting or in service at center 2. Since the number ($N$) of customers in the network is fixed, it is sufficient to record either only the number of customers waiting or in service at center 1 or the number waiting or in service at center 2. Let $X(t)$ be the number of customers waiting or in service at center 2 at time $t$, and take event $e_i$ = "service completion at center $i$." The stochastic process $\{X(t)\colon\ t \geq 0\}$ has finite state space $S = \{0, 1, \ldots, N\}$ and makes state transitions in accordance with the occurrence of events in the set $E = \{e_1, e_2\}$. Figure 1.2 shows a sample path of the process $\{X(t)\colon\ t \geq 0\}$ along with the underlying service times at centers 1 and 2. Observe that the process $\{X(t)\colon\ t \geq 0\}$ has piecewise constant sample paths.

The server at center 1 is busy at time $t$ ($A(t) = 1$) if and only if $X(t) < N$. Therefore,

$$A = \lim_{t\to\infty} E\left[\frac{1}{t}\int_0^t f(X(u))du\right] = \lim_{t\to\infty} P\left\{X(t) < N\right\},$$

where the real-valued function $f$ has domain $S$ and is such that $f(s) = 1$ if $0 \leq s < N$ and $f(s) = 0$ otherwise. The limiting expected value $Q$ of the workload at the repair facility can be defined as

$$Q \stackrel{\text{def}}{=} \lim_{t\to\infty} E\left[\frac{1}{t}\int_0^t X(u)du\right].$$

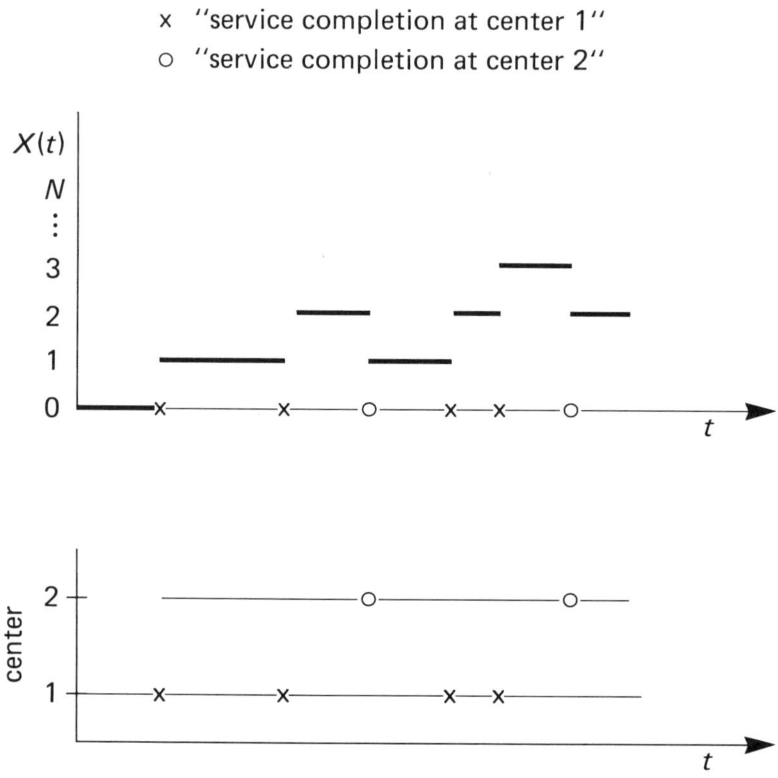


**Figure 1.2.** Sample path for machine availability.

Observe that there is nothing that prevents us from writing and executing a simulation program that reports "estimates" for quantities defined as limits that do not exist. (Of course the output of such a simulation isn't very interesting!) For example, we can simulate an unstable ("$\rho > 1$") single-server queue and report the sample mean of the first 1000 (or 10000) waiting times as an estimate of the "steady-state average waiting time."

**Example 2.2.** (Patrolling repairman) Consider a group of $N$ ($\geq 2$) machines numbered $1, 2, \ldots, N$ under the care of a single patrolling repairman who walks round in a strictly defined order: $1, 2, \ldots, N, 1, 2, \ldots$. The repairman repairs and restarts a machine that is stopped and passes a machine that is running. The time to repair and restart machine $j$ ($j = 1, 2, \ldots, N$) is distributed as a positive random variable $R_j$, and the walking time from machine $j$ to the next machine is a positive constant $W_j$. The times (lifetimes) between end of repair and the next stoppage of machine $j$ are i.i.d. as a positive random variable $L_j$ with absolutely continuous distribution so that, with probability 1, no two stoppages occur simultaneously. The assumption that each $L_j$ has an absolutely continuous distribution ensures that a stoppage and an arrival of repairman at machine cannot occur simultaneously and a stoppage and an end of repair cannot occur simultaneously.

Suppose that we wish to obtain estimates for the amounts of time that individual (stopped) machines wait for repair. Irrespective of the particular characteristics of these waiting times that are of interest, it is necessary to define the state of the system so that for each of the machines the random times at which a stoppage occurs and the random times at which a repair starts can be observed. Although at first glance it may appear that it is necessary to keep track of the exact location of the repairman in order to observe these random times, this is not the case. To see this, set

$$X(t) = (X_1(t), X_2(t), \ldots, X_N(t)), \tag{2.3}$$

where

$$X_j(t) = \begin{cases} 2 & \text{if machine } j \text{ is waiting for repair at time } t \\ 1 & \text{if machine } j \text{ is under repair at time } t \\ 0 & \text{if machine } j \text{ is running at time } t. \end{cases}$$

The stochastic process $\{X(t)\colon\ t \geq 0\}$ has finite state space

$$S = \Big\{(x_1, \ldots, x_N) \in \{0, 1, 2\}^N : \ x_j = 1 \text{ for at most one } j\Big\}.$$

According to this definition, machine 1 is stopped and waiting for repair and all other machines are running when the system is in state $(2, 0, \ldots 0)$. A state transition from $(2, x_2, \ldots, x_N) \in S$ to $(1, x_2, \ldots, x_N)$ corresponds to a start of repair for machine 1. Similarly, a state transition from $(0, x_2, \ldots, x_N) \in S$ to $(2, x_2, \ldots, x_N)$ corresponds to a stoppage of machine 1. Since the waiting times for repair start and end at state transitions of the stochastic process $\{X(t)\colon t \geq 0\}$, the random vector $X(t)$ records sufficient information to serve as the state of the system at time $t$.

Observe that the process $\{X(t)\colon t \geq 0\}$ does not maintain the location of the repairman explicitly. Some states of the process (for example, $(1, 0, \ldots, 0)$) reveal the location of the repairman. However, other states (for example, $(0, \ldots, 0)$) do not. Nevertheless, the process $\{X(t)\colon t \geq 0\}$ records sufficient information about the system so that the waiting time for repair experienced by any of the machines can be measured.

Let $\zeta_n$ be the time of the $n$th state transition, $n \geq 0$. To see that the state transitions and the state transition times determine the waiting times for repair experienced by machine $j$, set

$$\mathcal{S}_n(j) = \inf\{\zeta_k > \mathcal{T}_{n-1}(j)\colon\ X_j(\zeta_k) = 2, X_j(\zeta_{k-1}) = 0\}$$

and

$$\mathcal{T}_n(j) = \inf\{\zeta_k > \mathcal{S}_n(j)\colon\ X_j(\zeta_k) = 1, X_j(\zeta_{k-1}) = 2\}$$

for $n \geq 1$ ($\mathcal{T}_0(j) = 0$). Then the $n$th waiting time for repair experienced by machine $j$ is $\mathcal{T}_n(j) - \mathcal{S}_n(j)$.

**Exercises**

2.1. (Poisson shocks) Consider a system that experiences shocks that occur in accordance with a Poisson process of rate 1. Each shock causes a certain amount of damage. These damages are independent of the times at which the shocks occur and independent and identically distributed as a positive random variable $X$ with density function $g$ given by $g(x) = xe^{-x}$ for $x \geq 0$. Damages dissipate over time at exponential rate $\alpha$ ($> 0$); that is, a shock that occurs at time $t$ and causes initial damage $x$ has remaining damage $xe^{-\alpha u}$ at time $t + u$, $u \geq 0$. The damage values are cumulative. Thus, for example, if there have been exactly two shocks by time $t$ and these shocks occurred at times $t_1$ and $t_2$ with initial damages $x_1$ and $x_2$, the total damage at time $t$ is $\sum_{i=1}^{2} x_i e^{-\alpha(t-t_i)}$. The system fails when the total damage exceeds a fixed positive constant $K$.

Let $D(t)$ be the total damage at time $t$. Specify the state space $S$ of the process $\{D(t)\colon\ t \geq 0\}$ and the set $E$ of events associated with simulation of the process. Show that

$$D(T_k) = e^{-\alpha(T_k - T_{k-1})} D(T_{k-1}) + X_k,$$

where $X_k$ is the initial damage caused by the $k$th shock and $T_k$ is the time of the $k$th shock, $k \geq 0$. Using this result, describe how you would simulate the process $\{D(T_k)\colon\ k \geq 0\}$ to obtain an estimate for the expected time until the system fails. Specify your algorithm explicitly.

## 1.3. Time-Advance Procedures

The underlying stochastic process of a discrete-event simulation on a finite or countably infinite state space makes state transitions in accordance with the occurrence of *events* associated with the occupied state. Each of the several events associated with a state competes to "trigger" the next state transition, and each of these events has its own distribution for determining the next state. At each state transition of the process, new events may be scheduled. For each of these new events and according to an independent stochastic mechanism, a *clock* is set with a reading that indicates the time at which the event is scheduled to occur. If a scheduled event does not trigger a state transition but is associated with the next state, the corresponding clock (after possibly being reset) continues to run; if such an event is not associated with the next state, the event is cancelled and the corresponding clock reading is discarded.

Implementation of a discrete-event simulation requires a mechanism to ensure that the events occur at the proper simulated time and in the proper sequence. This mechanism is often referred to as the *time-flow* or *time-advance mechanism.* For systems with a finite or countably infinite state space, it is natural to use a *variable time-advance procedure*. An alternative is a *fixed time-advance procedure* that increases time in small increments, and at each time advance, executes all the events that are available for execution.

With a variable time-advance procedure, simulated time goes forward to the next time at which some event triggers a state transition, and all intervening time is skipped. Thus, the observed values of simulated time are the successive state transition times (times at which events occur). This time advance is accomplished by maintaining, usually in a "doubly linked" list, a *future-event set*. The "records" in this set are ordered according to the times at which the events are scheduled to occur. Other data structures are available for maintaining the future-event set. A record in the future-event set consists of the (simulated) *time* at which the event is scheduled to occur and the *type* of event. New events may be scheduled at the times at which events occur by adding records to the future-event set. The events scheduled may depend on the type of the trigger event and the new state of the system. Previously scheduled events may be cancelled at the times at which events occur by removing records from the future-event set.

For ease of exposition, we assume that, with probability 1, no two events occur simultaneously. The iterative step of a variable time-advance procedure can be expressed in terms of a *state transition mechanism* and an *event scheduling mechanism*. The procedure has the following general form.

**Algorithm 3.1.** *Variable time advance*

1. *At time $\zeta$ in state $s$, from the future-event set determine the time $\zeta + t^*$ of occurrence and type $e^*$ of the next trigger event (and delete the corresponding record from the future-event set).*
2. *According to the state transition mechanism, from $s$ and $e^*$ determine the next state $s'$.*
3. *According to the event scheduling mechanism, determine the (possibly empty) set of new events and the (possibly empty) set of old events. Also determine the time at which each event is scheduled to occur (and enter the corresponding record in the future-event set).*
4. *According to the event scheduling mechanism, determine the scheduled events (if any) that are to be cancelled (and delete the corresponding record from the future-event set).*
5. *Advance time from $\zeta$ to $\zeta + t^*$.*

Programming languages designed for simulation provide a *timing routine* that determines the time $(\zeta + t^*)$ and type $(e^*)$ of the next event, advances time (from $\zeta$ to $\zeta + t^*$), and transfers control to the appropriate subprogram for the type of event.

**Example 3.2.** (Single-server queue) Customers arrive at a queueing system one at a time. The 0th customer arrives at time $T_0 = 0$, finds the server free, and experiences a service time distributed as a positive random variable $V_0$; the $n$th customer arrives at time $T_n$ and experiences a service time distributed as $V_n$, $n \geq 1$. Denote the interarrival time $T_n - T_{n-1}$ by $U_n$ for $n \geq 1$. The queueing discipline is FCFS. The server is never idle when any customers are present in the system, and the waiting room has infinite capacity. Assume that the sequences $\{V_n\colon n \geq 0\}$ and $\{U_n\colon n \geq 1\}$ each consist of i.i.d. random variables and are themselves independent. Let $\lambda > 0$ be the arrival rate and $\mu > 0$ be the service rate so that $E[U_n] = \lambda^{-1}$ and $E[V_n] = \mu^{-1}$. The natural measure of congestion for the system is the traffic intensity $\rho = \lambda/\mu$.

Suppose that we wish to obtain estimates for characteristics of the queue length or waiting times experienced by customers. Let $Q(t)$ be the number of customers waiting or in service at time $t$, and let $W_n$ be the waiting time (exclusive of service time) of the $n$th customer, $n \geq 0$. The process $\{Q(t)\colon t \geq 0\}$ has countably infinite state space $S = \{0, 1, \ldots\}$. Observe that the state transitions and state transition times of $\{Q(t)\colon t \geq 0\}$ determine the waiting times.

To generate sample paths of the process $\{Q(t)\colon t \geq 0\}$ using a variable time-advance procedure, we view the process as making state transitions in accordance with the occurrence of "arrival" events and "service completion" (customer departure) events; see Figure 1.3. Assume that

(i) the interarrival times become available one at a time so that $U_{n+1}$ is determined at time $T_n$; and
(ii) the service times become available one at a time so that $V_n$ is determined when service for the $n$th customer begins.

Also assume that, with probability 1, no arrival and service completion occur simultaneously. The state transition and event scheduling mechanisms are as follows.

*State transition mechanism*

(i) if $e^* =$ "service completion," then $s' = s - 1$; and
(ii) if $e^* =$ "arrival," then $s' = s + 1$.

*Event scheduling mechanism*

(i) if $e^* =$ "arrival," let $u$ be a random sample from the distribution of interarrival time and schedule an "arrival"

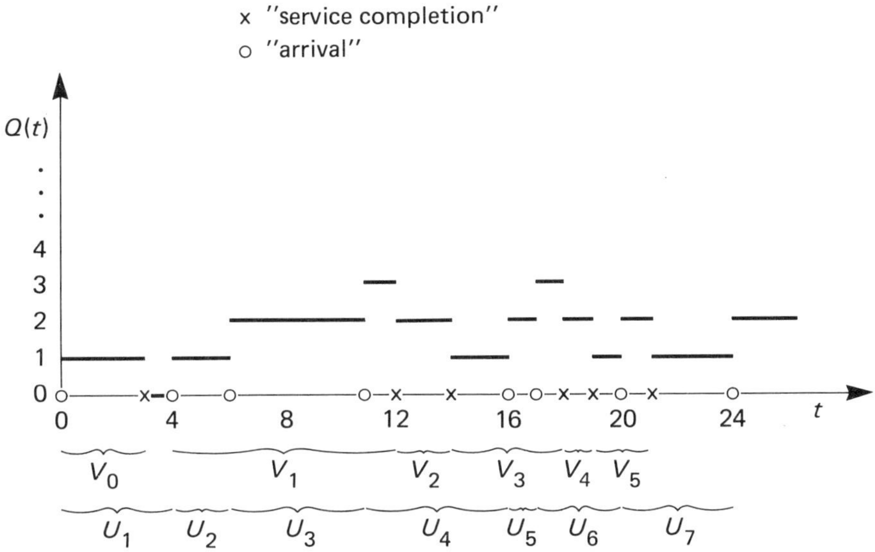


**Figure 1.3.** Sample path for single-server queue.

event at time $\zeta + t^* + u$. If $s' = 1$, let $v$ be a random sample from the distribution of service time and schedule a "service completion" event at time $\zeta + t^* + v$; and

(ii) if $e^*$ = "service completion" and $s' > 0$, let $v$ be a random sample from the distribution of service time and schedule a "service completion" event at time $\zeta + t^* + v$.

Suppose that $U_1 = 4$, $U_2 = 2$, $U_3 = 5$, $U_4 = 5$, $U_5 = 1$, $U_6 = 3$, $U_7 = 4$, $U_8 = 6, \ldots$, and $V_0 = 3$, $V_1 = 8$, $V_2 = 2$, $V_3 = 4$, $V_4 = 1$, $V_5 = 2$, $V_6 = 1$, $V_7 = 3, \ldots$ Beginning in state $s = 1$ at time 0, events then occur and time advances in the following manner.

$\zeta = 0$
$s = 1$

initialize: schedule "arrival" ($U_1 = 4$) and "service completion" ($V_0 = 3$)

| TIME | TYPE |
|---|---|
| 3 | "service completion" |
| 4 | "arrival" |

$\zeta = 0$
$s = 1$
$\zeta + t^* = 3$
$e^* =$ "service completion"
$s' = 0$
"service completion" and $s' = 0$: no new events scheduled

| TIME | TYPE |
|---|---|
| 4 | "arrival" |

$\zeta = 3$
$s = 0$
$\zeta + t^* = 4$
$e^* =$ "arrival"
$s' = 1$
"arrival" and $s' = 1$: schedule "arrival" ($U_2 = 2$) and "service completion" ($V_1 = 8$)

| TIME | TYPE |
|---|---|
| 6 | "arrival" |
| 12 | "service completion" |

$\zeta = 4$
$s = 1$
$\zeta + t^* = 6$
$e^* =$ "arrival"
$s' = 2$
"arrival": schedule "arrival" ($U_3 = 5$)

| TIME | TYPE |
|---|---|
| 11 | "arrival" |

| | |
|---|---|
| 12 | "service completion" |

$\zeta = 6$
$s = 2$
$\zeta + t^* = 11$; $e^*$ = "arrival"
$s' = 3$
"arrival": schedule "arrival" ($U_4 = 5$)

| TIME | TYPE |
|---|---|
| 12 | "service completion" |
| 16 | "arrival" |

$\zeta = 11$
$s = 3$
$\zeta + t^* = 12$; $e^*$ = "service completion"
$s' = 2$
"service completion" and $s' > 0$: schedule "service completion" ($V_2 = 2$)

| TIME | TYPE |
|---|---|
| 14 | "service completion" |
| 16 | "arrival" |

$\zeta = 12$
$s = 2$
$\zeta + t^* = 14$
$e^*$ = "service completion"
$s' = 1$
"service completion" and $s' > 0$: schedule "service completion" ($V_3 = 4$)

| TIME | TYPE |
|---|---|
| 16 | "arrival" |
| 18 | "service completion" |

$\zeta = 14$
$s = 1$
$\zeta + t^* = 16$
$e^*$ = "arrival"

$s' = 2$
"arrival": schedule "arrival" ($U_5 = 1$)

| TIME | TYPE |
|---|---|
| 17 | "arrival" |
| 18 | "service completion" |

$\zeta = 16$
$s = 2$
$\zeta + t^* = 17$
$e^* =$ "arrival"
$s' = 3$
"arrival": schedule "arrival" ($U_6 = 3$)

| TIME | TYPE |
|---|---|
| 18 | "service completion" |
| 20 | "arrival" |

$\zeta = 17$
$s = 3$
$\zeta + t^* = 18$
$e^* =$ "service completion"
$s' = 2$
"service completion" and $s' > 0$: schedule "service completion" ($V_4 = 1$)

| TIME | TYPE |
|---|---|
| 19 | "service completion" |
| 20 | "arrival" |

$\zeta = 18$
$s = 2$
$\zeta + t^* = 19$
$e^* =$ "service completion"
$s' = 1$
"service completion" and $s' > 0$: schedule "service completion" ($V_5 = 2$)

| TIME | TYPE |
|---|---|
| 20 | "arrival" |
| 21 | "service completion" |

$\zeta = 19$
$s = 1$
$\zeta + t^* = 20$
$e^* =$ "arrival"
$s' = 2$
"arrival": schedule "arrival" ($U_7 = 4$)

| TIME | TYPE |
|---|---|
| 21 | "service completion" |
| 24 | "arrival" |

**Example 3.3.** (Patrolling repairman) Assume that the walking time from machine $j$ to the next machine is a positive constant $W_j$ and the time to repair and restart machine $j$ is distributed as a positive random variable $R_j$. Also assume that the lifetimes for machine $j$ are i.i.d. as a positive random variable $L_j$ with absolutely continuous distribution so that, with probability 1, no two stoppages occur simultaneously. This assumption ensures that a stoppage and an arrival of repairman at machine cannot occur simultaneously and a stoppage and an end of repair cannot occur simultaneously.

To generate sample paths of the process $\{X(t): t \geq 0\}$ defined by (2.3) using a variable time-advance procedure, we view the process as making state transitions in accordance with the occurrence of "arrival of repairman at machine $j$," "end of repair for machine $j$," and "stoppage of machine $j$" events ($j = 1, 2, \ldots, N$); see Figure 1.4. Observe that the state of the process $\{X(t): t \geq 0\}$ alone does not determine the set of events that are scheduled to occur; that is, there is no mapping $(s \mapsto E(s))$ from the state space $S$ to the nonempty subsets of the set $E$ of events such that $E(s)$ is the set of all events that are scheduled to occur when the process is in state $s$. For

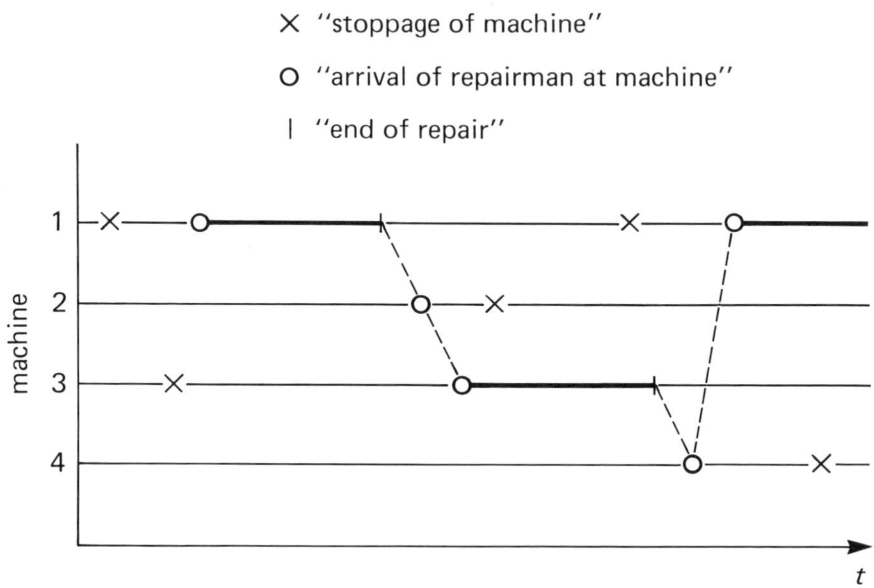


**Figure 1.4.** Time-line diagram for patrolling repairman.

example, if we know only that the process is in state $(0, \ldots, 0)$, we cannot determine which "arrival of repairman at machine $j$" event is scheduled to occur.

Suppose that an "end of repair for machine 1" event occurs at time 0 and all other machines are running: $X(0) = (0, \ldots, 0)$. Schedule an "arrival of repairman at machine 2" event at time $W_1$ and for each $j$ schedule a "stoppage of machine $j$" event at time $l_j$, where $l_j$ is a random sample of $L_j$. The state transition and event scheduling mechanisms are as follows. By convention, a reference to machine $N + 1$ is interpreted as a reference to machine 1.

*State transition mechanism*

(i) if event $e^* =$ "stoppage of machine $j$" and state $s = (x_1, \ldots, x_{j-1}, 0, x_{j+1}, \ldots, x_N)$, then state $s' = (x_1, \ldots,$

$x_{j-1}, 2, x_{j+1}, \ldots, x_N)$;

(ii) if event $e^*$ = "arrival of repairman at machine $j$" and state $s = (x_1, \ldots, x_{j-1}, 2, x_{j+1}, \ldots, x_N)$, then state $s' = (x_1, \ldots, x_{j-1}, 1, x_{j+1}, \ldots, x_N)$;

(iii) if event $e^*$ = "arrival of repairman at machine $j$" and state $s = (x_1, \ldots, x_{j-1}, 0, x_{j+1}, \ldots, x_N)$, then state $s' = (x_1, \ldots, x_{j-1}, 0, x_{j+1}, \ldots, x_N)$; and

(iv) if event $e^*$ = "end of repair for machine $j$" and state $s = (x_1, \ldots, x_{j-1}, 1, x_{j+1}, \ldots, x_N)$, then state $s' = (x_1, \ldots, x_{j-1}, 0, x_{j+1}, \ldots, x_N)$.

*Event scheduling mechanism*

(i) if event $e^*$ = "end of repair for machine $j$," state $s = (x_1, \ldots, x_{j-1}, 1, x_{j+1}, \ldots, x_N)$, and state $s' = (x_1, \ldots, x_{j-1}, 0, x_{j+1}, \ldots, x_N)$, then let $l_j$ be a random sample of $L_j$ and schedule an "arrival of repairman at machine $j+1$" event at time $\zeta + t^* + W_j$ and a "stoppage of machine $j$" event at time $\zeta + t^* + l_j$;

(ii) if event $e^*$ = "arrival of repairman at machine $j$," state $s = (x_1, \ldots, x_{j-1}, 2, x_{j+1}, \ldots, x_N)$ and state $s' = (x_1, \ldots, x_{j-1}, 1, x_{j+1}, \ldots, x_N)$, then let $r_j$ be a random sample of $R_j$ and schedule an "end of repair for machine $j$" event at time $\zeta + t^* + r_j$; and

(iii) if event $e^*$ = "arrival of repairman at machine $j$" and state $s = s' = (x_1, \ldots, x_{j-1}, 0, x_{j+1}, \ldots, x_N)$, then schedule an "arrival of repairman at machine $j + 1$" event at time $\zeta + t^* + W_j$.

**Exercises**

3.1. (Road traffic) Consider a single lane of one-way traffic proceeding along a main road. At a fixed merge point on the main road, there is a side road. Upon arriving at the merge point, side-road cars attempt to merge with the stream of traffic on the main road. Side-road cars arrive and join a waiting line at the merge point according to a Poisson process of rate $\lambda > 0$. Main-road cars arrive at the merge point according to a renewal process $\{A_n\colon n \geq 0\}$; that is, the $n$th main-road car arrives at the merge point at time $A_n$, and the interarrival times are i.i.d. as a positive random variable $U$ with absolutely continuous distribution.

There is a gap-acceptance function $L$ that determines the manner in which (instantaneous) merging takes place. Assume that

$$L(x) = \begin{cases} 1 & \text{if } x > \Delta_1 \\ 0 & \text{if } x \leq \Delta_1, \end{cases}$$

where $\Delta_1 > 0$ and $L(x)$ is the probability that the side-road car at the head of the line merges, given that the time until the next main-road car passes the merge point is $x$. When a side-road car merges, $\Delta_2$ time units elapse before another side-road car may merge.

What continuous time stochastic process $\{X(t)\colon t \geq 0\}$ would you simulate to estimate the delay (before merging) experienced by side-road cars? Using a variable time-advance procedure, describe how you would generate sample paths of the process $\{X(t)\colon t \geq 0\}$. Also show that the state transitions and state transition times determine the delays experienced by side-road cars.

3.2. (Baggage claim) An airplane lands with a fixed number $N$ of passengers aboard and reaches the arrival gate at time 0. The times for passengers that arrive to walk to the baggage claim area are mutually independent positive random variables having an exponential distribution with mean 1. Suitcases arrive at the baggage claim area according to the following stochastic process. An exponentially distributed (rate parameter $\mu$) time $A$ elapses after the airplane lands. Thereafter, at regular time intervals of length $L$ ($> 0$), suitcases arrrive one at a time at the baggage claim area. Thus, the $k$th suitcase arrives at the baggage claim area at time $Y + kL$ ($k = 1, 2, \ldots, N$).

Each passenger has one suitcase, and the arrival order of suitcases at the baggage claim area is random in the sense that all permutations are equally likely. A passenger at the baggage claim area removes his suitcase as soon as it arrives.

What continuous time stochastic process $\{X(t)\colon\ t \geq 0\}$ would you simulate to estimate the time until all passengers have obtained their baggage and the delay (waiting for baggage) experienced by passengers at the baggage claim area? Using a variable time-advance procedure, describe how you would generate sample paths of the process $\{X(t)\colon\ t \geq 0\}$. Also show that the state transitions and state transition times determine the elapsed time $T$ and the delay $D_k$ experienced by the $k$th passenger to arrive in the baggage claim area.

## 1.4. Generalized Semi-Markov Processes

The usual formal model for the underlying stochastic process of a discrete-event simulation on a finite or countably infinite state space is a "generalized semi-Markov process" (GSMP). Representation as a generalized semi-Markov process leads directly to an algorithm for generating sample paths of the underlying stochastic process of a simulation and is the basis for a rich theory of analysis of simulation output. This representation is particularly advantageous in connection with nonMarkovian systems. Formal definition of a GSMP is in terms of a general state space Markov chain (GSSMC) that describes the process at successive state transition times.

Let $S$ be a finite or countably infinite set of *states* and $E = \{e_1, e_2, \ldots, e_M\}$ be a finite set of *events*. We assume throughout that $S$ is a subset of $d$-dimensional Euclidean space $\Re^d$ for some $d \geq 1$. For $s \in S$, let $s \mapsto E(s)$ be a mapping from $S$ to the nonempty subsets of $E$; $E(s)$ denotes the set of all events that are scheduled to occur when the process is in state $s$. When the process is in state $s$, the occurrence of an event $e \in E(s)$ triggers a state transition to a state $s'$. Denote the probability that the new state is $s'$ given that event $e$ triggers a state transition in state $s$ by $p(s'; s, e)$. For each $s \in S$ and $e \in E(s)$, we assume that $p(\cdot\,; s, e)$ is a probability mass function. (If $e \in E(s)$ but, with probability 1, event $e$ does not trigger a state transition in state $s$, the probability mass function $p(\cdot\,; s, e)$ may be defined arbitrarily.) The actual event $e \in E(s)$ that triggers a state transition in state $s$ depends on clocks associated with the events in $E(s)$ and the speeds at which these clocks run. Associated with each event is a *clock* with a reading that records the remaining time until the event is scheduled to trigger a state transition. We denote the *speed* (finite, deterministic rate) at which the clock associated with event $e$ runs in state $s$ by $r(s, e)$ ($\geq 0$). By convention, for each $s \in S$, $r(s, e) = 0$ if $e \notin E(s)$ and we assume that $r(s, e) > 0$ for some $e \in E(s)$.

Typically in applications, all speeds $r(s, e)$ for scheduled events are equal to 1. However, there are models in which speeds other than 1 as well as state-dependent speeds are convenient. For example, positive speeds not equal to 1 are convenient for specification of queueing systems in which the service rate or arrival rate depends on the number in the system, and we use zero speeds for specification of queueing systems with service interruptions of the "preemptive-resume" type.

For $s \in S$, define the set $C(s)$ of possible clock-reading vectors in state $s$ as

$$\begin{aligned} C(s) \stackrel{\text{def}}{=} \bigl\{(c_1, \ldots, c_M)\colon\ & c_i \geq 0 \text{ with } c_i > 0 \text{ if and only if} \\ & e_i \in E(s);\ c_i/r(s, e_i) \neq c_j/r(s, e_j) \text{ for } i \neq j \\ & \text{with } c_i c_j r(s, e_i) r(s, e_j) > 0\bigr\}. \end{aligned} \tag{4.1}$$

The $i$th component of a clock-reading vector $c = (c_1, \ldots, c_M)$ is the clock reading associated with event $e_i$. The conditions in (4.1) ensure that no two events simultaneously trigger a state transition (as defined below). The clock reading $c_i$ and the event $e_i$ are said to be *active* in state $s$ if $e_i \in E(s)$. For $s \in S$ and $c \in C(s)$, set

$$t^* = t^*(s,c) = \min_{\{i:\ e_i \in E(s)\}} \{c_i/r(s,e_i)\}, \tag{4.2}$$

where we take $c_i/r(s,e_i)$ to be $+\infty$ when $r(s,e_i) = 0$. Also set

$$c_i^* = c_i^*(s,c) = c_i - t^*(s,c)r(s,e_i) \tag{4.3}$$

for $e_i \in E(s)$ and

$$i^* = i^*(s,c) = i \text{ such that } e_i \in E(s) \text{ and } c_i^*(s,c) = 0. \tag{4.4}$$

Beginning in state $s$ with clock-reading vector $c$, $t^*(s,c)$ is the time to the next state transition and $i^*(s,c)$ is the index of the unique *trigger event* $e^* = e^*(s,c) = e_{i^*(s,c)}$.

At a state transition from $s$ to $s'$ triggered by the occurrence of event $e^*$, a clock reading is generated for each new event $e' \in N(s';s,e^*) = E(s') - (E(s) - \{e^*\})$. Denote the *clock-setting distribution function* (distribution function of such a new clock reading) by $F(\,\cdot\,;s',e',s,e^*)$, and assume that $F(0;s',e',s,e^*) = 0$. For each old event $e' \in O(s';s,e^*) = E(s') \cap (E(s) - \{e^*\})$, the old clock reading is kept after the state transition. For $e' \in (E(s) - \{e^*\}) - e(s')$, event $e'$ is cancelled, and the corresponding clock reading is discarded; see Figure 1.5.

Next consider a GSSMC $\{(S_n, C_n)\colon\ n \geq 0\}$ taking values in the set

$$\Sigma = \bigcup_{s \in S} (\{s\} \times C(s))$$

and representing the state $(S_n)$ and clock-reading vector $(C_n)$ at successive state transition times. The transition kernel of the GSSMC $\{(S_n, C_n)\colon\ n \geq 0\}$ is given by

$$P((s,c),A) = p(s';s,e^*) \prod_{e_i \in N(s')} F(a_i;s',e_i,s,e^*) \prod_{e_i \in O(s')} 1_{[0,a_i]}(c_i^*) \tag{4.5}$$

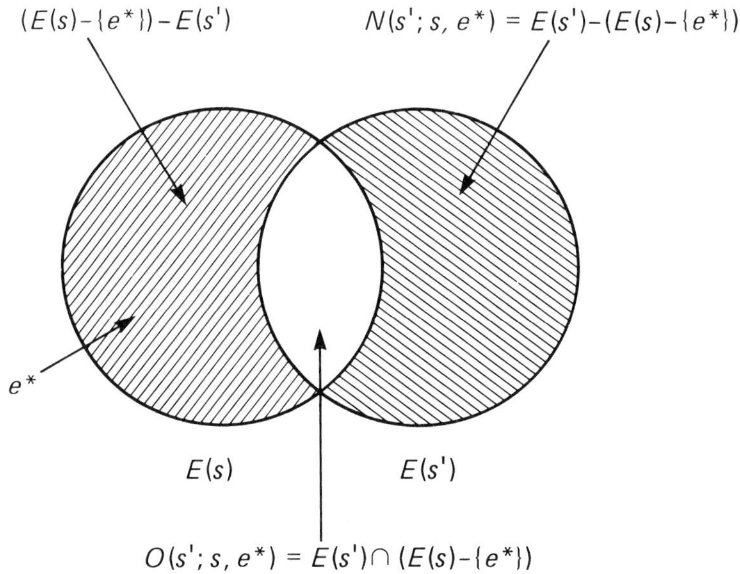


**Figure 1.5.** Sets of new and old events.

for all sets

$$A = \{s'\} \times \big\{(c_1', \ldots, c_M') \in C(s') \colon 0 \le c_i' \le a_i \text{ for } 1 \le i \le M\big\},$$

where $N(s') = N(s'; s, e^*)$ and $O(s') = O(s'; s, e^*)$. (The transition kernel in (4.5) is defined on the measurable space $(\Sigma, \mathcal{S})$, where $\mathcal{S}$ is the $\sigma$-field generated by sets of the form $\{s\} \times \big(C(s) \cap B\big)$ with $s \in S$ and $B$ a Borel subset of $\Re_+^M$.) The right side of (4.5) is the probability beginning in state $s$ with clock-reading vector $c$ that the GSSMC hits state $s'$ with the reading $c_i'$ on the clock associated with event $e_i$ set to a value in $[0, a_i]$. (We assume that the clock-setting distributions are such that $P\big((s,c), \Sigma\big) = 1$ for all $(s,c) \in \Sigma$. This assumption rules out simultaneous occurrence of events.) Denote the initial distribution of the GSSMC by $\mu$; that is, for any (measurable) subset $B \subseteq \Sigma$, $\mu(B)$ represents the probability that $(S_0, C_0) \in B$. For ease of exposition, we restrict attention to initial distributions $\mu$ satisfying

$$\mu(A) = q(s_0) \prod_{e_i \in E(s_0)} F_0\bigl(a_i; e_i, s_0\bigr)$$

for all sets

$$A = \{s_0\} \times \bigl\{(c_{0,1}, \ldots, c_{0,M}) \in C(s_0)\colon\ 0 \le c_{0,i} \le a_i \text{ for } 1 \le i \le M\bigr\},$$

where $q$ is a probability distribution on $S$ and $F_0(\,\cdot\,; e, s)$ is a proper distribution function for $s \in S$ and $e \in E(s)$. The form of the initial distribution amounts to selection of an initial state $s_0$ according to the probability mass function $q$ and then, for each active event $e_i \in E(s_0)$, selection of a corresponding clock reading $c_{0,i}$ according to the clock-setting distribution function $F_0(\,\cdot\,; e_i, s_0)$. We denote by $P_\mu$ the probability distribution for the GSSMC when the initial distribution is $\mu$.

Finally, construct a continuous time process with piecewise constant sample paths from the GSSMC $\{(S_n, C_n)\colon\ n \ge 0\}$ in the following manner. Set $\zeta_0 = 0$ and

$$\zeta_n = \sum_{k=0}^{n-1} t^*(S_k, C_k)$$

for $n \ge 1$, so that $\zeta_n$ is the time of the $n$th state transition. Let $\Delta \notin S$ and set

$$X(t) = \begin{cases} S_{N(t)} & \text{if } N(t) < \infty \\ \Delta & \text{if } N(t) = \infty, \end{cases} \tag{4.6}$$

where

$$N(t) = \max\{n \ge 0\colon\ \zeta_n \le t\}.$$

The process $\{X(t)\colon t \geq 0\}$ defined by (4.6) is a GSMP. We assume that

$$P_\mu\Big\{\sup_{n\geq 0}\zeta_n = \infty\Big\} = 1 \tag{4.7}$$

so that, with probability 1, $X(t) \neq \Delta$ for every finite time. It can be shown that (4.7) holds if $S$ is finite or if the speeds $r(s,e)$ and clock-setting distribution functions $F(\,\cdot\,; s', e', s, e^*)$ do not depend explicitly on the states $s'$ or $s$.

Observe that the process $\{X(t)\colon t \geq 0\}$ defined by (4.6) has right-continuous sample paths and limits from the left; that is, for $t \geq 0$

$$X(t) = \lim_{u\downarrow t} X(u)$$

and for all $t > 0$

$$X(t-) = \lim_{u\uparrow t} X(u)$$

exists with probability 1.

**Example 4.8.** (Machine availability) Assume that service times at center $i$ are identically distributed as a positive random variable $V_i$. Also assume that each distribution of service time is absolutely continuous so that, with probability 1, no two service completions occur simultaneously. Let $X(t)$ be the number of customers (components) waiting or in service at center 2 at time $t$. Take event $e_i$ = "service completion at center $i$." Formal specification of the process $\{X(t)\colon t \geq 0\}$ is as a GSMP with state space $S = \{0, 1, \ldots, N\}$ and event set $E = \{e_1, e_2\}$. For $s \in S$, event $e_1 \in E(s)$ if and only if $s < N$ and event $e_2 \in E(s)$ if and only if $s > 0$. All speeds $r(s,e)$ for active events are equal to 1.

If $e = e_1$ = "service completion at center 1", then the state transition probability $p(s'; s, e) = 1$ when

$$0 \leq s < N \text{ and } s' = s + 1.$$

If $e = e_2 =$ "service completion at center 2", then $p(s'; s, e) = 1$ when

$$0 < s \leq N \text{ and } s' = s - 1.$$

All other state transition probabilities $p(s'; s, e)$ are equal to 0. The set $N(s+1; s, e_1)$ of new events (when event $e_1$ triggers a state transition from $s$ to $s+1$) equals $\{e_1, e_2\}$ if $s = 0$, equals $\{e_1\}$ if $1 \leq s < N-1$, and equals $\emptyset$ if $s = N-1$. The set $N(s-1; s, e_2)$ equals $\{e_1, e_2\}$ if $s = N$, equals $\{e_2\}$ if $1 < s < N$, and equals $\emptyset$ if $s = 1$. The clock-setting distribution functions are as follows. If the new event $e' = e_i$, then $F(x; s', e', s, e^*) = P\{V_i \leq x\}$. (The set $O(s-1; s, e_2) = \{e_1\}$ if $s < N$ and equals $\emptyset$ if $s = N$. The set $O(s+1; s, e_1) = \{e_2\}$ if $s > 0$ and equals $\emptyset$ if $s = 0$.)

Suppose that at time 0 all customers are at center 2 and service to a customer has just started. Then the initial distribution $\mu$ is

$$\mu(A) = 1_{\{N\}}(s)\, P\{V_2 \leq a_2\}$$

for all sets

$$A = \{s\} \times \{(c_1, c_2)\colon\ 0 \leq c_i \leq a_i \text{ for } 1 \leq i \leq 2\}.$$

**Example 4.9.** (Patrolling repairman) Assume that the lifetimes for machine $j$ are identically distributed as a positive random variable $L_j$ and the time to repair and restart machine $j$ is distributed as a positive random variable $R_j$. The walking time from machine $j$ to the next machine is a positive constant $W_j$. Also assume that the distribution of each $L_j$ is absolutely continuous so that, with probability 1, no two stoppages occur simultaneously. Set

$$X(t) = (Z_1(t), \ldots, Z_N(t), M(t), N(t)),$$

where

$$Z_j(t) = \begin{cases} 1 & \text{if machine } j \text{ is waiting for repair at time } t \\ 0 & \text{otherwise,} \end{cases}$$

$$M(t) = \begin{cases} j & \text{if machine } j \text{ is under repair at time } t \\ 0 & \text{if no machine is under repair at time } t, \end{cases}$$

$N(t) = 1$ if machine $N$ is under repair or the repairman is walking to machine 1 at time $t$, and $N(t) = j$ if machine $j-1$ is under repair or the repairman is walking to machine $j$ at time $t$ $(j = 2, 3, \ldots, N)$.

Take event $e_{N+2}$ = "arrival of repairman at machine," $e_{N+1}$ = "end of repair," and $e_j$ = "stoppage of machine $j$" $(j = 1, 2, \ldots, N)$. Formal specification of the process $\{X(t)\colon\ t \geq 0\}$ is as a GSMP with finite state space

$$S = \big\{(z_1, \ldots, z_N, m, n) \in \{0,1\}^N \times \{0, 1, \ldots, N\} \times \\ \{1, 2, \ldots, N\}:\ \ n = m + 1 \text{ when } 0 < m < N;\ n = 1 \\ \text{when } m = N;\ m = j \text{ only if } z_j = 0 \text{ for } j = 1, 2, \ldots, N\big\}$$

and event set $E = \{e_1, e_2, \ldots, e_{N+2}\}$. For $s = (z_1, \ldots, z_N, m, n) \in S$, event $e_{N+2}$ = "arrival of repairman at machine" $\in E(s)$ if and only if $m = 0$ and $e_{N+1}$ = "end of repair" $\in E(s)$ if and only if $m > 0$. Event $e_j$ = "stoppage of machine $j$" $\in E(s)$ if and only if $z_j = 0$ and $m \neq j$. All speeds $r(s, e)$ for active events are equal to 1.

If $e = e_{N+2}$ = "arrival of repairman at machine," then the state transition probability $p(s'; s, e) = 1$ when

$$s = (z_1, \ldots, z_{n-1}, 1, z_{n+1}, \ldots, z_N, 0, n) \text{ with } n < N \\ \text{and } s' = (z_1, \ldots, z_{n-1}, 0, z_{n+1}, \ldots, z_N, n, n+1),$$

when

$$s = (z_1, \ldots, z_{N-1}, 1, 0, N) \text{ and } s' = (z_1, \ldots, z_{N-1}, 0, N, 1),$$

when

$$s = (z_1, \ldots, z_{n-1}, 0, z_{n+1}, \ldots, z_N, 0, n) \text{ with } n < N \\ \text{and } s' = (z_1, \ldots, z_{n-1}, 0, z_{n+1}, \ldots, z_N, 0, n+1),$$

and when

$$s = (z_1, \ldots, z_{N-1}, 0, 0, N) \quad \text{and } s' = (z_1, \ldots, z_{N-1}, 0, 0, 1).$$

If $e = e_{N+1} =$ "end of repair," then $p(s'; s, e) = 1$ when

$$s = (z_1, \ldots, z_N, m, m+1) \text{ with } 0 < m < N \\ \text{and } s' = (z_1, \ldots, z_N, 0, m+1)$$

and when

$$s = (z_1, \ldots, z_N, N, 1) \text{ and } s' = (z_1, \ldots, z_N, 0, 1).$$

If $e = e_j =$ "stoppage of machine $j$," then $p(s'; s, e) = 1$ when

$$s = (z_1, \ldots, z_{j-1}, 0, z_{j+1}, \ldots, z_N, m, n) \\ \text{and } s' = (z_1, \ldots, z_{j-1}, 1, z_{j+1}, \ldots, z_N, m, n).$$

All other state transition probabilities $p(s'; s, e)$ are equal to 0. The clock-setting distribution functions are as follows. If the new event $e' = e_{N+2}$ and $s' = (z_1, \ldots, z_N, m, n)$, then $F(x; s', e', s, e^*) = 1_{[W_{n-1}, \infty)}(x)$ if $n > 1$ and equals $1_{[W_N, \infty)}(x)$ if $n = 1$. If $e' = e_{N+1}$ and $s' = (z_1, \ldots, z_N, m, n)$, then $F(x; s', e', s, e^*) = P\{R_m \le x\}$. If $e' = e_j$, then $F(x; s', e', s, e^*) = P\{L_j \le x\}$.

Suppose that at time 0 the repairman has just arrived at machine 1 and all machines are stopped. Then the initial distribution $\mu$ is

$$\mu(A) = 1_{\{s_0\}}(s) P\{R_1 \le a_{N+1}\}$$

for all sets

$$A = \{s\} \times \{(c_1, \ldots, c_{N+2})\colon\ 0 \le c_i \le a_i \text{ for } 1 \le i \le N+2\},$$

where $s_0 = (0, 1, 1, \ldots, 1, 1, 2)$.

Zero speeds in a GSMP are convenient for specification of queueing systems with service interruptions of the preemptive-resume type.

**Example 4.10.** (Preemptive cyclic queues) Consider a network of queues with two single-server centers that provides service to a fixed number $N$ of customers labelled $1, 2, \ldots, N$. After completion of service at center 1, a customer moves instantaneously from center 1 to center 2, and after completion of service at center 2 moves instantaneously back to center 1. Suppose that the queueing discipline at each center is last-come, first-served (LCFS). Also suppose that service interruptions are of the preemptive-resume type (service continues from the point of interruption when the preempted job goes back into service). Customers waiting in queue appear in the order of preemption at the center, the most recently preempted customer at the head of the queue. Assume that all service times are mutually independent and the service times at center $i$ are identically distributed as a positive random variable $V_i$. Also assume that the distribution of each $V_i$ is absolutely continuous so that, with probability 1, no two service completions occur simultaneously.

Let $X(t)$ be the number of customers waiting or in service at center 2 at time $t$. Take event $e_j$ = "service completion at center 1 with $j$ customers at center 1" and event $e_{N+j}$ = "service completion at center 2 with $j$ customers at center 2" ($j = 1, 2, \ldots, N$). Formal specification of the process $\{X(t)\colon t \geq 0\}$ is as a GSMP with finite state space $S = \{0, 1, \ldots, N\}$ and event set $E = \{e_1, e_2, \ldots, e_{2N}\}$.

For $s \in S$, event $e_j \in E(s)$ if and only if $j \leq N - s$ and event $e_{N+j} \in E(s)$ if and only if $j \leq s$; thus, $E(0) = \{e_1, \ldots, e_N\}$, $E(N) = \{e_{N+1}, \ldots, e_{2N}\}$, and

$$E(s) = \{e_1, \ldots, e_{N-s}, e_{N+1}, \ldots, e_{N+s}\}$$

for $0 < s < N$. The speed $r(s, e_i)$ at which the clock associated with event $e_i$ runs in state $s$ is equal to 1 when $s = 0$ and $i = N$ and when $s = N$ and $i = 2N$. For $0 < s < N$, the speed $r(s, e_i)$ is equal to 1 when $i = N - s$ and when $i = N + s$. All other speeds $r(s, e_i)$ for active events are equal to 0.

If $e = e_j$ = "service completion at center 1 with $j$ customers at center 1," then the state transition probability $p(s'; s, e) = 1$ when

$$s = N - j \text{ and } s' = N - j + 1.$$

If $e = e_{N+j}$ = "service completion at center 2 with $j$ customers at center 2," then $p(s'; s, e) = 1$ when

$$s = j \text{ and } s' = j - 1.$$

All other state transition probabilities $p(s'; s, e)$ are equal to 0. The clock-setting distribution functions are as follows. If the new event $e' = e_j$, then $F(x; s', e', s, e^*) = P\{V_1 \leq x\}$. If $e' = e_{N+j}$, then $F(x; s', e', s, e^*) = P\{V_2 \leq x\}$. Observe that $N(j + 1; j, e_{N-j}) = \{e_{N+j+1}\}$ for $j = 0, 1, \ldots, N - 1$ and $N(j - 1; j, e_{N+j}) = \{e_{N-j+1}\}$ for $j = 1, 2, \ldots, N$.

**Example 4.11.** (Preemptive cyclic queues) Let $Z(t)$ be the number of customers waiting or in service at center 1 at time $t$. Define the position $N_i(t)$ of customer $i$ at time $t$: $N_i(t) = j$ if customer $i$ is waiting in queue at center 1 at time $t$ and there are $j - 1$ customers waiting at center 1 behind customer $i$ $(j = 1, 2, \ldots, N - 1)$; $N_i(t) = Z(t)$ if customer $i$ is in service at center 1; $N_i(t) = Z(t) + j$ if customer $i$ is waiting in queue at center 2 and there are $j - 1$ customers waiting at center 2 behind customer $i$; and $N_i(t) = N$ if customer $i$ is in service at center 2. Set

$$X(t) = (Z(t), N_1(t), N_2(t), \ldots, N_N(t)).$$

Take event $e_i$ = "service completion to customer $i$." Formal specification of the process $\{X(t)\colon t \geq 0\}$ is as a GSMP with finite state space

$$S = \big\{(z_1, n_1, \ldots, n_N) \in \{0, 1, \ldots, N\} \times \{1, 2, \ldots, N\}^N : \\ n_i \neq n_j \text{ when } i \neq j \text{ for } i, j = 1, 2, \ldots, N\big\}$$

and event set $E = \{e_1, e_2, \ldots, e_N\}$. The event set $E(s) = E$ for all $s \in S$. The speed $r(s, e)$ at which the clock associated with event $e$ runs in state $s = (z_1, n_1, \ldots, n_N)$ is equal to 1 when $n_i = z_1$ and when $n_i = N$; all other speeds $r(s, e)$ for active events are equal to 0.

If $e = e_i$ = "service completion to customer $i$," then the state transition probability $p(s'; s, e_i) = 1$ when

$$s = (z_1, n_1, \ldots, n_{i-1}, z_1, n_{i+1}, \ldots, n_N) \text{ with } 0 < z_1 \leq N \\ \text{and } s' = (z_1 - 1, n_1', \ldots, n_{i-1}', N, n_{i+1}', \ldots, n_N'),$$

where $n_j'$ equals $n_j$ if $n_j < z_1$ and equals $n_j - 1$ if $n_j > z_1$ $(j \neq i)$. The state transition probability $p(s'; s, e_i) = 1$ when

$$s = (z_1, n_1, \ldots, n_{i-1}, N, n_{i+1}, \ldots, n_N) \text{ with } 0 \leq z_1 < N \\ \text{and } s' = (z_1 + 1, n_1', \ldots, n_{i-1}', z_1 + 1, n_{i+1}', \ldots, n_N'),$$

where $n_j'$ equals $n_j$ if $n_j \leq z_1$ and equals $n_j + 1$ if $n_j > z_1$. All other state transition probabilities $p(s'; s, e)$ are equal to 0.

**Example 4.12.** (Particle counter) Particles arrive, one at a time, at a counter. A particle arrives at time 0 and locks the counter for an interval of fixed length $T$. If there are no arrivals in $(0, T]$, the counter becomes unlocked at time $T$; the next particle gets registered and the counter is locked again for an interval of length $T$. A particle that arrives when the counter is locked does not get registered, but extends the locked interval so that the counter remains locked for an interval of length $T$

after the arrival. Assume that the successive interarrival times for particles are i.i.d. as a positive random variable $U$ with absolutely continuous distribution.

Set

$$Z(t) = \begin{cases} 1 & \text{if the counter is locked at time } t \\ 0 & \text{otherwise.} \end{cases}$$

Take event $e_1$ = "arrival of particle" and $e_2$ = "end of locked interval." The process $\{Z(t): t \geq 0\}$ cannot be specified as a GSMP with state space $\{0,1\}$ and event set $\{e_1, e_2\}$. The difficulty is in a GSMP the clock associated with an "old event" cannot be reset. Observe that event $e_1$ = "arrival of particle" is scheduled to occur in state 0 and in state 1 and event $e_2$ = "end of locked interval" is scheduled to occur in state 1, but is not scheduled to occur in state 0. Also observe that, with probability 1, there is a state transition to state 1 when event $e_1$ occurs in state 0. Likewise, with probability 1, there is a state transition to state 1 when event $e_1$ occurs in state 1, as well as a state transition to state 0 when event $e_2$ occurs in state 1. Since a particle that arrives when the counter is locked extends the locked interval for an interval of length $T$ after the arrival, it is necessary to reset the clock associated with event $e_2$ when event $e_1$ triggers a state transition from state 1 to state 1. For such a state transition, event $e_2$ is an old event. However, in a GSMP the clock associated with an old event continues to run after the state transition.

To specify the process $\{Z(t): t \geq 0\}$, we use one "arrival of particle" event along with two copies of an "end of locked interval" event. The idea is that the two "end of locked interval" events alternate. In effect, we reset the clock associated with an old "end of locked interval" event by cancelling the event and scheduling a new "end of locked interval" event. Take event $e_1$ = "arrival of particle," $e_2$ = "(copy 1) end of locked interval," and $e_3$ = "(copy 2) end of locked interval." Set

$$X(t) = (Z(t), Y(t)),$$

where

$$Y(t) = \begin{cases} 3 & \text{if event } e_3 \text{ is active at time } t \\ 2 & \text{if event } e_2 \text{ is active at time } t \\ 0 & \text{if neither event } e_2 \text{ nor } e_3 \text{ is active at time } t\ . \end{cases}$$

Formal specification of the process $\{X(t)\colon\ t \geq 0\}$ is as a GSMP with state space $S = \{(0,0),(1,2),(1,3)\}$ and event set $E = \{e_1, e_2, e_3\}$. For $s = (x,y) \in S$, event $e_2 \in E(s)$ if and only if $y = 2$ and event $e_3 \in E(s)$ if and only if $y = 3$. Event $e_1 \in E(s)$ for all $s \in S$. All speeds $r(s,e)$ for active events are equal to 1.

If $e = e_1$, then the state transition probability $p(s'; s, e) = 1$ when

$$s = (0,0) \text{ and } s' = (1,2),$$

when

$$s = (1,2) \text{ and } s' = (1,3),$$

and when

$$s = (1,3) \text{ and } s' = (1,2).$$

If $e = e_2$, then $p(s'; s, e) = 1$ when

$$s = (1,2) \text{ and } s' = (0,0).$$

If $e = e_3$, then $p(s'; s, e) = 1$ when

$$s = (1,3) \text{ and } s' = (0,0).$$

All other state transition probabilities $p(s'; s, e)$ are equal to 0. The clock-setting distribution functions are as follows. If the new event $e' = e_1$, then $F(x; s', e', s, e^*) = P\{U \leq x\}$. If $e' = e_2$ or if $e' = e_3$, then $F(x; s', e', s, e^*) = 1_{[T,\infty)}(x)$.

Theorem 4.14 provides conditions under which a GSMP $\{X(t)\colon\ t \geq 0\}$ with a finite state space has a limiting distribution; that is, there exists a random variable $X$ such that

$$\lim_{t \to \infty} P\{X(t) \leq x\} = P\{X \leq x\}$$

for all $x$ at which the right side is continuous. We denote this convergence in distribution by "$X(t) \Rightarrow X$ as $t \to \infty$." Portions of the proof require methods of advanced probability.

**Definition 4.13.** *A generalized semi-Markov process with state space $S$ and event set $E$ is said to be irreducible if for each pair $s, s' \in S$ there exist states $s = s_0, s_1, \dots, s_{n-1}, s_n = s' \in S$ and events $e_{i_0}, e_{i_1}, \dots, e_{i_{n-1}} \in E$ such that*

$$p(s_1; s_0, e_{i_0})r(s_0, e_{i_0})p(s_2; s_1, e_{i_1})r(s_1, e_{i_1}) \cdots \\ p(s_n; s_{n-1}, e_{i_{n-1}})r(s_{n-1}, e_{i_{n-1}}) > 0.$$

**Theorem 4.14.** *Let $\{X(t)\colon t \geq 0\}$ be an irreducible generalized semi-Markov process with finite state space $S$ and event set $E$. Assume that for $s, s' \in S$, $e^* \in E$, and $e' \in N(s'; s, e^*)$, the clock-setting distribution function $F(\cdot\,; s', e', s, e^*)$ has finite mean and a density function that is continuous and positive on $(0, +\infty)$. Then $X(t) \Rightarrow X$ as $t \to \infty$.*

This result provides a means of establishing the existence of a limiting distribution for the underlying stochastic process of a discrete-event simulation on a finite state space. It is sufficient to show that the process can be specified as an irreducible GSMP in which all clock-setting distribution functions have finite mean and a density function that is continuous and positive on $(0, +\infty)$.

**Example 4.15.** (Telephone system) Consider a telephone system with $N$ ($\geq 2$) telephones connected to a switchboard by lines numbered $1, 2, \dots, N$. The switchboard has $K$ ($< N$) links numbered $1, 2, \dots, K$, each of which can connect any two lines subject to the restriction that only one connection at a

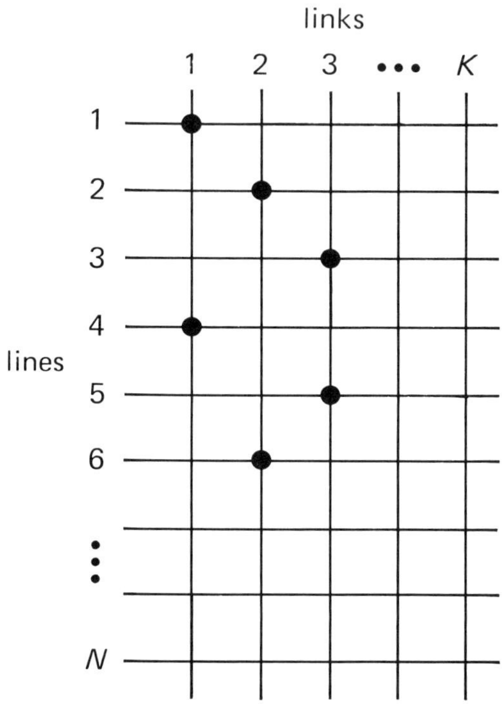


**Figure 1.6.** Telephone system.

time can be made to each line. If more than one link is available and the called line is not in use, a placed call is connected (instantaneously) on the lowest numbered available link; see Figure 1.6. Assume that the system is a lost-call system in the sense that any call is immediately lost if no connection can be made when it is placed. A call is lost if there is at least one link available but the called line is in use (a *busy call*), and a call is lost if no link is available (a *blocked call*). Figure 1.7 shows a time-line diagram for the telephone system with $N = 6$ lines and $K = 2$ links. The initial call (placed at line

1 to line 3) is connected on link 1 and the next call (placed at line 2 to line 1) is a busy call. The call placed at line 5 (to line 4) is connected on link 2 and the call placed at line 6 (to line 2) is a blocked call. In the figure a circled number represents the link on which a call is connected and a number displayed above an $\times$ represents the destination of a call.

Assume that the duration of a call placed at line $i$ is distributed as a positive random variable $L_i$. Also assume that calls are placed randomly in the sense that the time from the end of a call placed or received at line $i$ to the next call placed at line $i$ is distributed as a positive random variable $A_i$. When a call is placed, with independent probability $p_{ij}$ the called line is line $j$ ($\neq i$). After a lost call placed at line $i$, the time to the next call placed at line $i$ is distributed as $A_i$. Assume that the distribution of each $L_i$ and $A_i$ is absolutely continuous.

Set

$$X(t) = (X_1(t), X_2(t), \ldots, X_K(t)), \tag{4.16}$$

where $X_k(t)$ equals $(i,j)$ if a call placed at line $i$ is connected to line $j$ on link $k$ at time $t$ and equals $(0,0)$ if there is no call connected on link $k$ at time $t$. Take event $e_i$ = "call placed at line $i$" ($i = 1, 2, \ldots, N$) and event $e_{N+k}$ = "completion of call on link $k$" ($k = 1, 2, \ldots, K$). Formal specification of the process $\{X(t)\colon\ t \geq 0\}$ is as a GSMP with finite state space

$$\begin{aligned} S = \big\{(i_1, j_1, \ldots, i_K, j_K) \in \{0, 1, \ldots, N\}^{2K} &: \ i_m = j_m \text{ only if} \\ i_m = j_m = 0; i_m \neq i_n, j_m \neq j_n, i_m \neq j_n, j_m \neq i_n &\text{ or} \\ i_m = i_n = j_m = j_n = 0 \text{ if } m \neq n \text{ with } m, n &= 1, 2, \ldots, K\big\} \end{aligned}$$

and event set $E = \{e_1, e_2, \ldots, e_N, e_{N+1}, \ldots, e_{N+K}\}$. For $s = (i_1, j_1, \ldots, i_K, j_K) \in S$, event $e_i \in E(s)$ if and only if $i_k \neq i$ and $j_k \neq i$ for $k = 1, 2, \ldots, K$ and $e_{N+k} \in E(s)$ if and only if $i_k > 0$. All speeds $r(s,e)$ for active events are equal to 1.

If $e = e_i$ = "call placed at line $i$," then the state transition

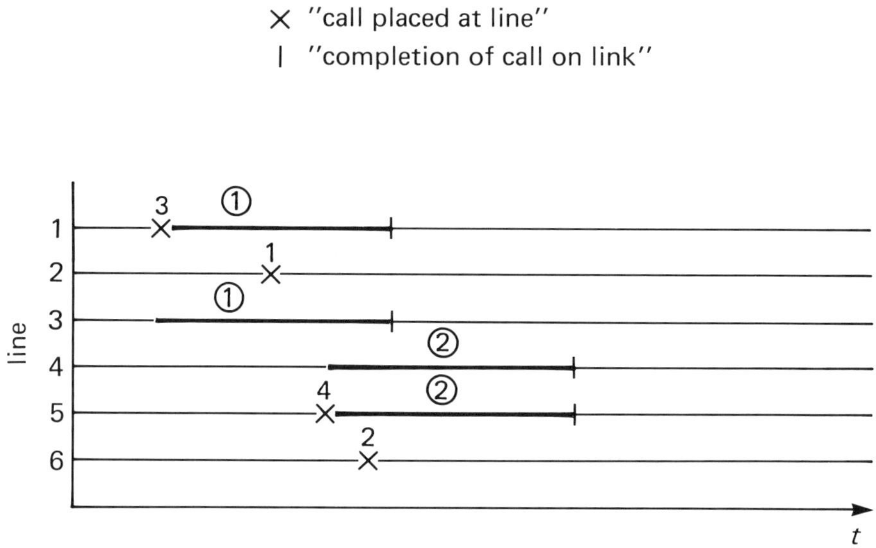


**Figure 1.7.** Time-line diagram for telephone system.

probability $p(s'; s, e) = p_{ij}$ when (for a "successful call")

$$\begin{aligned} s = (i_1, j_1, \ldots, i_K, j_K) &\text{ with } i_1, i_2, \ldots, i_m > 0 \\ &\text{and } \; i_{m+1} = 0 \text{ for some } m < K, \\ &(i_1 - j) \cdots (i_K - j)(j_1 - j) \cdots (j_K - j) \neq 0 \\ \text{and } s' &= (i_1, j_1, \ldots, i_{m-1}, j_{m-1}, i, j, i_{m+1}, \ldots, j_K), \end{aligned}$$

$p(s'; s, e) = \sum_{j:(i_1-j)\cdots(i_K-j)(j_1-j)\cdots(j_K-j)=0} p_{ij}$ when (for a "busy call")

$$\begin{aligned} s = (i_1, j_1, \ldots, i_K, j_K) &\text{ with } i_1, i_2, \ldots, i_m > 0 \text{ and } i_{m+1} = 0 \\ &\text{for some } m < K, (j_1 - j) \cdots (j_K - j) = 0 \\ \text{and } s' &= s \end{aligned}$$

and $p(s'; s, e) = 1$ when (for a "blocked call")

$$s = (i_1, j_1, \ldots, i_K, j_K) \text{ with } i_1, i_2, \ldots, i_K > 0 \text{ and } s' = s.$$

If $e = e_{N+k} =$ "completion of call on link $k$," then $p(s'; s, e) = 1$ when

$$s = (i_1, j_1, \ldots, i_K, j_K)$$
$$\text{and } s' = (i_1, j_1, \ldots, i_{k-1}, j_{k-1}, 0, 0, i_{k+1}, j_{k+1}, \ldots, i_K, j_K).$$

All other state transition probabilities $p(s'; s, e)$ are equal to 0. The clock-setting distribution functions are as follows. If the new event $e' = e_i =$ "call placed at line $i$," then $F(x; s', e', s, e^*) = P\{A_i \le x\}$. If $e' = e_{N+k} =$ "completion of call on link $k$" and $s' = (i_1, j_1, \ldots, i_K, j_K)$, then $F(x; s', e', s, e^*) = P\{L_{i_k} \le x\}$.

The state transitions and state transition times determine whether a call is successful, busy, or blocked. A state transition from $s = (i_1, j_1, \ldots, i_{m-1}, j_{m-1}, 0, 0, i_{m+1}, j_{m+1}, \ldots, i_K, j_K) \in S$ to $s' = (i_1, j_1, \ldots, i_{m-1}, j_{m-1}, i, j, i_{m+1}, j_{m+1}, \ldots, i_K, j_K)$ with $i, j > 0$ corresponds to a successful call. A state transition from some $s = (i_1, j_1, \ldots, i_K, j_K) \in S$ with $i_1 i_2 \cdots i_K = 0$ to $s' = s$ corresponds to a busy call. A state transition from some $s = (i_1, j_1, \ldots, i_K, j_K) \in S$ with $i_1 i_2 \cdots i_K > 0$ to $s' = s$ corresponds to a blocked call.

It is straightforward to show that the GSMP is irreducible. For example, let $s = (i_1, j_1, \ldots, i_{m+1}, j_{m+1}, 0, 0, \ldots, 0, 0) \in S$ with $i_1, i_2, \ldots, i_{m+1} > 0$ and $s' = (i_1', j_1', 0, 0, \ldots, 0, 0) \in S$. Take $e_{l_{m+1}} = e_{i_1'}$, $e_{l_{k-1}} = e_{N+k}$, and $s_k = (0, 0, \ldots, 0, 0, i_{k+1}, j_{k+1}, \ldots, i_{m+1}, j_{m+1}, 0, 0, \ldots, 0, 0)$ for $k = 1, 2, \ldots, m+1$. Then $p(s_1; s, e_{l_0}) r(s, e_{l_0}) p(s_2; s_1, e_{l_1}) r(s_1, e_{l_1}) \cdots p(s'; s_{m+1}, e_{l_{m+1}})$ $r(s_{m+1}, e_{l_{m+1}}) > 0$. Theorem 4.14 implies that $X(t) \Rightarrow X$ as $t \to \infty$ provided that each clock-setting distribution has finite mean and a density function that is continuous and positive on $(0, +\infty)$.

Formal specification as a GSMP leads to an algorithm for generating sample paths of the underlying stochastic process of a simulation. Since the event-set mapping, state transition probabilities, speeds, and clock-setting distribution functions determine the transition kernel of the associated GSSMC, sample paths of the GSMP can be generated directly and usually without enumeration of the state space.

The form of the transition kernel in (4.5) leads to the following algorithm for generating sample paths of the underlying GSSMC $\{(S_n, C_n)\colon\ n \geq 0\}$ of the GSMP.

**Algorithm 4.17.** *Sample path generation*

1. *Set $\zeta = 0$. Select an initial state $s \in S$ according to the initial probability distribution $q$. For each active event $e_i \in E(s)$, generate a corresponding clock reading $c_i$ according to the clock-setting distribution function $F_0(\,\cdot\,; e_i, s)$. Set $c_i = 0$ for each $e_i \notin E(s)$.*
2. *Determine the index $i^* = i^*(s,c)$ of the next trigger event $e^* = e_{i^*(s,c)}$ as the index $i^*$ such that $c_{i^*}/r(s, e_{i^*}) < c_i/r(s, e_i)$ for all $i \neq i^*$. Also determine the time $t^*(s,c)$ to the next state transition according to $t^*(s,c) = c_{i^*}/r(s, e_{i^*})$.*
3. *Generate the new state $s'$ according to the probability mass function $p(\,\cdot\,; s, e^*)$.*
4. *For each event $e_i \in N(s'; s, e^*) = E(s') - (E(s) - e^*)$, generate a new clock reading $c_i'$ according to the distribution function $F(\,\cdot\,; s', e_i, s, e^*)$.*
5. *For each event $e_i \in O(s'; s, e^*) = E(s') \cap \big(E(s) - e^*\big)$, set $c_i' = c_i - t^*(s,c) r(s, e_i)$.*
6. *For each event $e_i \in \big(E(s) - \{e^*\}\big) - e(s')$, set $c_i' = 0$.*
7. *Return to Step 2 and iterate with $s'$ playing the role of $s$ and $c'$ the role of $c$.*

A sample path of the GSMP can be obtained from a sample path of the GSSMC. Recall that by construction, the GSMP takes values in the set $S \cup \{\Delta\}$ and has piecewise constant, right-continuous sample paths. A sample path of the GSMP can be represented as a sequence $\{(X(\zeta_n), \zeta_n)\colon n \geq 0\}$, where $\zeta_0 = 0$ and $\zeta_n = \sum_{k=0}^{n-1} t^*(S_k, C_k)$ for $n \geq 1$. In this representation, $\zeta_n$ is the time of the $n$th state transition and $X(\zeta_n)$ is the state of the GSMP at time $\zeta_n$. The following algorithm produces a sample path of the GSMP.

**Algorithm 4.18.** *Sample path generation*

1. *Set* $n = 0$ *and* $\zeta_n = 0$.
2. *Set* $X(\zeta_n) = S_n$ *and* $\zeta_{n+1} = \zeta_n + t^*(S_n, C_n)$.
3. *Increment* $n$ *by 1 and go to Step 2.*

**Exercises**

4.1. (Generalized semi-Markov process with resettable old clocks) In a "GSMP with resettable old clocks," the clock associated with an old event can be reset after a state transition. At a state transition from $s$ to $s'$ triggered by event $e^*$, clock readings are generated for each new event $e' \in N(s'; s, e^*) = E(s') - (E(s) - \{e^*\})$. Denote the clock-setting distribution function (distribution function of such a new clock reading) by $F(\,\cdot\,; s', e', s, e^*)$. The resettable old clocks are specified by a (possibly empty) subset $R(s'; s, e^*)$ of $O(s'; s, e^*) = E(s') \cap (E(s) - \{e^*\})$. For $e' \in R(s'; s, e^*) \subseteq O(s'; s, e^*)$, the clock associated with event $e'$ is reset after the state transition, and the distribution function of the new clock reading is $G(\,\cdot\,; s', e', s, e^*)$. For each old event $e' \in O(s'; s, e^*) - R(s'; s, e^*)$, the old clock reading is kept after the state transition. For $e' \in (E(s) - \{e^*\}) - e(s')$, event $e'$ is cancelled, and the corresponding clock reading is discarded. Formal definition of a GSMP with resettable old clocks is in terms of a GSSMC that describes the state of the system at successive state transition times. Specify the transition kernel of the GSSMC. Also show that there is no loss of generality to assume that clocks associated with old events in a GSMP are not resettable.

4.2. (Random polling system) Consider a random polling system with a fixed number of stations, numbered $1, 2, \ldots, N$. At each station, customers arrive according to a renewal process and join the tail of a queue. A single-server switches from station to station according to a stochastic mechanism and serves the customers queued at a station in the order of their arrival at the station. The number of customers served at a station before the server "polls" (switches to) another station depends upon the fixed service policy of the polling system. Suppose that the server polls station $j$ and finds at least one customer queued for service. Under a "limited service policy" exactly one customer is served. Upon completion of service at station $j$, the server switches to station $k$ ($\neq j$) with probability $p_{jk}$; the time required for this switch is distributed as a positive random variable $R_j$. When the server arrives at station $j$ and finds no customers queued for service, the server immediately switches to another station. Assume that the interarrival time for customers at station $j$ is distributed as a positive random variable $W_j$ and the service time for a customer at station $j$ is distributed as a positive random variable $L_j$. All interarrival times, service times, and times required to switch are mutually independent.

What continuous time stochastic process would you simulate to estimate the number of stations that have a customer queued for service and the waiting times (times from arrival of a customer until the start of service) experienced by customers as an individual station? Show that the state transitions and state transition times determine the start times and termination times of the waiting times at center $j$.

4.3. (Road traffic) Consider the road traffic system of Exercise 3.1. Set

$$X(t) = (Q(t), G(t), B(t)),$$

where $Q(t)$ is the number of cars waiting on the side road at time $t$,

$$G(t) = \begin{cases} 2 & \text{if } A_n \leq t < A_{n+1},\ A_{n+1} - A_n > \Delta_1, \\ & \quad A_{n+1} - t > \Delta_1 \\ 1 & \text{if } A_n \leq t < A_{n+1},\ A_{n+1} - A_n > \Delta_1, \\ & \quad A_{n+1} - t \leq \Delta_1 \\ 0 & \text{if } A_n \leq t < A_{n+1},\ A_{n+1} - A_n \leq \Delta_1, \end{cases}$$

and

$$B(t) = \begin{cases} 0 & \text{if } G(t) = 2 \text{ and a car has merged in } [t - \Delta_2, t) \\ 1 & \text{otherwise .} \end{cases}$$

Take event $e_1$ = "arrival of main-road car at the merge point," $e_2$ = "arrival of side-road car," $e_3$ = "end of time interval of length $\Delta_2$," and $e_4$ = "end of time interval when side-road car may merge." Specify the process $\{X(t)\colon t \geq 0\}$ as a GSMP with a countably infinite state space $S$ and event set $E = \{e_1, e_2, e_3, e_4\}$.

4.4. (Telephone system) Consider the telephone system of Example 4.15. Assume that the duration of a call placed at line $i$ is distributed as a positive random variable $L_i$. Also assume that calls are placed randomly in the sense that the time from the end of a call (placed or received) at line $i$ to the next call placed at line $i$ is distributed as a positive random variable $A_i$. When a call is placed, with independent probability $p_{ij}$ the called line is line $j$ ($\neq i$). After a lost call placed at line $i$, the time to the next call placed at line $i$ is distributed as $A_i$. Assume that the distribution of each $L_i$ and each $A_i$ is absolutely continuous.

Set

$$X(t) = (X_1(t), \ldots, X_N(t)),$$

where $X_i(t)$ equals $k$ if line $i$ is connected (to some other line) on link $k$ at time $t$ and equals 0 if line $i$ is not connected at time $t$. Take event $e_i$ = "call placed at line $i$" ($i = 1, 2, \ldots, N$) and event $e_{N+k}$ = "completion of call on link $k$" ($k = 1, 2, \ldots, K$). Specify the process $\{X(t)\colon\ t \geq 0\}$ as a GSMP with a finite state space $S$ and event set $E = \{e_1, e_2, \ldots, e_N, e_{N+1}, \ldots, e_{N+K}\}$. Show that the state transitions and state transition times determine the calls that are successful, the calls that are busy, and the calls that are blocked.

4.5. (Operating unit failure) Consider a system consisting of $N$ identical operating units that are subject to failure. A failed unit is sent to a repair facility that operates as a single-server queue: units are restored to a like-new condition one at a time on a first-come, first-served basis. When a unit in "like-new" condition is placed in service, the unit operates for a random time distributed as a positive random variable $L$ and then fails. Assume that the time required for the repairman to restore a failed unit is exponentially distributed. Restoration time has mean 1 if the repairman is idle when the unit fails, and has mean 2 if the repairman is busy when the unit fails. After being restored, the unit goes back into operation immediately.

Set

$$X(t) = (N(t), M_1(t), \ldots, M_N(t)),$$

where $N(t)$ is the number of units at the repair facility at time $t$, $M_i(t)$ equals $j+1$ if unit $i$ is at the repair facility at time $t$ and there are $j$ units waiting for repair behind unit $i$ ($j = 0, 1, \ldots, N-1$), and $M_i(t)$ equals 0 if unit $i$ is not at the repair facility at time $t$. Take event $e_i$ = "failure of unit $i$" and event $e_{N+i}$ = "end of repair for unit $i$." Specify the process $\{X(t)\colon\ t \geq 0\}$ as a GSMP with a finite state space $S$ and event set $E = \{e_1, e_2, \ldots, e_{2N}\}$.

4.6. (Shuttle train with finite waiting room) Consider a train that shuttles back and forth between two stations and has seats for $K$ ($\geq 1$) passengers. Passengers that wish to board the train and ride to the other station arrive at station $i$ according to a renewal process, queue, and subsequently board the train in the order in which they arrive at the station. Assume that passengers board the train instantaneously and all passengers that board at station $i$ disembark instantaneously when the train arrives at the other station. At each station passengers disembark before any passengers board the train. The waiting room at station $i$ has finite capacity $B_i$ ($\geq 1$). A passenger that wishes to board the train at station $i$ and arrives when there are already $B_i$ passengers waiting to board is turned away. The time required for the train to travel from one station to the other is distributed as a positive random variable $L$ with finite mean. Travel times are independent of interarrival times, and the arrival process of passengers for boarding at station 1 is independent of the arrival process of passengers for boarding at station 2. Assume that distribution of interarrival times for passengers at station $i$ is absolutely continuous.

What continuous time stochastic process $\{X(t)\colon\ t \geq 0\}$ would you simulate to obtain estimates for the number of passengers waiting to board the train at station $i$, the number of occupied seats on the train, and the waiting times (before boarding) experienced by passengers? Specify the state space $S$ of the process $\{X(t)\colon\ t \geq 0\}$ and the set $E$ of events associated with simulation of the process. Also specify the process $\{X(t)\colon\ t \geq 0\}$ as a GSMP with state space $S$ and event set $E$. Show that the state transitions and state transition times determine the occurrence of trips from station 1 to station 2 in which all seats are occupied and the waiting times experienced by passengers that board the train at station $i$.

4.7. (Airport shuttle) Consider an airport shuttle that provides transportation service to $N$ stations numbered $1, 2, \ldots, N$. The shuttle has seats for $K$ ($\geq 1$) passengers and moves from station to station in a strictly defined order: $1, 2, \ldots, N, N-1, \ldots, 1, 2, \ldots$. Passengers that wish to board the shuttle at station $i$ in the "northbound" (from station 1 to station $N$)

direction arrive at station $i$ according to a Poisson process of rate $\lambda_i$ and disembark at station $j$ ($> i$) with probability $p_{i,j}$ ($i = 1, 2, \ldots, N-1$). Passengers that wish to board the shuttle at station $i+1$ in the "southbound" (from station $N$ to station 1) direction arrive at station $i+1$ according to a Poisson process of rate $\lambda_{2N-(i+1)}$ and disembark at station $j$ ($< i+1$) with probability $p_{2N-(i+1),j}$ ($i = 1, 2, \ldots, N-1$). Passengers that arrive for boarding at station $i$ in each direction queue (and subsequently board the shuttle) in the order in which they arrive at the station. At each station passengers disembark before any waiting passengers board the shuttle. Assume that passengers board the shuttle instantaneously and passengers disembark instantaneously. The time required for the shuttle to travel (in the northbound direction) from station $i$ to station $i+1$ is distributed as a positive random variable $L_i$ with finite first and second moment. The time required for the shuttle to travel (in the southbound direction) from station $i+1$ to station $i$ is distributed as a positive random variable $L_{2N-(i+1)}$ with finite first and second moment. Travel times, interarrival times, and destinations are mutually independent.

What continuous time stochastic process $\{X(t)\colon t \geq 0\}$ would you simulate to obtain estimates for the number of occupied seats on the shuttle and the number of passengers waiting to board at station $i$? Specify the state space $S$ of the process $\{X(t)\colon t \geq 0\}$ and the set $E$ of events associated with simulation of the process. Also specify the process $\{X(t)\colon t \geq 0\}$ as a GSMP with state space $S$ and event set $E$. Show that the state transitions and state transition times determine the starts of trips from station $i$ (to station $i+1$) in the northbound direction with all seats occupied.

4.8. (Flexible manufacturing system) Consider a flexible manufacturing system that produces two types of parts and has three machines numbered 1, 2, and 3. Parts of type 1 require processing first by machine 1 and then by machine 2.

There are two processes that can produce parts of type 2. The first process consists of a fast intervention by machine 1 followed by a refinement performed by machine 2. The second process, performed entirely by machine 3, is much slower but produces finished parts. The duration of the refinement operation performed by machine 2 on parts processed by machine 1 is independent of the part type. Finished parts are unloaded (instantaneously) from the system and replaced by raw ones three at a time; thus, there are exactly three parts in the system at any time. Machine 2 processes parts in the order of their arrival in its input queue, as does machine 3. Machine 1 processes parts of the same type in the order of their arrival, but parts of type 2 have nonpreemptive priority over parts of type 1 for processing by machine 1.

The identification of raw parts loaded (instantaneously) into the system as being of type 1 or of type 2 is probabilistic, and raw parts loaded into the system move immediately to the input queue of machine 1 or to the input queue of machine 3 according to a fixed (probabilistic) scheduling mechanism. With independent probability $p$ ($0 < p < 1$) a raw part is of type 1, and with probability $1 - p$ a part is of type 2. Raw parts of type 2 move to the input queue of machine 1 with probability $q$ ($0 < q < 1$) and to the input queue of machine 3 with probability $1 - q$. The time required by machine 1 to process a part of type $i$ is distributed as a positive random variable $L_{1i}$. The time required by machine 2 to process a part is distributed as a positive random variable $L_2$, and the time required by machine 3 to process a part (of type 2) is distributed as a positive random variable $L_3$. All processing times are mutually independent.

What continuous time stochastic process $\{X(t)\colon\ t \geq 0\}$ would you simulate to estimate the time to produce a part of type $i$ and the number of parts that are finished but not unloaded? Specify the state space $S$ of the process $\{X(t)\colon\ t \geq 0\}$ and the set $E$ of events associated with simulation of the process. Also specify the process $\{X(t)\colon\ t \geq 0\}$ as a GSMP with state space $S$ and event set $E$.

4.9. (Finite-capacity queue with processor sharing) Consider an open queueing system with one center. Jobs arrive from an external source, receive service at the center, and depart to an external sink. A job that arrives when the system already contains $B-1$ ($\geq 0$) jobs causes the arrival process to shut down. The arrival process remains shut down until the first subsequent departure of a job. The server is never idle when any jobs are present in the system. All jobs in the system share the server equally; that is, when there are $q$ ($> 0$) jobs in the system, the remaining service for each job decreases at rate $1/q$. Assume that the successive interarrival times are i.i.d. as a positive random variable $U$ with absolutely continuous distribution and the successive service requirements for jobs are i.i.d. as a positive random variable $V$ with absolutely continuous distribution.

Take event $e_{B+1}$ = "arrival of job" and $e_i$ = "(copy $i$) service completion" ($i = 1, 2, \ldots, B$). Set

$$X(t) = (Q(t), K_1(t), \ldots, K_B(t)),$$

where $Q(t)$ is the number of jobs in the system at time $t$ and

$$K_i(t) = \begin{cases} 1 & \text{if event } e_i \text{ is active at time } t \\ 0 & \text{if event } e_i \text{ is not active at time } t. \end{cases}$$

Specify the process $\{X(t)\colon\ t \geq 0\}$ as a GSMP with a finite state space $S$ and event set $E = \{e_1, e_2, \ldots, e_B, e_{B+1}\}$.

## Chapter 2

# Regenerative Stochastic Processes

A regenerative stochastic process has the characteristic property that there exists a sequence of random times, referred to as "regeneration points," at which the process probabilistically restarts. The essence of regeneration is that, between any two successive regeneration points, the evolution of the process is a probabilistic replica of the process in any other such "cycle". Under mild regularity conditions, the regenerative property guarantees the existence of a limiting distribution for the process provided that the expected time between regeneration points is finite. Moreover, the behavior of a regenerative process in a cycle determines the limiting distribution of the process as a ratio of expected values. These results have important implications for the analysis of simulation output.

## 2.1. The Regenerative Property

Let $\{X(t)\colon\ t \geq 0\}$ be a continuous time stochastic process with a state space $S$ that is a subset of $d$-dimensional Euclidean space $\Re^d$ for some $d \geq 1$. Formal definition of a regenerative process is in terms of "stopping times" for a stochastic process. Recall that a *stopping time* for a stochastic process $\{X(t)\colon\ t \geq 0\}$ is a random variable $T$ taking values in $[0, +\infty)$ and such that for every finite $t \geq 0$, the history $\{X(u)\colon\ u \leq t\}$ of the process up to time $t$ determines the occurrence or nonoccurrence of the event $\{T \le t\}$. The definition of a regenerative process in discrete time is analogous.

**Definition 1.1.** *The stochastic process $\{X(t)\colon t \ge 0\}$ is said to be a regenerative process in continuous time provided that:*

*(i) there exists an increasing sequence $\{T_k\colon k \ge 0\}$ of stopping times such that $\{T_{k+1} - T_k\colon k \ge 0\}$ are independent and identically distributed; and*
*(ii) for every sequence of times $0 \le t_1 < t_2 < \cdots < t_m$ $(m \ge 1)$, $n \ge 1$, and $k \ge 0$, the vectors $\{X(T_0 + t_1), \ldots, X(T_0 + t_m), T_1 - T_0, \ldots, T_n - T_{n-1}\}$ and $\{X(T_k + t_1), \ldots, X(T_k + t_m), T_{k+1} - T_k, \ldots, T_{k+n} - T_{k+n-1}\}$ are identically distributed, and the processes $\{X(t)\colon 0 \le t < T_k;\ T_l - T_{l-1},\ 1 \le l \le k\}$ and $\{X(T_k + t)\colon t \ge 0;\ T_l - T_{l-1},\ l \ge k+1\}$ are independent.*

The random times $\{T_k\colon k \ge 0\}$ are said to be *regeneration points* (or *regeneration times*) for the process $\{X(t)\colon t \ge 0\}$, and $\{X(t)\colon T_{k-1} \le t < T_k\}$ is said to be the $k$th *cycle* of the process. The requirement that the $\{T_k\colon k \ge 0\}$ be stopping times for $\{X(t)\colon t \ge 0\}$ means that for any fixed $t$ the occurrence of a regeneration point prior to time $t$ (that is, $T_1 \le t$) may depend on the evolution of the process in the time interval $(0, t]$ but not beyond time $t$. Often in the context of discrete-event simulation when $\{X(t)\colon t \ge 0\}$ can be specified as a GSMP, $T_k$ is the time of the $\beta_k$th state transition for some random index $\beta_k$ $(T_k = \zeta_{\beta_k})$. We then require the random indices $\{\beta_k\colon k \ge 0\}$ to be stopping times for the underlying GSSMC $\{(S_n, C_n)\colon n \ge 0\}$.

The most familiar example of a regenerative process in continuous time is an irreducible positive recurrent continuous time Markov chain (CTMC) with a finite or countably infinite state space $S$. Recall that the given data of a CTMC is the infinitesimal generator $\boldsymbol{Q} = \{q_{ij}\colon i, j \in S\}$. The holding time in state $i \in S$ is exponentially distributed with rate parameter $q_i = -q_{ii}$, and the elements of the jump matrix $\boldsymbol{R} = \{r_{ij}\colon i, j \in S\}$ of the CTMC are

$$r_{ij} = \begin{cases} q_{ij}/q_i & \text{if } j \neq i \\ 0 & \text{if } j = i. \end{cases}$$

The element $r_{ij}$ of the matrix $\boldsymbol{R}$ represents the probability that the CTMC makes a state transition from $i$ to $j$. Also recall that an irreducible positive recurrent CTMC hits each state infinitely often with probability 1 and an irreducible CTMC with finite state space is necessarily positive recurrent. The successive random times at which there is a state transition to any fixed state $i \in S$ form a sequence of regeneration points for the process, and the expected time between regeneration points is finite.

**Example 1.2.** (Single-server queue) Assume that the sequence $\{(V_{n-1}, U_n)\colon n \geq 1\}$ of (service time, interarrival time) pairs consists of i.i.d. random vectors with finite mean. Also assume that the traffic intensity $\rho = \lambda/\mu$ is less than 1, where $E[U_n] = \lambda^{-1}$ and $E[V_n] = \mu^{-1}$. Let $Q(t)$ be the number of customers waiting or in service at time $t$. The successive random times at which there is an arrival to an empty system form a sequence of regeneration points for the process $\{Q(t)\colon t \geq 0\}$. This assertion says that, if $\rho < 1$, there exists an infinite sequence of random times at which the process $\{Q(t)\colon t \geq 0\}$ makes a state transition from 0 to 1 and the process $\{Q(t)\colon t \geq 0\}$ probabilistically restarts at any such time.

Now let $V(t)$ be the virtual waiting time at time $t$; that is, $V(t)$ is the waiting time of a hypothetical customer that arrives just after time $t$. Observe that the process $\{V(t)\colon t \geq 0\}$ does not have piecewise constant sample paths. The successive random times at which there is an arrival to an empty system form a sequence of regeneration points for the process $\{V(t)\colon t \ge 0\}$. Finally, let $W_n$ be the waiting time (exclusive of service time) of the $n$th customer, $n \ge 0$. The process $\{W_n\colon n \ge 0\}$ is a regenerative process in discrete time. The successive random indices for customers that experience waiting time equal to zero form a sequence of regeneration points for the process $\{W_n\colon n \ge 0\}$.

**Example 1.3.** (Single-server queue with scheduled arrivals) Customers arrive at a queueing system one at a time. The 0th customer arrives at time $T_0 = 0$, finds the server free, and experiences a service time distributed as a positive random variable $V_0$. The $n$th customer arrives at time $T_n = n + L_n$, where the random variables $\{L_n\colon n \ge 1\}$ are independent and have a common uniform distribution on $(-1/2, 1/2)$. The service time of the $n$th customer is distributed as a positive random variable $V_n$, and the random variables $\{V_n\colon n \ge 0\}$ are i.i.d. with $1/2 < E[V_0] < 1$. Denote by $W_n$ the waiting time experienced by customer $n$, $n \ge 0$. Then

$$\begin{aligned} W_n &= [W_{n-1} + V_{n-1} - (T_n - T_{n-1})]^+ \\ &= [W_{n-1} + V_{n-1} - (1 + L_n - L_{n-1})]^+ , \end{aligned}$$

where $[x]^+$ equals $x$ if $x \ge 0$ and equals 0 if $x < 0$. Since $W_{n+1}$ and $W_n$ both depend on $L_n$, the process $\{W_n\colon n \ge 0\}$ is not a regenerative process in discrete time.

**Example 1.4.** (Machine availability) Assume that service time at center $i$ is exponentially distributed (with rate parameter $\lambda_i$) as a random variable $V_i$. Let $X(t)$ be the number of customers waiting or in service at center 2 at time $t$. The process $\{X(t)\colon t \ge 0\}$ is a CTMC with state space $S = \{0, 1, \ldots, N\}$. The nonzero elements of the jump matrix $\boldsymbol{R}$ are $r_{01} = 1$, $r_{NN-1} = 1$, and

$$r_{ii+1} = 1 - r_{ii-1} = \frac{\lambda_1}{(\lambda_1 + \lambda_2)}$$

for $i = 1, 2, \ldots, N-1$. The elements of the vector $\boldsymbol{q} = (q_0, \ldots, q_N)$ of rate parameters for exponentially distributed holding times are $q_0 = \lambda_1$, $q_N = \lambda_2$, and $q_i = \lambda_1 + \lambda_2$ for $i = 1, 2, \ldots, N-1$.

Observe that in state 0 all $N$ customers are at center 1 and after an exponentially distributed (rate parameter $\lambda_1$) amount of time, there is a state transition to 1. Similarly, in state $N$ all customers are at center 2, and after an exponentially distributed (rate parameter $\lambda_2$) amount of time, there is a state transition to $N-1$. In state $i$ there is a customer in service at center 1 and a customer in service at center 2 ($1 \leq i \leq N-1$). The process $\{X(t)\colon\ t \geq 0\}$ makes a state transition from $i$ to $i+1$ if there is a service completion at center 1 before there is a service completion at center 2. Thus,

$$r_{ii+1} = 1 - r_{ii-1} = P\{V_1 \leq V_2\} = \frac{\lambda_1}{(\lambda_1 + \lambda_2)}.$$

The holding time in state $i$ is distributed as $\min(V_1, V_2)$. Since $V_1$ and $V_2$ are exponentially distributed and

$$\min(V_1, V_2) > t \text{ if and only if } V_1 > t \text{ and } V_2 > t$$

for all $t > 0$, the holding time in state $i$ is exponentially distributed with rate parameter $\lambda_1 + \lambda_2$.

The CTMC $\{X(t)\colon\ t \geq 0\}$ is irreducible positive recurrent since $S$ is finite and any state of the embedded jump chain $\{X(\zeta_n)\colon\ n \geq 0\}$ can be reached from any other state in some finite number of state transitions. It follows that the successive random times at which there is a state transition to any fixed state $s_0 \in S$ form a sequence of regeneration points for the process and the expected time between regeneration points is finite.

**Example 1.5.** (Machine availability) Assume that service time at center 1 is exponentially distributed as a random variable $V_1$ and service time at center 2 is distributed as a positive random variable $V_2$ with finite mean but otherwise arbitrary distribution. Let $X(t)$ be the number of customers waiting or in service at center 2 at time $t$. Observe that since service time at center 2 has an arbitrary, not necessarily exponential, distribution the process $\{X(t)\colon\ t \geq 0\}$ is not a CTMC. However, $\{X(t)\colon\ t \geq 0\}$ is a regenerative process in continuous time, and the expected time between regeneration points is finite.

Let $T_n$ be the time of the $n$th service completion at center 2, $n \geq 0$. The discrete time stochastic process $\{X(T_n)\colon\ n \geq 0\}$ records the state of the process $\{X(t)\colon\ t \geq 0\}$ at the random times at which there is a service completion at center 2. Fix $s_0'$ $(0 \leq s_0' < N)$. The successive random times at which $\{X(T_n)\colon\ n \geq 0\}$ hits state $s_0'$ form a sequence of regeneration points for the process $\{X(t)\colon\ t \geq 0\}$. To see this, consider the random times at which event $e_2$ = "service completion at center 2" occurs and there is a state transition from $s_0' + 1$ to $s_0'$. Following such a state transition, there is a start of service at center 1 if $s_0' = N - 1$. If $s_0' < N - 1$, as a consequence of the "memoryless property" of the exponential distribution the clock associated with the service in progress at center 1 can be viewed as having been reset probabilistically; no matter when the clock associated with event $e_1$ = "service completion at center 1" was set, the remaining time until event $e_1$ triggers a state transition is exponentially distributed with the same parameter. Likewise, if $s_0' > 0$, there is a start of service at center 2. Thus, the process $\{X(t)\colon\ t \geq 0\}$ probabilistically restarts whenever $\{X(T_n)\colon\ n \geq 0\}$ hits state $s_0'$.

We must show that $\{X(T_n)\colon\ n \geq 0\}$ hits state $s_0'$ infinitely often (i.o.) with probability 1:

$$P\{X(T_n) = s_0' \text{ i.o.}\} = 1.$$

The assumption that service time at center 1 is exponentially distributed implies that the process $\{X(T_n)\colon\ n \geq 0\}$ is a time-homogeneous discrete time Markov chain (DTMC) with finite state space $S - \{N\} = \{0, 1, \ldots, N-1\}$; that is,

$$\begin{aligned}P\{X(T_n) = s' | X(T_{n-1}), X(T_{n-2}), \ldots, X(T_0)\} \\ = P\{X(T_n) = s' | X(T_{n-1})\}\end{aligned}$$

for all $n \geq 0$ and $s' \in S - \{N\}$. Moreover, this "embedded Markov chain" is irreducible. Then, since the state space $S - \{N\}$ is finite, $\{X(T_n)\colon\ n \geq 0\}$ hits state $s_0'$ infinitely often with probability 1.

To show that the expected time between regeneration points is finite, observe that the process $\{(X(T_n), T_n)\colon\ n \geq 0\}$ is a Markov renewal process (MRP) with finite state space $S - \{N\}$; that is,

$$\begin{aligned}P\{X(T_{n+1}) = s', T_{n+1} - T_n \leq t | X(T_n), X(T_{n-1}), \ldots, X(T_0); \\ T_n, T_{n-1}, \ldots, T_0\} \\ = P\{X(T_{n+1}) = s', T_{n+1} - T_n \leq t | X(T_n)\}\end{aligned}$$

for all $n \geq 0$, $s' \in S - \{N\}$, and $t \geq 0$. The process $\{Y(t)\colon\ t \geq 0\}$ defined by

$$Y(t) = X(T_n) \text{ for } T_n \leq t < T_{n+1}$$

is the associated semi-Markov process (SMP). The key observation is that the mean recurrence time for state $s_0'$ is equal to the expected time between regeneration points for the process $\{X(t)\colon\ t \geq 0\}$. The mean recurrence time for state $s_0'$ is finite since the SMP is irreducible and has a finite state space.

**Example 1.6.** (Machine availability) Assume that service time at center $i$ is distributed as a positive random variable $V_i$ with finite mean and absolutely continuous but otherwise arbitrary distribution. Let $X(t)$ be the number of customers waiting or in service at center 2 at time $t$. Take event $e_i =$

"service completion at center $i$." The process $\{X(t)\colon t \geq 0\}$ probabilistically restarts whenever (event $e_2$ occurs and) there is a state transition from $N$ to $N-1$. Observe that whenever event $e_2$ triggers a state transition from $N$ to $N-1$ there are no old events ($O(N-1; N, e_2) = \emptyset$) and there is a start of service at center 1 and at center 2 ($N(N-1; N, e_2) = \{e_1, e_2\}$).

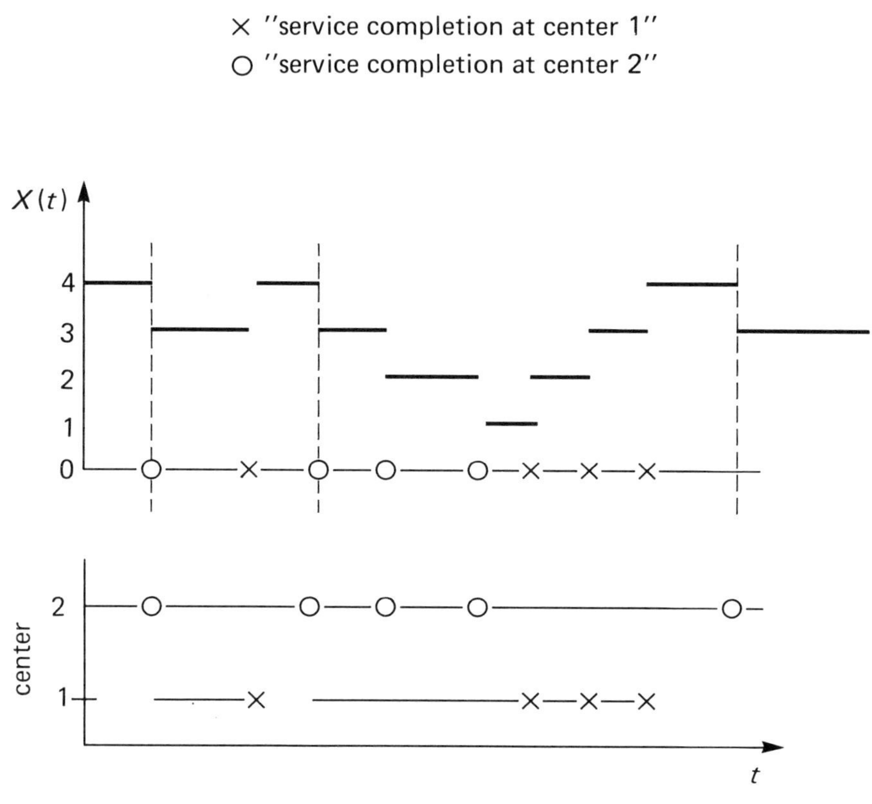


**Figure 2.1.** Regenerative cycle for machine availability.

Let $T_n$ be the time of the $n$th service completion at center 2, $n \geq 0$. The successive random times at which $\{X(T_n)\colon n \geq 0\}$ hits state $N-1$ form a sequence of regeneration points for the process $\{X(t)\colon t \geq 0\}$, provided that $\{X(T_n)\colon n \geq 0\}$ hits state $N-1$ infinitely often with probability 1 ($P\{X(T_n) = N-1 \text{ i.o.}\} = 1$); see Figure 2.1. Additional conditions on the service times are needed to ensure that $P\{X(T_n) = N-1 \text{ i.o.}\} = 1$. To see this, suppose that $P\{V_1 > V_2\} = 1$. Then, beginning in any state and with any clock-reading vector, $\{X(T_n)\colon n \geq 0\}$ hits state $N-1$ only a finite number of times.

**Example 1.7.** (Patrolling repairman) Assume that the walking time from machine $j$ to the next machine is a positive constant $W_j$ and the time to repair and restart machine $j$ is distributed as a positive random variable $R_j$ with finite mean. Also assume that the lifetime for machine $j$ is exponentially distributed as a random variable $L_j$. Set

$$X(t) = (Z_1(t), \ldots, Z_N(t), M(t), N(t)), \tag{1.8}$$

where

$$Z_j(t) = \begin{cases} 1 & \text{if machine } j \text{ is waiting for repair at time } t \\ 0 & \text{otherwise,} \end{cases}$$

$$M(t) = \begin{cases} j & \text{if machine } j \text{ is under repair at time } t \\ 0 & \text{if no machine is under repair at time } t, \end{cases}$$

$N(t) = 1$ if machine $N$ is under repair or the repairman is walking to machine 1 at time $t$, and $N(t) = j$ if machine $j-1$ is under repair or the repairman is walking to machine $j$ at time $t$ ($j = 2, 3, \ldots, N$). Take event $e_{N+2}$ = "arrival of repairman at machine," $e_{N+1}$ = "end of repair," and $e_j$ = "stoppage of machine $j$" ($j = 1, 2, \ldots, N$). Recall from Example 4.9 of Chapter 1 that formal specification of the process $\{X(t)\colon t \geq 0\}$ is as a GSMP with finite state space

$$\begin{aligned} S = \big\{&(z_1, \ldots, z_N, m, n) \in \{0,1\}^N \times \{0, 1, \ldots, N\} \times \\ &\{1, 2, \ldots, N\}\colon\ n = m+1 \text{ when } 0 < m < N;\ n = 1 \\ &\text{when } m = N;\ m = j \text{ only if } z_j = 0 \text{ for } j = 1, 2, \ldots, N\big\} \end{aligned}$$

and event set $E = \{e_1, e_2, \ldots, e_{N+2}\}$.

The process $\{X(t)\colon t \geq 0\}$ is a regenerative process in continuous time, and the expected time between regeneration points is finite. To see this, let $T_n$ be the time of the $n$th state transition triggered by event $e_{N+2}$ = "arrival of repairman at machine," $n \geq 0$. Take $s_0 = (1, 0, \ldots, 0, 0, 1)$ and $s_0' = (0, \ldots, 0, 1, 2)$. The successive random times at which (event $e_{N+2}$ occurs and) there is a state transition from $s_0$ to $s_0'$ form a sequence of regeneration points for the process $\{X(t)\colon t \geq 0\}$. These random times correspond to starts of repair for machine 1 with machines $2, 3, \ldots, N$ running. Observe that when event $e_{N+2}$ occurs and there is a state transition from $s_0$ to $s_0'$, the set $N(s_0'; s_0, e_{N+2}) = \{e_{N+1}\}$, the clock associated with new event $e_{N+1}$ is set according to $F(\,\cdot\,; s_0', e_{N+1}, s_0, e_{N+2})$, and the old clock reading is kept for each event $e' \in O(s_0'; s_0, e_{N+2}) = \{e_2, e_3, \ldots, e_N\}$. By the memoryless property of the exponential distribution, each of the clocks running at time $T_n-$ can be viewed as having been reset probabilistically at time $T_n$; no matter when the clock associated with event $e' \in O(s_0'; s_0, e_{N+2})$ was set, the remaining time until event $e'$ is scheduled to trigger a state transition is exponentially distributed with the same parameter. Thus, the process $\{X(t)\colon t \geq 0\}$ probabilistically restarts every time $\{X(T_n)\colon n \geq 0\}$ hits state $s_0'$. The assumption that each $L_j$ is exponentially distributed implies that $\{X(T_n)\colon n \geq 0\}$ is an irreducible DTMC with a finite state space, and therefore $P\{X(T_n) = s_0' \text{ i.o.}\} = 1$. An argument using a Markov renewal process as in Example 1.5 shows that the expected time between regeneration points is finite.

The successive random times at which event $e_{N+1}$ = "end of repair" triggers a state transition from $s_0 = (0, \ldots, 0, N, 1)$ to $s_0' = (0, \ldots, 0, 1)$ also form a sequence of regeneration points for the process $\{X(t)\colon t \geq 0\}$. These random times correspond to ends of repair for machine $N$ with all other machines running.

**Example 1.9.** (Shuttle train) Consider a train that shuttles back and forth between two stations and has seats for $K$ ($\geq 1$) passengers. Passengers that wish to board the train and ride to the other station arrive at station $i$ according to a Poisson process of rate $\lambda_i$, queue, and subsequently board the train in the order in which they arrive at the station. Assume that passengers board the train instantaneously and all passengers that board at station $i$ disembark instantaneously when the train arrives at the other station. The time required for the train to travel from station $i$ to the other station is distributed as a positive random variable $L_i$ with finite mean. Travel times are independent of interarrival times of passengers, and the arrival process of passengers for boarding at station 1 is independent of the arrival process of passengers for boarding at station 2. Assume that $(\lambda_1 + \lambda_2)(E[L_1] + E[L_2]) < K$.

Set

$$X(t) = (K_1(t), K_2(t), B(t), D(t)),$$

where $K_i(t)$ is the number of passengers waiting to board the train at station $i$ at time $t$, $B(t)$ is the number of passengers on board at time $t$, and

$$D(t) = \begin{cases} 1 & \text{if the train is moving to station 1 at time } t \\ 2 & \text{if the train is moving to station 2 at time } t. \end{cases}$$

Take event $e_3$ = "arrival of train at station" and $e_i$ = "arrival of passenger for boarding at station $i$" ($i = 1, 2$). Formal specification of the process $\{X(t)\colon t \geq 0\}$ is as a GSMP with a countably infinite state space $S$ and event set $E = \{e_1, e_2, e_3\}$.

The process $\{X(t)\colon t \geq 0\}$ is a regenerative process in continuous time, and the expected time between regeneration points is finite. Let $T_n$ be the time of the $n$th departure of the the train from station 1, $n \geq 0$. Take $s_0' = (0, 0, b_0, 2) \in S$. The successive random times at which $\{X(T_n)\colon n \geq 0\}$ hits state $s_0'$ form a sequence of regeneration points for the process $\{X(t)\colon t \geq 0\}$. To see this, consider the random times at which the train departs from station 1 and there is a state transition to $s_0'$. The only clocks that are active at such a time can be viewed as just having been reset probabilistically since $O(s_0'; s^*, e_3) = \{e_1, e_2\}$ for all $s^* \in S$; no matter when the clock associated with event $e' \in O(s_0'; s^*, e_3)$ was set, the remaining time until event $e'$ is scheduled to trigger a state transition is exponentially distributed with the same parameter. The joint distribution of the state just after event $e_3$ occurs and the clock (associated with event $e_3$) set at this time depends on the past history of $\{X(t)\colon\ t \geq 0\}$ only through $s_0'$, the previous state $s^*$, and the trigger event $e_3$. Since the new event and clock-setting distribution are the same for all $s^*$, the process $\{X(t)\colon\ t \geq 0\}$ probabilistically restarts whenever event $e_3$ triggers a state transition to $s_0'$.

To see that $P\{X(T_n) = s_0' \text{ i.o.}\} = 1$, observe that the interarrival times for passengers for boarding at station $i$ are exponentially distributed with rate parameter $\lambda_i$. These assumptions imply that $\{X(T_n)\colon\ n \geq 0\}$ is an irreducible DTMC with countably infinite state space $G = \{0, 1, \ldots, \}^2 \times \{0, 1, \ldots, K\} \times \{2\}$. Foster's criteria can be used to show that this chain $\{X(T_n)\colon\ n \geq 0\}$ is positive recurrent. Set

$$G_0 = \{(k_1, k_2, b, 2) \in G\colon\ 0 \leq k_1 < K, 0 \leq k_2 < K\}$$

and observe that $G_0$ is a finite subset of $G$. It is sufficient to show that for some nonnegative real-valued function $g$ with domain $G$ and some $\epsilon > 0$,

$$E[g(X(T_{n+1}))|X(T_n) = s] < \infty$$

for $s \in G_0$ and

$$E[g(X(T_{n+1}))|X(T_n) = s] \leq g(s) - \epsilon$$

for $s \in G - G_0$. Set

$$g(s) = \max(k_1, k_2)$$

for $s = (k_1, k_2, b, 2) \in G$. Denote by $N_i(L)$ the (random) number of passengers that arrive for boarding at station $i$ in a time interval of random length $L$, where $L$ is a positive random variable. Take $\epsilon = K - E[\max(N_1(L_1 + L_2), N_2(L_1 + L_2))]$, and observe that

$$\begin{aligned}\epsilon &= K - E[\max(N_1(L_1 + L_2), N_2(L_1 + L_2)]\\ &\geq K - E[N_1(L_1 + L_2) + N_2(L_1 + L_2)]\\ &= K - (\lambda_1 + \lambda_2)(E[L_1] + E[L_2])\\ &> 0\end{aligned}$$

by assumption. If $s \in G_0$, then

$$\begin{aligned}E[g(X(T_{n+1}))|X(T_n) = s] \leq E[\max(N_1(L_1 + L_2),\\ N_2(L_1 + L_2))] < \infty.\end{aligned}$$

If $s \in G - G_0$ with $k_1, k_2 \geq K$, then

$$\begin{aligned}&E[g(X(T_{n+1}))|X(T_n) = s]\\ &\quad = E[\max(k_1 - K + N_1(L_1 + L_2), k_2 - K + N_2(L_1 + L_2))]\\ &\quad \leq E[\max(\max(k_1, k_2) - K + N_1(L_1 + L_2), \max(k_1, k_2)-\\ &\qquad\quad K + N_2(L_1 + L_2))]\\ &\quad = E[\max(k_1, k_2) + \max(N_1(L_1 + L_2), N_2(L_1 + L_2)) - K]\\ &\quad = g(s) - \epsilon.\end{aligned}$$

If $s \in G - G_0$ with $k_1 < K$ and $k_2 \geq K$, then

$$\begin{aligned}&E[X(T_{n+1}))|X(T_n) = s]\\ &\quad = E[\max((k_1 - K + N_1(L_1 + L_2))^+, k_2 - K+\\ &\qquad\quad N_2(L_1 + L_2))]\\ &\quad \leq E[\max(\max(k_1, k_2) - K + N_1(L_1 + L_2), \max(k_1, k_2)-\\ &\qquad\quad K + N_2(L_1 + L_2))]\\ &\quad = E[\max(k_1, k_2) + \max(N_1(L_1 + L_2), N_2(L_1 + L_2)) - K]\\ &\quad = g(s) - \epsilon.\end{aligned}$$

Finally, if $s \in G - G_0$ with $k_1 \geq K$ and $k_2 < K$, then

$$\begin{aligned}
&E[X(T_{n+1}))|X(T_n) = s] \\
&\quad = E[\max(k_1 - K + N_1(L_1 + L_2), (k_2 - K + N_2(L_1))^+ \\
&\qquad + N_2(L_2))] \\
&\quad \leq E[\max(k_1 - K + N_1(L_1 + L_2), k_1 - K + N_2(L_1) \\
&\qquad + N_2(L_2))] \\
&\quad \leq E[\max(k_1, k_2) + \max(N_1(L_1 + L_2), N_2(L_1 + L_2)) - K] \\
&\quad = g(s) - \epsilon.
\end{aligned}$$

Then, by Foster's criteria, the chain $\{X(T_n)\colon\ n \geq 0\}$ is positive recurrent, and $P\{X(T_n) = s_0' \text{ i.o.}\} = 1$.

We now show that the expected time between regeneration points is finite. Let $T_{\beta_k}$ be the $k$th time at which the process $\{X(T_n)\colon\ n \geq 0\}$ hits state $s_0'$. Set $\alpha_k = \beta_k - \beta_{k-1}$, and observe that $\alpha_k$ is the number of times that the train travels from station $i$ to the other station in the time interval $[T_{\beta_{k-1}}, T_{\beta_k})$. Denote by $L_{i,m,k}$ the $m$th such time for the train to travel from station $i$ to the other station. Also observe that $E[\alpha_k] < \infty$ since the chain $\{X(T_n)\colon\ n \geq 0\}$ is positive recurrent. Then

$$T_{\beta_k} - T_{\beta_{k-1}} = \sum_{m=1}^{\alpha_k} (L_{1,m,k} + L_{2,m,k})$$

and, applying Wald's identity,

$$\begin{aligned}
E[T_{\beta_k} - T_{\beta_{k-1}}] &= E\Big[\sum_{m=1}^{\alpha_k} (L_{1,m,k} + L_{2,m,k})\Big] \\
&= E[\alpha_k](E[L_1] + E[L_2]) \\
&< \infty.
\end{aligned}$$

It can be difficult to determine assumptions on clock-setting distributions that are sufficient to ensure that a GSMP hits a fixed state infinitely often with probability 1.

**Example 1.10.** (Patrolling repairman) Assume that the walking time from machine $j$ to the next machine is a positive constant $W_j$ and the time to repair and restart machine $j$ is distributed as a positive random variable $R_j$ with finite mean. Also assume that the lifetimes for machine $j$ are i.i.d. as a positive random variable $L_j$ with finite mean and absolutely continuous distribution so that, with probability 1, no two stoppages occur simultaneously. Set $X(t) = (Z_1(t), \ldots, Z_N(t), M(t), N(t))$ as in (1.8).

Let $T_n$ be the time of the $n$th arrival of the repairman at machine 1, $n \geq 0$. Take $s_0' = (0, 1, \ldots, 1, 1, 2)$, so that $X(T_n) = s_0'$ if machines $2, 3, \ldots, N$ are stopped and the repairman starts repair of machine 1 at time $T_n$. Observe that there is a state transition to $s_0'$ when event $e_{N+2}$ = "arrival of repairman at machine" occurs in state $s_0 = (1, \ldots, 1, 0, 1)$. The process $\{X(t)\colon t \geq 0\}$ probabilistically restarts whenever (event $e_{N+2}$ occurs and) there is a state transition from $s_0$ to $s_0'$. The set $N(s_0'; s_0, e_{N+2}) = \{e_{N+1}\}$, the clock for new event $e_{N+1}$ is set according to $F(x; s_0', e_{N+1}, s_0, e_{N+2}) = P\{R_1 \leq x\}$, and there are no old clocks running ($O(s_0'; s_0, e_{N+2}) = \emptyset$). The successive random times at which $\{X(T_n)\colon n \geq 0\}$ hits state $s_0'$ form a sequence of regeneration points for the process $\{X(t)\colon t \geq 0\}$ provided that $P\{X(T_n) = s_0' \text{ i.o.}\} = 1$.

Additional conditions on the lifetimes, repair times and walking times are needed to ensure that $P\{X(T_n) = s_0' \text{ i.o.}\} = 1$. To see this, suppose that the time $R_j$ to repair machine $j$ is a positive constant, the lifetimes for machine $j$ are identically distributed as a positive random variable $L_j$ with finite mean, and the walking time from machine $j$ to the next machine is a positive constant $W_j$. Also suppose that $P\big\{W_j < L_j < W_j + R_{j+1}\big\} = 1$. Then, starting from state $s_0'$ at time $T_0$, $X(T_n) = (0, 1, 1, \ldots, 1, 0, 1, 2)$ for all $n \geq 1$. The following events occur in sequence: "end of repair," "arrival of repairman at machine," "stoppage of machine 1," "end of repair," "arrival of repairman at machine," "stoppage of machine 2," ..., "end of repair," "arrival of repairman at machine," "stoppage of machine $N$," "end of repair," "arrival of repairman at machine," "stoppage of machine 1," .... The resulting sequence of states is $(0, 1, 1, \ldots, 1, 0, 2)$, $(0, 0, 1, \ldots, 1, 2, 3)$, $(1, 0, 1, \ldots, 1, 2, 3)$, $(1, 0, 1, \ldots, 1, 0, 3)$, $(1, 0, 0, 1, \ldots, 1, 3, 4)$, $(1, 1, 0, 1, \ldots, 1, 3, 4)$, $(1, 1, 0, 1, \ldots, 1, 0, 4), \ldots,$ $(1, 1, \ldots, 1, 0, 0, 1)$, $(0, 1, 1, \ldots, 1, 0, 1, 2)$, $(0, 1, 1, \ldots, 1, 1, 2)$, $(0, 1, 1, \ldots, 1, 0, 2)$, $(0, 0, 1, \ldots, 1, 2, 3) \ldots$.

**Example 1.11.** (Producer/consumer with infinite buffers) Consider a system consisting of two producers, two consumers, and two buffers, each numbered 1 and 2. The producers share a single channel for transmission, one at a time, of items to consumers. Producer $i$ creates items for consumer $i$ one at a time; items created, but not yet transmitted, are placed in buffer $i$ for transmission. Items created by producer $i$ are transmitted in the order in which they are created. Assume that producer/consumer pair 1 has nonpreemptive priority over producer/consumer pair 2 for use of the channel. Also assume that the times required by producer $i$ to create an item are independent and exponentially distributed with rate parameter $\lambda_i$ and the time to transmit an item from buffer $i$ to consumer $i$ is distributed as a positive random variable $L_i$ with finite mean. All times to create an item and times to transmit an item are mutually independent.

Set

$$X(t) = \big(B_1(t), B_2(t), M(t)\big),$$

where $B_i(t)$ is the number of items in buffer $i$ at time $t$ and

$$M(t) = \begin{cases} 2 & \text{if producer 2 is transmitting at time } t \\ 1 & \text{if producer 1 is transmitting at time } t \\ 0 & \text{if neither producer is transmitting at time } t \,. \end{cases}$$

Take event $e_3$ = "end of transmission to consumer" and $e_i$ = "creation of item by producer $i$" $(i = 1, 2)$. Formal specification of the process $\{X(t)\colon\ t \geq 0\}$ is as a GSMP with countably infinite state space

$$S = \{(b_1, b_2, m) \in \{0, 1, \ldots\}^2 \times \{0, 1, 2\} : \ m = 0 \text{ only if } b_1 = b_2 = 0\}$$

and event set $E = \{e_1, e_2, e_3\}$. For $s = (b_1, b_2, m) \in S$, event $e_3$ = "end of transmission to consumer" $\in E(s)$ if and only if $m > 0$; event $e_i$ = "creation of item by producer $i$" $\in E(s)$ for all $s$. All speeds $r(s, e)$ for active events are equal to 1.

If $e = e_3$ = "end of transmission to consumer," then the state transition probability $p(s'; s, e) = 1$ when

$$s = (b_1, b_2, m) \text{ with } b_1 > 0 \text{ and } s' = (b_1 - 1, b_2, 1),$$

when

$$s = (0, b_2, m) \text{ with } b_2 > 0 \text{ and } s' = (0, b_2 - 1, 2),$$

and when

$$s = (0, 0, m) \text{ and } s' = (0, 0, 0).$$

If $e = e_1$ = "creation of item by producer 1," then $p(s'; s, e) = 1$ when

$$s = (0, 0, 0) \text{ and } s' = (0, 0, 1)$$

and when

$$s = (b_1, b_2, m) \text{ with } m > 0 \text{ and } s' = (b_1 + 1, b_2, m).$$

If $e = e_2$ = "creation of item by producer 2," then $p(s'; s, e) = 1$ when

$$s = (0, 0, 0) \text{ and } s' = (0, 0, 2)$$

and when

$$s = (b_1, b_2, m) \text{ with } m > 0 \text{ and } s' = (b_1, b_2 + 1, m).$$

All other state transition probabilities are equal to 0.

For $s = (b_1, b_2, m) \in S$, the set $N(s'; s, e_3)$ of new events when event $e_3$ triggers a state transition from $s$ to $s'$ equals $\{e_3\}$ if $b_1 > 0$ or $b_2 > 0$ and equals $\emptyset$ otherwise. The set $N(s'; s, e_1)$ equals $\{e_3, e_1\}$ if $s = (0,0,0)$ and equals $\{e_1\}$ otherwise. The set $N(s'; s, e_2)$ equals $\{e_3, e_2\}$ if $s = (0,0,0)$ and equals $\{e_2\}$ otherwise. The clock-setting distribution functions are as follows. If the new event $e' = e_3 =$ "end of transmission to consumer" and $s' = (b_1', b_2', i)$, then $F(x; s', e', s, e^*) = P\{L_i \le x\}$ for $x \ge 0$. If $e' = e_i =$ "creation of item by producer $i$", then $F(x; s', e', s, e^*) = 1 - e^{-\lambda_i x}$ for $x \ge 0$.

Let $T_n$ be the time of the $n$th state transition triggered by event $e_3 =$ "end of transmission to consumer," $n \ge 0$. Select $s_0' \in S - \{(b_1, b_2, 2) \in S\colon\ b_1 \ge 1\}$, and consider the successive random times at which $\{X(T_n)\colon\ n \ge 0\}$ hits state $s_0'$. For example, take $s_0' = (0,0,0)$. When event $e_3$ occurs in (0,0,1) or in (0,0,2), the process $\{X(t)\colon\ t \ge 0\}$ makes a state transition to $s_0'$ $(X(T_n) = s_0')$. Following the state transition, both buffers are empty. When $X(T_n) = s_0'$, events $e_1$ and $e_2$ are old events and event $e_3$ is a new event (unless $s_0' = (0,0,0)$, in which case there are no new events). Since the clock-setting distribution functions $F(\,\cdot\,; s', e_1, s, e)$ and $F(\,\cdot\,; s', e_2, s, e)$ are exponential with rate parameter independent of $s'$, $s$, and $e$ and the clock-setting distribution function $F(\,\cdot\,; s_0', e_3, s, e_3) = F(\,\cdot\,; s_0', e_3)$ for all $s$, the process $\{X(t)\colon\ t \ge 0\}$ probabilistically restarts whenever $\big\{X(T_n)\colon\ n \ge 0\big\}$ hits state $s_0'$. Additional conditions on the times to create an item and the times to transmit an item are needed to ensure that $P\{X(T_n) = s_0' \text{ i.o.}\} = 1$.

A GSMP need not be a regenerative process in continuous time in order to exhibit stable limiting behavior.

**Example 1.12.** (Two-server cyclic queues) Consider a queueing system with two centers, each having two servers, that provides service to a fixed number $N$ $(\ge 2)$ of customers. After service completion at center 1, a customer moves instantaneously from center 1 to center 2. After service completion at center 2, a customer moves back to center 1. If servers 1 and 2 at center $i$ are idle when a customer joins the queue at center $i$, server 1 serves the customer. The queueing discipline at each center is FCFS. Assume that all service times are mutually independent and service times at center $i$ are identically distributed as a positive random variable $V_i$. Also assume that the distribution of each $V_i$ has finite mean and a density function that is continuous and positive on $(0, +\infty)$.

Set

$$X(t) = (Z_1(t), S_{11}(t), S_{12}(t), S_{21}(t), S_{22}(t)),$$

where $Z_1(t)$ is the number of customers waiting or in service at center 1 at time $t$ and

$$S_{ij}(t) = \begin{cases} 1 & \text{if server } j \text{ at center } i \text{ is busy at time } t \\ 0 & \text{if server } j \text{ at center } i \text{ is idle at time } t. \end{cases}$$

Take event $e_{ij}$ = "service completion by server $j$ at center $i$." Formal specification of the process $\{X(t)\colon\ t \geq 0\}$ is as a GSMP with finite state space

$$S = \{(z_1, s_{11}, s_{12}, s_{21}, s_{22}) \in \{0, 1, \ldots, N\} \times \{0, 1\}^4 : \\ s_{11} + s_{12} = 2 \wedge z_1; s_{21} + s_{22} = 2 \wedge (N - z_1)\}$$

and event set $E = \{e_{11}, e_{12}, e_{21}, e_{22}\}$. (Recall that $x \wedge y$ equals $x$ if $x \leq y$ and equals $y$ otherwise.) For $s = (z_1, s_{11}, s_{12}, s_{21}, s_{22}) \in S$, event $e_{ij} \in E(s)$ if and only if $s_{ij} = 1$. All speeds for active events are equal to 1.

If $e = e_{12}$, then the state transition probability $p(s'; s, e) =$ 1 when

$$s = (z_1, s_{11}, 1, 0, s_{22}) \text{ with } z_1 = 1 \text{ or } 2 \\ \text{and } s' = (z_1 - 1, s_{11}, 0, 1, s_{22}),$$

when

$$s = (z_1, s_{11}, 1, 0, s_{22}) \text{ with } z_1 \geq 3 \quad \text{and } s' = (z_1 - 1, s_{11}, 1, 1, s_{22}),$$

when

$$s = (z_1, s_{11}, 1, 1, s_{22}) \text{ with } z_1 = 1 \text{ or } 2 \quad \text{and } s' = (z_1 - 1, s_{11}, 0, 1, 1),$$

and when

$$s = (z_1, s_{11}, 1, 1, s_{22}) \text{ with } z_1 \geq 3 \quad \text{and } s' = (z_1 - 1, s_{11}, 1, 1, 1).$$

If $e = e_{11}$, then $p(s'; s, e) = 1$ when

$$s = (z_1, 1, s_{12}, 0, s_{22}) \text{ with } z_1 = 1 \text{ or } 2 \quad \text{and } s' = (z_1 - 1, 0, s_{12}, 1, s_{22}),$$

when

$$s = (z_1, 1, s_{12}, 0, s_{22}) \text{ with } z_1 \geq 3 \quad \text{and } s' = (z_1 - 1, 1, s_{12}, 1, s_{22}),$$

when

$$s = (z_1, 1, s_{12}, 1, s_{22}) \text{ with } z_1 = 1 \text{ or } 2 \quad \text{and } s' = (z_1 - 1, 0, s_{12}, 1, 1),$$

and when

$$s = (z_1, 1, s_{12}, 1, s_{22}) \text{ with } z_1 \geq 3 \quad \text{and } s' = (z_1 - 1, 1, s_{12}, 1, 1).$$

If $e = e_{22}$, then $p(s'; s, e) = 1$ when

$$s = (z_1, 0, s_{12}, s_{21}, 1) \text{ with } z_1 \geq N - 2 \quad \text{and } s' = (z_1 + 1, 1, s_{12}, s_{21}, 0),$$

when

$$s = (z_1, 0, s_{12}, s_{21}, 1) \text{ with } z_1 < N - 2 \quad \text{and } s' = (z_1 + 1, 1, s_{12}, s_{21}, 1),$$

when

$$s = (z_1, 1, s_{12}, s_{21}, 1) \text{ with } z_1 \geq N - 2$$
$$\text{and } s' = (z_1 + 1, 1, 1, s_{21}, 0),$$

and when

$$s = (z_1, 1, s_{12}, s_{21}, 1) \text{ with } z_1 < N - 2$$
$$\text{and } s' = (z_1 + 1, 1, 1, s_{21}, 1).$$

If $e = e_{21}$, then $p(s'; s, e) = 1$ when

$$s = (z_1, 0, s_{12}, 1, s_{22}) \text{ with } z_1 \geq N - 2$$
$$\text{and } s' = (z_1 + 1, 1, s_{12}, 0, s_{22}),$$

when

$$s = (z_1, 0, s_{12}, 1, s_{22}) \text{ with } z_1 < N - 2$$
$$\text{and } s' = (z_1 + 1, 1, s_{12}, 1, s_{22}),$$

when

$$s = (z_1, 1, s_{12}, 1, s_{22}) \text{ with } z_1 \geq N - 2$$
$$\text{and } s' = (z_1 + 1, 1, 1, 0, s_{22}),$$

and when

$$s = (z_1, 1, s_{12}, 1, s_{22}) \text{ with } z_1 < N - 2$$
$$\text{and } s' = (z_1 + 1, 1, 1, 1, s_{22}).$$

All other state transition probabilities $p(s'; s, e)$ are equal to 0.

For $s, s' \in S$ and $e^* \in E(s)$, the sets $N(s'; s, e^*)$ of new events when event $e^*$ triggers a state transition from $s$ to $s'$ are as follows. Let $s = (z_1, s_{11}, s_{12}, s_{21}, s_{22}) \in S$ and set $z_2 = N - z_1$. For $e^* = e_{i1}$ $(i = 1, 2)$ and $k = 3 - i$, the event $e_{i1} \in N(s'; s, e^*)$ if and only if $z_i \geq 3$; $e_{k1} \in N(s'; s, e^*)$ if and only if $s_{k1} = 0$; and $e_{k2} \in N(s'; s, e^*)$ if and only if $s_{k1} = 1$ and $s_{k2} = 0$. For $e^* = e_{i2}$ and $k = 3 - i$, the event $e_{i2} \in N(s'; s, e^*)$ if and only if $z_i \geq 3$; $e_{k1} \in N(s'; s, e^*)$ if and only if $s_{k1} = 0$; and $e_{k2} \in N(s'; s, e^*)$ if and only if $s_{k1} = 1$ and $s_{k2} = 0$. The clock-setting distribution functions are as follows. If the new event $e' = e_{ij}$, then $F(x; s', e', s, e^*) = P\{V_i \leq x\}$.

Theorem 4.14 of Chapter 1 implies that $X(t) \Rightarrow X$ as $t \to \infty$. However without further restrictions on the distributions of service time, the process $\{X(t)\colon\ t \geq 0\}$ is not a regenerative process in continuous time. Since $N \geq 2$ in any state there are at least two active events; that is, there are always at least two clocks running. The assumptions on the distributions of service time imply that, with probability 1, no two events occur simultaneously. It follows that for any increasing sequence $\{T_k\colon\ k \geq 0\}$ of stopping times the processes $\{X(t)\colon\ t < T_k\}$ and $\{X(T_k + t)\colon\ t \geq 0\}$ are not independent.

**Exercises**

1.1. (Shuttle train with finite waiting room) Consider a train that shuttles back and forth between two stations and has seats for $K$ ($\geq 1$) passengers. Passengers that wish to board the train and ride to the other station arrive at station $i$ according to a renewal process, queue, and subsequently board the train in the order in which they arrive at the station. Assume that passengers board the train instantaneously and all passengers that board at station $i$ disembark instantaneously when the train arrives at the other station. At each station passengers disembark before any passengers board the train. The waiting room at station $i$ has finite capacity $B_i$ ($\geq 1$). A passenger that wishes to board the train at station $i$ and arrives when there are already $B_i$ passengers waiting to board is turned away. The time required for the train to travel from one station to the other is distributed as a positive random variable $L$ with finite mean. The successive interarrival times at station $i$ are i.i.d. as a positive random variable $A_i$ with finite mean and absolutely continuous distribution. Travel times are independent of interarrival times of passengers, and the arrival process of passengers for boarding at station 1 is independent of the arrival process of passengers for boarding at station 2.

Set

$$X(t) = (K_1(t), K_2(t), B(t), D(t)),$$

where $K_i(t)$ is the number of passengers waiting to board the train at station $i$ at time $t$, $B(t)$ is the number of passengers on board at time $t$, and

$$D(t) = \begin{cases} 1 & \text{if the train is moving to station 1 at time } t \\ 2 & \text{if the train is moving to station 2 at time } t. \end{cases}$$

Is $\{X(t)\colon t \geq 0\}$ a regenerative process in continuous time? Explain your answer.

## 2.2. Limit Theorems

The "regenerative method" for analysis of simulation output discussed in Chapter 3 rests on structural properties of regenerative stochastic processes. Let $\{X(t)\colon t \geq 0\}$ be a regenerative process in continuous time with state space $S$ as in Definition 1.1 and $\{T_k\colon k \geq 0\}$ be a sequence of regeneration points. Set $\tau_k = T_k - T_{k-1}$ for $k \geq 1$, and denote the distribution function of $\tau_1$ by $F$. Also, for a real-valued measurable function $f$ with domain $S$, set $T_{-1} = 0$ and

$$Y_k(f) = \int_{T_{k-1}}^{T_k} f\bigl(X(u)\bigr)\, du$$

for $k \geq 0$. For a regenerative process in discrete time, set

$$Y_k(f) = \sum_{n=T_{k-1}}^{T_k - 1} f(X_n).$$

**Theorem 2.1.** *The sequence* $\{(Y_k(f), \tau_k)\colon\ k \geq 1\}$ *consists of independent and identically distributed random vectors.*

This result follows directly from Definition 1.1. Theorems 2.2 and 2.3 give a law of large numbers and a central limit theorem (CLT) for regenerative processes.

**Theorem 2.2.** *Assume that* $E\big[\tau_1\big] < \infty$. *Then*

$$\lim_{t\to\infty} \frac{1}{t} \int_0^t f\big(X(u)\big)\, du = \frac{E\big[Y_1(f)\big]}{E\big[\tau_1\big]} \stackrel{\text{def}}{=} r(f) \text{ a.s.}$$

*for any real-valued measurable function* $f$ *such that* $Y_0(|f|) < \infty$ *a.s. and* $E[Y_1(|f|)] < \infty$.

Denote the indicator function of a set $A$ by $1_A$; $1_A(x)$ equals 1 if $x \in A$ and equals 0 otherwise. Theorem 2.2 implies that

$$\frac{1}{t} \int_0^t 1_A\big(X(u)\big)\, du \to \frac{E\left[\int_{T_0}^{T_1} 1_A\big(X(u)\big)\, du\right]}{E\big[T_1 - T_0\big]} \quad \text{a.s.}$$

for all $A \subseteq S$ provided that $E\big[T_1 - T_0\big] < \infty$.

**Theorem 2.3.** *Assume that* $E[\tau_1] < \infty$. *If Var* $[Y_1(f)] < \infty$ *and Var* $[\tau_1] < \infty$, *then*

$$\frac{\int_0^t f(X(u))du - r(f)t}{\sqrt{t}\,\sigma(f)/\sqrt{E[\tau_1]}} \Rightarrow N(0,1)$$

*as* $t \to \infty$, *where*

$$\sigma^2(f) \stackrel{\text{def}}{=} \textit{Var}\,[Y_1(f)] - 2r(f)\ \textit{Cov}\,[Y_1(f), \tau_1] + r^2(f)\ \textit{Var}\,[\tau_1].$$

Theorem 2.5 shows that under mild regularity conditions a regenerative process has a limiting distribution provided that the expected time between regeneration points is finite.

**Definition 2.4.** *The random variable $\tau_1$ (and distribution function $F$) is said to be periodic with period $d$ if $d$ is the largest real number such that, with probability 1, $\tau_1$ assumes values in the set $\{0, d, 2d, \ldots\}$. If there is no such number, then $\tau_1$ (and distribution function $F$) is said to be aperiodic.*

**Theorem 2.5.** *Assume that $\tau_1$ is aperiodic with $E[\tau_1] < \infty$. Also assume that $\{X(t)\colon\ t \geq 0\}$ has right-continuous sample paths. Then $X(t) \Rightarrow X$ as $t \to \infty$ and*

$$E[f(X)] = \frac{E[Y_1(f)]}{E[\tau_1]}$$

*for any real-valued measurable function $f$ such that $E[|f(X)|] < \infty$ or $E[Y_1(|f|)] < \infty$. Moreover,*

$$\lim_{t\to\infty} E\big[f(X(t))\big] = E\big[f(X)\big]$$

*for all bounded functions $f$ such that $P\{X \in D(f)\} = 0$, where $D(f)$ is the set of discontinuity points of $f$.*

**Example 2.6.** (Machine availability) Assume that service time at center 1 is exponentially distributed as a random variable $V_1$ and service time at center 2 is distributed as a positive random variable $V_2$ with finite mean but otherwise arbitrary distribution. Let $X(t)$ be the number of customers waiting or in service at center 2 at time $t$. Also let $T_n$ be the time of the $n$th service completion at center 2, $n \geq 0$. Take $s_0' \in S - \{N\} = \{0, 1, \ldots, N-1\}$, and recall from Example 1.6 that the successive random times at which $\{X(T_n)\colon\ n \geq 0\}$ hits state $s_0'$ form a sequence of regeneration points for the process $\{X(t)\colon\ t \geq 0\}$. The time $\tau_1$ between regeneration points is aperiodic since the density function of service time at center 1 is positive on $(0, +\infty)$, and the argument in Example 1.5 shows that $E[\tau_1]$ is finite.

Let $f$ be the identity function: $f(s) = s$ for $s \in S$. Theorem 2.5 implies that $X(t) \Rightarrow X$ as $t \to \infty$ and

$$E[X] = \frac{E[Y_1(f)]}{E[\tau_1]}.$$

This "ratio formula" asserts that the expected value of the limiting number of customers at center 2 is equal to the expected time-average queue length in a regenerative cycle divided by the expected length of the cycle.

Now let $f$ be the indicator function of the set $S - \{N\}$, and observe that $Y_1(f)$ is the total amount of time in the first cycle that the number of customers at center 2 is less than $N$. Theorem 2.5 implies that

$$E[f(X)] = P\{X < N\} = \frac{E[Y_1(f)]}{E[\tau_1]}.$$

This ratio formula asserts that the limiting probability that there are fewer than $N$ customers at center 2 is equal to the expected amount of time that the server at center 1 is busy during a cycle divided by the expected length of the cycle.

## 2.3. Regenerative Generalized Semi-Markov Processes

Recall that a GSMP with state space $S$ and event set $E$ is said to be irreducible if for each pair $s, s' \in S$ there exist $s = s_0, s_1, \ldots, s_{n-1}, s_n = s' \in S$ and $e_{i_0}, e_{i_1}, \ldots, e_{i_{n-1}} \in E$ such that

$$\begin{aligned} p(s_1; s_0, e_{i_0})r(s_0, e_{i_0})p(s_2; s_1, e_{i_1})r(s_1, e_{i_1}) \cdots \\ p(s_n; s_{n-1}, e_{i_{n-1}})r(s_{n-1}, e_{i_{n-1}}) > 0, \end{aligned}$$

where $p(s'; s, e)$ is the probability that the new state is $s'$ given that event $e$ triggers a state transition in state $s$ and $r(s, e)$

is the speed at which the clock associated with event $e$ runs in state $s$. Also recall that $\zeta_n$ is the time of the $n$th state transition, $n \geq 0$.

Proposition 3.1 provides conditions on the clock-setting distributions which ensure that an irreducible GSMP with finite state space hits each state infinitely often with probability 1. We assume that the initial distribution $\mu$ of the GSMP is fixed and suppress $\mu$ in the notation.

**Proposition 3.1.** *Let $\{X(t)\colon t \geq 0\}$ be an irreducible generalized semi-Markov process with a finite state space $S$ and event set $E$. Assume that for all $s, s' \in S$, $e^* \in E$, and $e' \in N(s'; s, e^*)$ the clock-setting distribution function $F(\,\cdot\,; s', e', s, e^*)$ has finite mean and a density function that is continuous and positive on $(0, +\infty)$. Then $P\{X(\zeta_n) = s_0 \text{ i.o.}\} = 1$ for all $s_0 \in S$.*

The idea of the proof is as follows. It is sufficient to show that the GSSMC $\{(S_n, C_n)\colon n \geq 0\}$ associated with state transitions of the GSMP hits the set $\{s_0\} \times C(s_0)$ infinitely often, where $C(s_0)$ is the set of possible clock-reading vectors in state $s_0$. It follows immediately that $P\{X(\zeta_n) = s_0 \text{ i.o.}\} = 1$. Methods of advanced probability can be used to show that the GSSMC hits the set $\{s_0\} \times C(s_0)$ infinitely often with probability 1, provided that

(i) the GSMP is irreducible;
(ii) each clock-setting distribution has finite mean and a density function that is continuous and positive on $(0, +\infty)$; and
(iii) a "recurrence measure" assigns positive measure to the set $\{s\} \times C(s)$ for all states $s$ of the process.

Theorem 3.2 provides conditions under which a GSMP is a regenerative process in continuous time and the expected time between regeneration points is finite.

**Theorem 3.2.** *Let $\{X(t)\colon t \geq 0\}$ be an irreducible generalized semi-Markov process with a finite state space $S$ and event set $E$. Assume that for $s, s' \in S$, $e^* \in E$, and $e' \in N(s'; s, e^*)$ the clock-setting distribution function $F(\,\cdot\,; s', e', s, e^*)$ has finite mean and a density function that is continuous and positive on $(0, +\infty)$. Also assume that there exist $s_0, s_0' \in S$ and $e \in E$ such that for $e^* \in E(s_0)$,*

*(i)* $O(s_0'; s_0, e^*) = \emptyset$;
*(ii)* $N(s_0'; s_0, e^*) = N(s_0'; s_0, e)$; *and*
*(iii)* $F(\,\cdot\,; s_0', e', s_0, e^*) = F(\,\cdot\,; s_0', e', s_0, e)$ *for all events* $e' \in N(s_0'; s_0, e^*)$.

*Then $\{X(t)\colon t \geq 0\}$ is a regenerative process in continuous time, and the expected time between regeneration points is finite.*

The conclusion of Theorem 3.2 also holds if Condition (i) is replaced by:

(i′) $O(s_0'; s_0, e^*) \neq \emptyset$ and for any $e' \in O(s_0'; s_0, e^*)$ the clock-setting distribution function $F(\,\cdot\,; s', e', s, e)$ is exponential with rate parameter $\lambda$ independent of $s$, $s'$, and $e$.

Condition (i′) ensures that irrespective of when the clock associated with event $e' \in O(s_0'; s_0, e^*)$ was set, the remaining time until event $e'$ triggers a state transition is exponentially distributed with rate parameter $\lambda$. Observe that since each clock-setting distribution has a density function that is positive on $(0, +\infty)$, the time $\tau_1$ between regeneration points is aperiodic, and $X(t) \Rightarrow X$ as $t \to \infty$ by Theorem 2.5.

The idea is to show that the successive random times at which there is a state transition from $s_0$ to $s_0'$ form a sequence of regeneration points for the process $\{X(t)\colon\ t \geq 0\}$. Let $\zeta_n$ be the time of the $n$th state transition, $n \geq 0$. Also let $L(t)$ be the last state occupied by $\{X(t)\colon\ t \geq 0\}$ before jumping to the state occupied at time $t$, and set

$$V(t) = (L(t), X(t)).$$

The process $\{X(t)\colon\ t \geq 0\}$ makes a state transition from $s_0$ to $s_0'$ infinitely often with probability 1. To see this, observe that the process $\{V(t)\colon\ t \geq 0\}$ can be represented as an irreducible GSMP with finite state space

$$G = \{(s, s') \subseteq S \times S\colon\ p(s'; s, e) > 0 \text{ for some } e \in E(s)\}.$$

Each clock-setting distribution has finite mean and a density function that is continuous and positive on $(0, +\infty)$. Set $v_0' = (s_0, s_0')$. Then $P\{V(\zeta_n) = v_0' \text{ i.o.}\} = 1$ by Proposition 3.1.

Next observe that if the process makes a state transition from $s_0$ to $s_0'$ at time $\zeta_n$ (that is, if $V(\zeta_n) = v_0'$), the only clocks that are active have just been set since $O(s_0'; s_0, e^*) = \emptyset$ for all $e^* \in E(s_0)$. The joint distribution of $X(\zeta_n)$ and the clocks set at time $\zeta_n$ depends on the past history of $\{X(t)\colon\ t \geq 0\}$ only through $s_0'$, the previous state $s_0$, and the trigger event $e^*$. Since the new events and clock-setting distributions are the same for all $e^*$, the process $\{X(t)\colon\ t \geq 0\}$ probabilistically restarts whenever there is a state transition from $s_0$ to $s_0'$. The hypotheses of the theorem ensure that the expected number of state transitions in a cycle is finite. Then, since $S$ is finite and each clock-setting distribution has finite mean, Wald's identity implies that the expected time between regeneration points is finite.

**Example 3.3.** (Machine availability) Assume that all service times are mutually independent and service time at center $i$ has finite mean and a density function that is continuous and positive on $(0, +\infty)$. Let $X(t)$ be the number of customers waiting or in service at center 2 at time $t$. Take event $e_i$ = "service completion at center $i$." Recall from Example 4.8 of Chapter 1 that formal specification of the process $\{X(t)\colon t \geq 0\}$ is as a GSMP with finite state space $S = \{0, 1, \ldots, N\}$ and event set $E = \{e_1, e_2\}$.

To see that the GSMP is irreducible, let $s, s' \in S$. If $s' = s$, then either $p(s+1; s, e_1)p(s; s+1, e_2) > 0$ or $p(s-1; s, e_2)p(s; s-1, e_1) > 0$. If $s' > s$, then $p(s+1; s, e_1)p(s+2; s+1, e_1) \cdots p(s'; s'-1, e_1) > 0$. If $s' < s$, then $p(s-1; s, e_2)p(s-2; s-1, e_2) \cdots p(s'; s'+1, e_2) > 0$.

Take $s_0' = N - 1$ and $s_0 = N$, and observe that there is a state transition from $N$ to $N - 1$ only if the trigger event is $e_2$. Also observe that $O(N-1; N, e_2) = \emptyset$ and $N(N-1; N, e_2) = \{e_1, e_2\}$. By Theorem 3.2, the successive random times at which there is a state transition from $s_0$ to $s_0'$ form a sequence of regeneration points for the process $\{X(t)\colon t \geq 0\}$, and the expected time between regeneration points is finite.

**Example 3.4.** (Telephone system) Assume that the duration of a call placed at line $i$ is distributed as a positive random variable $V_i$ with finite mean and a density function that is continuous and positive on $(0, +\infty)$. Also assume that calls are placed randomly in the sense that the time from the end of a call placed or received at line $i$ to the next call placed at line $i$ is exponentially distributed as a random variable $A_i$; with independent probability $p_{ij}$ the called line is line $j$ ($\neq i$). After a lost call placed at line $i$, the time to the next call placed at line $i$ is distributed as $A_i$.

Set

$$X(t) = (X_1(t), \ldots, X_K(t)),$$

where $X_k(t) = (i, j)$ if a call placed at line $i$ is connected to line $j$ on link $k$ at time $t$ and equals $(0, 0)$ if there is no call connected on link $k$ at time $t$. Take event $e_i$ = "call placed at line $i$" ($i = 1, 2, \ldots, N$) and $e_{N+k}$ = "completion of call on link

$k$" ($k = 1, 2, \ldots, K$). Recall from Example 4.15 of Chapter 1 that formal specification of the process $\{X(t)\colon\ t \geq 0\}$ is as an irreducible GSMP with finite state space

$$S = \{(i_1, j_1, \ldots, i_K, j_K) \in \{0, 1, \ldots, N\}^{2K} : i_m = j_m \text{ only if } i_m = j_m = 0; i_m \neq i_n, j_m \neq j_n, i_m \neq j_n, j_m \neq i_n \text{ or } i_m = i_n = j_m = j_n = 0 \text{ if } m \neq n \text{ with } m, n = 1, 2, \ldots, K\}$$

and event set $E = \{e_1, e_2, \ldots, e_N, e_{N+1}, \ldots, e_{N+K}\}$. All speeds for active events are equal to 1.

Select $i_1, j_1 \in \{1, 2, \ldots, N\}$ with $i_1 \neq j_1$. Take $s_0 = (0, 0, \ldots, 0, 0) \in S$ and $s_0' = (i_1, j_1, 0, \ldots, 0) \in S$. Observe that $E(s_0) = \{e_1, e_2, \ldots, e_N\}$ and $E(s_0') = \{e_1, e_2, \ldots, e_N, e_{N+1}\} - \{e_{i_1}, e_{j_1}\}$, so that $O(s_0'; s_0, e_{i_1}) = \{e_1, e_2, \ldots, e_N\} - \{e_{i_1}, e_{j_1}\}$ and $N(s_0'; s_0, e_{i_1}) = \{e_{i_1}\}$. (Event $e_{j_1}$ is cancelled when there is a state transition from $s_0$ to $s_0'$.) Also observe that the process $\{X(t)\colon\ t \geq 0\}$ makes a state transition from $s_0$ to $s_0'$ only if event $e^* = e_{i_1} =$ "call placed at line $i_1$" is the trigger event. By Theorem 3.2, the successive random times at which there is a state transition from $s_0$ to $s_0'$ form a sequence of regeneration points for the process $\{X(t)\colon\ t \geq 0\}$, and the expected time between regeneration points is finite. Figure 2.2 shows a regenerative cycle for the telephone system with $N = 6$ lines and $K = 2$ links. (Select $i_1 = 1$ and $j_1 = 3$ so that $s_0' = (1, 3, 0, 0)$.)

The state transitions and state transition times determine whether a call is successful, busy, or blocked. A state transition from $s = (i_1, j_1, \ldots, i_{m-1}, j_{m-1}, 0, 0, i_{m+1}, j_{m+1}, \ldots, i_K, j_K) \in S$ to $s' = (i_1, j_1, \ldots, i_{m-1}, j_{m-1}, i, j, i_{m+1}, j_{m+1}, \ldots, i_K, j_K)$ with $i, j > 0$ corresponds to a successful call. A state transition from some $s = (i_1, j_1, \ldots, i_K, j_K) \in S$ with $i_1 i_2 \cdots i_K = 0$ to $s' = s$ corresponds to a busy call. A state transition from some $s = (i_1, j_1, \ldots, i_K, j_K) \in S$ with $i_1 i_2 \cdots i_K > 0$ to $s' = s$ corresponds to a blocked call.

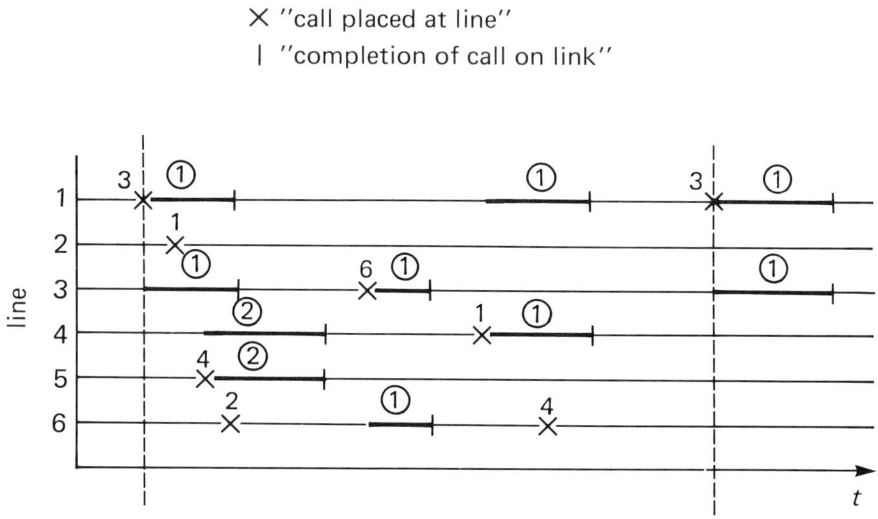


**Figure 2.2.** Regenerative cycle for telephone system.

Let $L(t)$ be the last state occupied by $\{X(t)\colon\ t \geq 0\}$ before jumping to the state occupied at time $t$, and set $V(t) = (L(t), X(t))$. The process $\{V(t)\colon\ t \geq 0\}$ has state space

$$G = \{(s, s') \subseteq S \times S\colon\ p(s'; s, e) > 0 \text{ for some } e \in E(s)\}\,.$$

Let $T_n$ be the time at which the $n$th call is placed, and set $V_n = V(T_n)$ for $n \geq 0$. The successive random times at which there is a state transition to $v_0 = (s_0, s_0')$ form a sequence of regeneration points for the process $\{V_n\colon\ n \geq 0\}$, and the expected time between regeneration points is finite. It follows that $V_n \Rightarrow V$ as $n \to \infty$. Define the limiting probability that a call is successful as $P\{V \in H\} = E[1_H(V)]$, where

$$H = \big\{(i_1, j_1, \ldots, i_K, j_K, i_1', j_1', \ldots, i_K', j_K') \in G\colon \\ i_k = 0 \text{ and } i_k' \neq 0 \text{ for some } k \text{ with } 1 \leq k \leq K\big\}.$$

Now let $\alpha_k$ be the length (that is, the number of state transitions) of the $k$th cycle of the discrete time process $\{V_n\colon\ n \geq 0\}$. According to this definition, $\alpha_k$ is the total number of calls placed irrespective of line in the $k$th cycle. Also set

$$Y_1(1_H) = \sum_{j=0}^{\alpha_1 - 1} 1_H(V_j).$$

Denote by $Y_k(1_H)$ the analogous quantity for the $k$th cycle, so that $Y_k(1_H)$ is the number of successful calls in the $k$th cycle. The sequence $\{(Y_k(1_H), \alpha_k)\colon\ k \geq 1\}$ consists of i.i.d. random vectors and

$$E[1_H(V)] = \frac{E[Y_1(1_H)]}{E[\alpha_1]}.$$

Lemma 3.5 is a special case of a generalized Borel-Cantelli lemma. This result provides a means of showing that a GSMP hits a fixed state infinitely often with probability 1. The elementary proof given below uses a "geometric trials" argument.

**Lemma 3.5.** *Let $\{Y_n\colon\ n \geq 0\}$ be a sequence of random variables defined on a probability space $(\Omega, \mathcal{F}, P)$ and taking values in a set $S$. Let $s' \in S$. Assume that there exists $\delta > 0$ such that*

$$P\{Y_n = s' | Y_{n-1}, \ldots, Y_0\} \geq \delta \text{ a.s.} \tag{3.6}$$

*for all $n \geq 1$. Then $P\{Y_n = s' \text{ i.o.}\} = 1$.*

*Proof.* Define a sequence of random indices by $I_0 = 0$ and

$$I_k = \inf\{n > I_{k-1}\colon\ Y_n = s'\}$$

for $k \geq 1$. According to this definition, $I_k$ is the index of the $k$th time that $\{Y_n\colon\ n \geq 0\}$ hits state $s'$. Then $P\{Y_n = s' \text{ i.o.}\} =$

1 if and only if $P\{I_k < \infty \text{ for } k \geq 0\} = 1$. Observe that $I_0$ is a.s. finite by definition, and assume for induction that $I_k$ is a.s. finite. Using (3.6) it follows that

$$\begin{aligned}
&P\{I_{k+1} - I_k > n, I_k = j\} \\
&\quad = P\{Y_{j+n} \neq s', \ldots, Y_{j+1} \neq s', I_k = j\} \\
&\quad = E\Big[P\{Y_{j+n} \neq s', \ldots, Y_{j+1} \neq s', I_k = j \\
&\qquad\qquad |Y_{j+n-1}, \ldots, Y_0\}\Big] \\
&\quad = E\Big[1_{\{Y_{j+n-1} \neq s', \ldots, Y_{j+1} \neq s', I_k = j\}} \\
&\qquad\qquad\qquad P\{Y_{j+n} \neq s' | Y_{j+n-1}, \ldots, Y_0\}\Big] \\
&\quad \leq E\big[1_{\{Y_{j+n-1} \neq s', \ldots, Y_{j+1} \neq s', I_k = j\}}(1-\delta)\big] \\
&\quad = (1-\delta) P\{I_{k+1} - I_k > n-1, I_k = j\}
\end{aligned}$$

so that

$$P\{I_{k+1} - I_k > n, I_k = j\} \leq (1-\delta)^n P\{I_k = j\}$$

for $n \geq 1$ and $j \geq 0$. Since $P\{I_k < \infty\} = 1$ by the induction hypothesis, we can sum over $j$ to obtain

$$P\{I_{k+1} - I_k > n\} \leq (1-\delta)^n$$

for $n \geq 1$, so that $I_{k+1} - I_k$ and hence $I_{k+1}$ is a.s. finite. Using the Bonferroni inequality, we have

$$\begin{aligned}
P\{Y_n = s' \text{ i.o.}\} &= P\{I_k < \infty \text{ for } k \geq 0\} \\
&\geq 1 - \sum_{k=0}^{\infty} P\{I_k = \infty\} \\
&= 1
\end{aligned}$$

and the desired result follows. □

Now let $\{X(t)\colon t \geq 0\}$ be a GSMP with state space $S$ and event set $E$. Define a finite-valued increasing sequence $\{T_n\colon n \geq 0\}$ of state transition times as follows. For fixed $e^* \in E$ and $S^* \subseteq S$, set $T_0 = 0$ and

$$\begin{aligned} T_n = \inf\bigl\{t > T_{n-1}\colon \ &\text{at time } t \text{ event } e^* \text{ triggers} \\ &\text{a state transition in some state } s^* \in S^*\bigr\} \end{aligned} \tag{3.7}$$

for $n \geq 1$. Thus, $T_n$ is the time of the $n$th occurrence of event $e^*$ when the underlying GSSMC $\{(S_k, C_k)\colon k \geq 0\}$ is in some state in the set $S^*$. Lemma 3.8 is an immediate consequence of Lemma 3.5 (with $Y_n = X(T_n)$). This geometric trials criterion for recurrence avoids the "positive density on the half line" assumptions used in Proposition 3.1.

**Lemma 3.8.** *(Geometric trials lemma) Let $\{T_n\colon n \geq 0\}$ be a finite-valued increasing sequence of state transition times as in (3.7). Let $s_0' \in S$ and assume that*

$$P\{X(T_n) = s_0' | X(T_{n-1}), \ldots, X(T_0)\} \geq \delta \text{ a.s.}$$

*for some $\delta > 0$ and all $n \geq 1$. Then $P\{X(T_n) = s_0' \text{ i.o.}\} = 1$.*

This result can be used in conjunction with "new better than used" assumptions on clock-setting distributions to show that a GSMP hits a fixed state infinitely often with probability 1.

**Definition 3.9.** *The distribution of a positive random variable $A$ is said to be new better than used (NBU) if*

$$P\{A > x + y | A > y\} \leq P\{A > x\}$$

*for all $x, y \geq 0$.*

Observe that every distribution with increasing failure rate (IFR) is NBU. Also observe that if $A$ and $B$ are independent random variables with distributions that are NBU, then the distributions of $A + B$, $\min(A, B)$, and $\max(A, B)$ are NBU. A distribution that is NBU has finite moments of all orders.

**Example 3.10.** (Patrolling repairman) Assume that the walking time from machine $j$ to the next machine is a positive constant $W_j$ and the time to repair and restart machine $j$ is distributed as a positive random variable $R_j$ with finite mean. Also assume that the distribution of each $L_j$ is NBU and

$$\delta_j \stackrel{\text{def}}{=} P\{L_j \leq W_N\} > 0 \tag{3.11}$$

for $j = 1, 2, \ldots, N$. Set $X(t) = (Z_1(t), \ldots, Z_N(t), M(t), N(t))$ as in (1.8). Take event $e^* = e_{N+2} =$ "arrival of repairman at machine" and $S^* = \{(z_1, \ldots, z_N, m, n) \in S\colon\ m = 0,\ n = 1\}$, so that $T_n$ defined by (3.7) is the time of the $n$th arrival of the repairman at machine 1. Observe that for $s_0' = (0, 1, \ldots, 1, 1, 2)$, $X(T_n) = s_0'$ if machines $2, 3, \ldots, N$ are stopped and repair of machine 1 starts at time $T_n$. Then $P\{X(T_n) = s_0' \text{ i.o.}\} = 1$. To see this, first observe that $T_n < \infty$ a.s. since

$$E[T_n - T_{n-1}] \leq W_1 + \cdots + W_N + \sum_{j=1}^{N} E[R_j] < \infty \tag{3.12}$$

for all $n \geq 1$. Now set

$$\delta = \prod_{j=1}^{N} \delta_j = \prod_{j=1}^{N} P\{L_j \leq W_N\}\,.$$

By (3.11), $\delta > 0$. Let $T_{n-1}^{+}$ be the first time after $T_{n-1}$ that the repairman leaves machine $N$, and set $S^+ = S^*$. Observe that $X(T_n) = s_0'$ if there is a stoppage of each machine that is running at time $T_{n-1}^{+}$ while the repairman is walking from machine $N$ to machine 1. For $s^+ = (z_1^+, \ldots, z_N^+, m^+, n^+) \in S^+$, set

$$K^+(s^+) = \{k\colon\ z_k^+ = 0\}.$$

When $X(T_{n-1}^+) = s^+$, for $k \in K(s^+)$ denote the latest time less than or equal to $T_{n-1}^+$ at which the clock associated with event $e_k$ was set by $S_{n,k}(s^+)$ and the setting on the clock at time $S_{n,k}(s^+)$ by $L_{n,k}(s^+)$. Then, using the fact that the distribution of each $L_j$ is NBU, we have

$$\begin{aligned}
&P\{X(T_n) = s_0', X(T_{n-1}) = x_{n-1}, \ldots, X(T_0) = x_0\} \\
&\quad = \sum_{s^+ \in S^+} P\{X(T_n) = s_0', X(T_{n-1}^+) = s^+, X(T_{n-1}) = x_{n-1}, \\
&\qquad\qquad \ldots, X(T_0) = x_0\} \\
&\quad \geq \sum_{s^+ \in S^+} P\{S_{n,k}(s^+) + L_{n,k}(s^+) \leq T_{n-1}^+ + W_N, k \in K(s^+); \\
&\qquad\qquad X(T_{n-1}^+) = s^+, X(T_{n-1}) = x_{n-1}, \ldots, X(T_0) = x_0\} \\
&\quad \geq \sum_{s^+ \in S^+} \delta P\{X(T_{n-1}^+) = s^+, X(T_{n-1}) = x_{n-1}, \ldots, \\
&\qquad\qquad X(T_0) = x_0\} \\
&\quad = \delta P\,\{X(T_{n-1}) = x_{n-1}, \ldots, X(T_0) = x_0\}
\end{aligned}$$

for all $x_0, x_1, \ldots, x_{n-1} \in S$ so that

$$P\,\{X(T_n) = s_0' | X(T_{n-1}), \ldots, X(T_0)\} \geq \delta \text{ a.s.}$$

and $P\,\{X(T_n) = s_0' \text{ i.o.}\} = 1$ by Lemma 3.8.

Theorem 3.13 prescribes conditions under which a GSMP is a regenerative process in continuous time and the expected time between regeneration points is finite.

**Theorem 3.13.** *Let $\{T_n \colon n \geq 0\}$ be a finite-valued increasing sequence of stopping times for the underlying general state space Markov chain $\{(S_n, C_n) \colon n \geq 0\}$ that are state transition times as in (3.7). Assume that there exist $s, s'_0 \in S$ and $\delta > 0$ such that*

$$P\{X(T_n) = s'_0 | X(T_{n-1}), \ldots, X(T_0)\} \geq \delta \text{ a.s.} \tag{3.14}$$

*for all $n \geq 1$ and for $s^* \in S^*$,*

*(i)* $O(s'_0; s^*, e^*) = \emptyset$;
*(ii)* $N(s'_0; s^*, e^*) = N(s'_0; s, e^*)$; *and*
*(iii)* $F(\,\cdot\,; s'_0, e', s^*, e^*) = F(\,\cdot\,; s'_0, e', s, e^*)$ *for all events* $e' \in N(s'_0; s^*, e^*)$.

*Then $\{X(t) \colon t \geq 0\}$ is a regenerative process in continuous time. Moreover, if $\sup_{n \geq 0} E[T_{n+1} - T_n] \leq b < \infty$, then the expected time between regeneration points is finite.*

Observe that Theorem 3.13 does not guarantee that the time $\tau_1$ between regeneration points is aperiodic.

*Proof.* Using Lemma 3.8, (3.14) implies that event $e^*$ triggers a state transition to $s'_0$ infinitely often with probability 1: $P\{X(T_n) = s'_0 \text{ i.o.}\} = 1$. Furthermore, at such a time $T_{\beta_n}$ the only clocks that are active have just been set since $O(s'_0; s^*, e^*) = \emptyset$ for all $s^* \in S$. The joint distribution of $X(T_{\beta_n})$ and the clocks set at time $T_{\beta_n}$ depends on the past history of $\{X(t) \colon t \geq 0\}$ only through $s'_0$, the previous state $s^*$, and the trigger event $e^*$. Since the new events and clock-setting distributions are the same for all $s^*$, the process $\{X(t) \colon t \geq 0\}$ probabilistically restarts whenever event $e^*$ triggers a state transition to $s'_0$.

We now show that the expected time between regeneration points is finite. Assume for convenience that $X(T_0) = X(0) = s'_0$. Set $X_n = X(T_n)$ and $D_n = T_{n+1} - T_n$ for $n \geq 0$. Observe that the random indices $\{\beta_n\colon\ n \geq 0\}$ form a sequence of regeneration points for the discrete time process $\{(X_n, D_n)\colon\ n \geq 0\}$. This assertion follows from that the fact $\{D_n\colon\ n \geq 0\}$ starts from scratch whenever $X(T_{\beta_n}) = s_0'$. Set $\alpha_n = \beta_n - \beta_{n-1}$, and observe that the $\alpha_n$ are i.i.d. as $\alpha_1$. Using (3.14), the argument in the proof of Lemma 3.5 shows that

$$P\{\alpha_1 > k\} \leq (1 - \delta)^k$$

so that the random variable $\alpha_1$ is stochastically dominated by a random variable having a geometric distribution with parameter $\delta$. It follows that $E[\alpha_1] < \infty$ and the expected time between regeneration points for the process $\{(X_n, D_n)\colon\ n \geq 0\}$ is finite. Since $\alpha_1$ is aperiodic by (3.14) and $E[\alpha_1] < \infty$, $(X_n, D_n) \Rightarrow (X, D)$ as $n \rightarrow \infty$. The continuous mapping theorem implies that $D_n \Rightarrow D$ as $n \rightarrow \infty$, and since $D_n \geq 0$ and $E[D_n] \leq b < \infty$,

$$E[D] \leq \liminf_{n \rightarrow \infty} E[|D_n|] = E[D_n] \leq b < \infty.$$

Since $\alpha_1$ is aperiodic and $E[\alpha_1] < \infty$,

$$E[D] = \frac{E[\sum_{j=0}^{\alpha_1 - 1} D_j]}{E[\alpha_1]}.$$

Therefore, $E[\sum_{j=0}^{\alpha_1 - 1} D_j] < \infty$, and the expected time between regeneration points for $\{X(t)\colon\ t \geq 0\}$ is finite. □

Observe (for fixed $S^* \subseteq S$, $e^* \in E$, and $s_0' \in S$) that if Condition (i) of Theorem 3.13 holds for every initial distribution $\mu$, then the limiting distribution of the GSMP is independent of $\mu$. This assertion is a consequence of the fact that for any initial distribution $\mu$ the random times $\{T_{\beta_n} : n \geq 0\}$ form a sequence of regeneration points for the process $\{X(t)\colon\ t \geq 0\}$.

Also observe that the conclusion of Theorem 3.13 also holds if Condition (i) is replaced by:

(i′) $O(s'_0; s^*, e^*) \neq \emptyset$ and for any $e' \in O(s'_0; s^*, e^*)$ the clock-setting distribution functic.n $F(\,\cdot\,; s', e', s, e)$ is exponential with rate parameter $\lambda$ independent of $s, s'$, and $e$.

Condition (i′) ensures that no matter when the clock associated with event $e' \in O(s'_0; s^*, e^*)$ was set, the remaining time until event $e'$ triggers a state transition is exponentially distributed with rate parameter $\lambda$.

The state transition times $\{T_n\colon\ n \geq 0\}$ of (3.7) are necessarily stopping times for $\{X(t)\colon\ t \geq 0\}$ if

$$p(s^*; s^*, e^*) = 0 \tag{3.15}$$

for all $s^* \in S^*$ and

$$e = e^* \text{ whenever } p(s; s^*, e) > 0 \text{ and } p(s; s^*, e^*) > 0 \tag{3.16}$$

for all $s^* \in S^*$ and $s \in S$. Conditions (3.15) and (3.16) imply that the sample paths of $\{X(t)\colon\ t \geq 0\}$ determine every occurrence of event $e^*$ in a state $s^* \in S^*$, and hence determine the state transition times $\{T_n\colon\ n \geq 0\}$. Also observe that since the expected time between regeneration points is finite, $X(t) \Rightarrow X$ as $t \to \infty$ provided that the time between regeneration points is aperiodic.

**Example 3.17.** (Patrolling repairman) Assume as in Example 3.10 that the distribution of each $L_j$ is NBU and $\delta_j = P\{L_j \leq W_N\} > 0$ for $j = 1, 2, \ldots, N$. Take event $e^* = e_{N+2} =$ "arrival of repairman at machine" and $S^* = \{(z_1, \ldots, z_N, m, n) \in S\colon\ m = 0,\ n = 1\}$, so that $T_n$ defined by (3.7) is the time of the $n$th arrival of the repairman at machine 1, $n \geq 0$. Also take $s'_0 = (0, 1, \ldots, 1, 1, 2)$. By the argument in Example 3.10, $P\{X(T_n) = s'_0 | X(T_{n-1}), \ldots, X(T_0)\} \geq \delta$ a.s., where $\delta = \prod_{j=1}^{N} \delta_j$. The process $\{X(t): t \geq 0\}$ hits state $s_0'$ when event $e^*$ is the trigger event only if the event occurs in state $s_0 = (1, \ldots, 1, 0, 1)$, and in this case there are no old clocks running ($O(s_0'; s_0, e^*) = \emptyset$). It follows from (3.12) and Theorem 3.13 that the successive random times at which $\{X(T_n): n \geq 0\}$ hits state $s_0'$ form a sequence of regeneration points for the process $\{X(t): t \geq 0\}$ and the expected time between regeneration points is finite; see Figure 2.3.

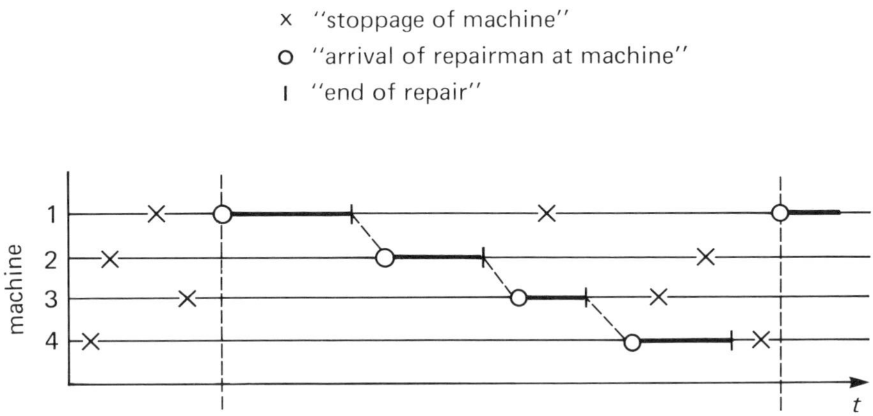


**Figure 2.3.** Regenerative cycle for patrolling repairman.

## Exercises

3.1. (Machine repair) Consider a group of $N$ ($\geq 1$) machines numbered $1, 2, \ldots, N$ under the care of a single repairman. The repairman provides service to (repairs) the machines one at a time in the order in which they fail. Successive repair times for machine $j$ are i.i.d. as a positive random variable $R_j$ with finite mean and a density function that is continuous and positive on $(0, +\infty)$. The machines stop randomly in the sense that the times (lifetimes) between end of repair and the next stoppage of machine $j$) are i.i.d. as a positive random variable $L_j$ with finite mean and a density function that is continuous and positive on $(0, +\infty)$. Lifetimes are independent of repair times, and quantities associated with distinct machines are independent.

Set

$$X(t) = (X_1(t), X_2(t), \ldots, X_N(t)),$$

where

$$X_j(t) = \begin{cases} 0 & \text{if machine } j \text{ is running at time } t \\ 1 & \text{if machine } j \text{ is under repair at time } t \\ k & \text{if machine } j \text{ is waiting for repair at time } t \\ & \text{with } k-1 \text{ machines ahead at the repair station.} \end{cases}$$

Show that $\{X(t)\colon\ t \geq 0\}$ is a regenerative process in continuous time and the expected time between regeneration points is finite.

3.2. (Telephone system) Consider a telephone system with $N$ $(\geq 2)$ telephones connected to a switchboard by lines numbered $1, 2, \ldots, N$. The switchboard has $K$ $(< N)$ links numbered $1, 2, \ldots, K$, each of which can connect any two lines subject to the restriction that only one connection at a time can be made to each line. If more than one link is available and the called line is not in use, a placed call is connected (instantaneously) on the lowest numbered available link. Assume that the system is a lost-call system in the sense that any call is immediately lost if no connection can be made when it is placed. A call is lost if there is at least one link available but the called line is in use (a *busy call*) and a call is lost if no link is available (a *blocked call*).

Assume that the duration of a call placed at line $i$ is distributed as a positive random variable $V_i$ with finite mean and a density function that is continuous and positive on $(0, +\infty)$.

Also assume that calls are placed randomly in the sense that the time from the end of a call placed or received at line $i$ to the next call placed at line $i$ is exponentially distributed as a random variable $A_i$. When a call is placed, with independent probability $p_{ij}$ the called line is line $j$ ($\neq i$). After a lost call placed at line $i$, the time to the next call placed at line $i$ is distributed as $A_i$.

Set

$$X(t) = (X_1(t), \ldots, X_K(t)),$$

where $X_k(t) = (i, j)$ if a call placed at line $i$ is connected to line $j$ on link $k$ at time $t$ and equals $(0, 0)$ otherwise. Take event $e_i$ = "call placed at line $i$" ($i = 1, 2, \ldots, N$) and event $e_{N+k}$ = "completion of call on link $k$" ($k = 1, 2, \ldots, K$). Formal specification of the process $\{X(t)\colon t \geq 0\}$ is as a GSMP with a finite state space $S$ and event set $E = \{e_1, \ldots, e_N, e_{N+1}, \ldots, e_{N+K}\}$.

Take $s_0 = (0, 0, \ldots, 0, 0) \in S$, and let $T_n$ be the time of the $n$th completion of a call, $n \geq 0$. Show that the successive random times at which $\{X(T_n)\colon n \geq 0\}$ hits state $s_0$ form a sequence of regeneration points for the process $\{X(t)\colon t \geq 0\}$.

3.3. (Telephone system) Consider the telephone system in Exercise 3.2. Set

$$X(t) = (X_1(t), \ldots, X_N(t)),$$

where $X_i(t)$ equals $k$ if line $i$ is connected (to some other line) on link $k$ at time $t$ and equals 0 if line $i$ is not connected at time $t$. Take event $e_i$ = "call placed at line $i$" ($i = 1, 2, \ldots, N$) and event $e_{N+k}$ = "completion of call on link $k$" ($k = 1, 2, \ldots, K$). Formal specification of the process $\{X(t)\colon t \geq 0\}$ is as a GSMP with a finite state space $S$ and event set $E = \{e_1, \ldots, e_N, e_{N+1}, \ldots, e_{N+K}\}$.

Take $s_0 = (0, 0, \ldots, 0, 0) \in S$ and $s_0' = (1, 1, 0, \ldots, 0) \in S$. Do the successive random times at which there is a state transition from $s_0$ to $s_0'$ form a sequence of regeneration points for the process $\{X(t)\colon\ t \geq 0\}$? Explain your answer.

3.4. (Operating unit failure) Consider a system consisting of $N$ identical operating units that are subject to failure. A failed unit is sent to a repair facility that operates as a single-server queue: units are restored to a like-new condition one at a time on a first-come, first-served basis. When a unit in like-new condition is placed in service, it functions for a random amount of time that is exponentially distributed with mean $N$ and then fails. Assume that the time required for the repairman to restore a failed unit is exponentially distributed. The time to restore a failed unit has mean 1 if the repairman is idle when the unit fails and has mean 2 if the repairman is busy when the unit fails. After being restored, the unit goes back into operation immediately. Let $X(t)$ be the number of units at the repair facility at time $t$. Show that the process $\{X(t)\colon\ t \geq 0\}$ is not a continuous time Markov chain. Also show that $\{X(t)\colon\ t \geq 0\}$ is a regenerative process in continuous time.

3.5. (Airport shuttle) Consider an airport shuttle that provides transportation service to $N$ stations numbered $1, 2, \ldots, N$. The shuttle has seats for $K$ ($\geq 1$) passengers and moves from station to station in a strictly defined order: $1, 2, \ldots, N, N-1, \ldots, 1, 2, \ldots$. Passengers that wish to board the shuttle at station $i$ in the "northbound" (from station 1 to station $N$) direction arrive at station $i$ according to a Poisson process of rate $\lambda_i$ and disembark at station $j$ ($> i$) with probability $p_{i,j}$ ($i = 1, 2, \ldots, N-1$). Passengers that wish to board the shuttle at station $i+1$ in the "southbound" (from station $N$ to station 1) direction arrive at station $i+1$ according to a Poisson process of rate $\lambda_{2N-(i+1)}$ and disembark at station $j$ ($< i+1$) with probability $p_{2N-(i+1),j}$ ($i = 1, 2, \ldots, N-1$). Passengers that arrive for boarding at station $i$ in each direction queue and subsequently board the shuttle in the order in which they arrive at the station. At each station passengers disembark before any waiting passengers board the shuttle. Assume that passengers board the shuttle instantaneously and passengers disembark instantaneously. The time required for the shuttle to travel (in the northbound direction) from station $i$ to station $i+1$ is distributed as a positive random variable $L_i$ with finite first and second moments, and the time required for the shuttle to travel (in the southbound direction) from station $i+1$ to station $i$ is distributed as a positive random variable $L_{2N-(i+1)}$ with finite first and second moment. Travel times, interarrival times, and destinations are mutually independent. Assume that $\sum_{k=1}^{2(N-1)} \lambda_k \sum_{k=1}^{2(N-1)} E[L_k] < K$.

Set

$$X(t) = (D_1(t), \ldots, D_N(t), B_1(t), \ldots, B_{2(N-1)}(t), N(t)),$$

where

$$N(t) = \begin{cases} i & \text{if the shuttle is moving from station } i \text{ to station } i+1 \text{ in the northbound direction at time } t \\ 2N-(i+i) & \text{if the shuttle is moving from station } i+1 \text{ to station } i \text{ in the southbound direction at time } t, \end{cases}$$

$D_i(t)$ is the number of passengers on board the shuttle at time $t$ that wish to disembark at station $i$ ($i = 1, 2, \ldots, N$), $B_i(t)$ is the number of passengers waiting at time $t$ to board the shuttle at station $i$ in the northbound direction, and $B_{2N-(i+1)}(t)$ is the number of passengers waiting at time $t$ to board the shuttle at station $i+1$ in the southbound direction ($i = 1, 2, \ldots, N-1$).

Let $T_n$ be the time of the $n$th departure of the shuttle from station 1, $n \geq 0$. Show that $\{X(T_n)\colon n \geq 0\}$ is an irreducible positive recurrent DTMC. Also show that $\{X(t)\colon t \geq 0\}$ is a regenerative process in continuous time and the expected time between regeneration points is finite.

3.6 (Regeneration theorem for generalized semi-Markov processes) Let $\{X(t)\colon t \geq 0\}$ be a GSMP with state space $S$ and event set $E$. Also let $\{T_n\colon n \geq 0\}$ be a finite-valued increasing sequence of state transition times such that for some $e^* \in E$ and $S^* \subseteq S$: $T_0 = 0$ and

$$T_n = \inf\big\{t > T_{n-1}\colon \text{ at time } t \text{ event } e^* \text{ triggers } \\ \text{a state transition in some state } s^* \in S^*\big\}$$

for $n \geq 1$. Assume that there exists $s_0 \in S^*$ such that

(i) $P\{X(T_n-) = s_0 | X(T_{n-1}-), \ldots, X(T_1-)\} \geq \delta$ a.s. for some $\delta > 0$ and all $n \geq 1$; and
(ii) the set $O(s; s_0, e^*) = \emptyset$ for all $s \in S$ with $p(s; s_0, e^*) > 0$.

Also assume that $\sup_{n \geq 0} E[T_{n+1} - T_n] \leq b < \infty$.

Show that $\{X(t)\colon t \geq 0\}$ is a regenerative process in continuous time and the expected time between regeneration points is finite.

## Chapter 3

# Regenerative Simulation

In the presence of mild regularity conditions, a regenerative stochastic process has a limiting distribution provided that the expected time between regeneration points is finite. Moreover, the behavior of the process within a regenerative cycle determines the limiting distribution of the process as a ratio of expected values. Strongly consistent point estimates and asymptotic confidence intervals for "steady-state means" can be obtained by observation of a finite portion of a single sample path of the regenerative process. The idea is to observe a finite number of regenerative cycles and to measure quantities within individual cycles.

When applicable, the regenerative method for analysis of simulation output is attractive because it provides valid point and interval estimates. The classical alternative entails selection of initial conditions for the underlying stochastic process of the simulation and observation of the process for some initial time period. We then discard this "initial transient" and observe the process for a subsequent time period to obtain point estimates. In general, no confidence interval is available, nor is there any guidance on the selection of the initial conditions. Moreover, the determination of a suitable initial time period and a suitable subsequent time period is often nontrivial and likely to require sophisticated statistical techniques. With the regenerative method, these difficulties are generally avoidable.

## 3.1. The Regenerative Method

According to Definition 1.1 of Chapter 2, a real (possibly vector-valued) stochastic process $\{X(t)\colon t \geq 0\}$ with state space $S \subseteq \Re^d$ for some $d \geq 1$ is a regenerative process in continuous time provided that: (i) there exists an increasing sequence $\{T_k\colon k \geq 0\}$ of stopping times such that $\{T_{k+1} - T_k\colon k \geq 0\}$ are independent and identically distributed; and (ii) for every sequence of times $0 \leq t_1 < t_2 < \cdots < t_m$ $(m \geq 1)$, $n \geq 1$, and $k \geq 0$, the vectors $\{X(T_0 + t_1), \ldots, X(T_0 + t_m), T_1 - T_0, \ldots, T_n - T_{n-1}\}$ and $\{X(T_k + t_1), \ldots, X(T_k + t_m), T_{k+1} - T_k, \ldots, T_{k+n} - T_{k+n-1}\}$ are identically distributed, and the processes $\{X(t)\colon 0 \leq t < T_k; T_l - T_{l-1}, 1 \leq l \leq k\}$ and $\{X(T_k + t)\colon t \geq 0; T_l - T_{l-1}, l \geq k+1\}$ are independent. For a real-valued measurable function $f$ with domain $S$, set $T_{-1} = 0$ and

$$Y_k(f) = \int_{T_{k-1}}^{T_k} f(X(u))du$$

for $k \geq 0$. Also set $\tau_k = T_k - T_{k-1}$ for $k \geq 1$. Recall that the definition of regenerative process ensures that the sequence $\{(Y_k(f), \tau_k)\colon k \geq 1\}$ consists of i.i.d. random vectors. Also recall from Theorem 2.5 of Chapter 2 that if the process $\{X(t)\colon t \geq 0\}$ has right-continuous sample paths, $\tau_1 = T_1 - T_0$ is aperiodic, and $E[\tau_1] < \infty$, then $X(t) \Rightarrow X$ as $t \to \infty$. Moreover,

$$E\big[f(X)\big] = \frac{E\big[Y_1(f)\big]}{E\big[\tau_1\big]} \stackrel{\text{def}}{=} r(f) \tag{1.1}$$

for any real-valued measurable function $f$ such that $E[|f(X)|] < \infty$ or $E[Y_1(|f|)] < \infty$.

Suppose that the goal of the simulation is the estimation of the quantity $r(f)$. Thus, we wish to estimate the ratio of the expected values of two dependent random variables from a sequence of independent observations of the pair of random variables. To obtain an estimator that is strongly consistent for $r(f)$ along with an asymptotic confidence interval, set

$$Z_k(f) = Y_k(f) - r(f)\tau_k$$

and observe that the $k$th cycle of the process $\{X(t)\colon t \geq 0\}$ determines the quantity $Z_k(f)$. The sequence $\{Z_k(f)\colon k \geq 1\}$ of random variables are i.i.d., and the ratio formula of (1.1) implies that $E[Z_k(f)] = 0$. Set

$$\sigma^2(f) \stackrel{\text{def}}{=} \text{Var}\,[Z_1(f)] = \text{Var}\,[Y_1(f) - r(f)\tau_1]$$

and write

$$E[Z_1^2(f)] = E\big[\{\big(Y_1(f) - E[Y_1(f)]\big) - r(f)\big(\tau_1 - E[\tau_1]\big)\}^2\big].$$

It follows that

$$\sigma^2(f) = \text{Var}\,[Y_1(f)] - 2r(f)\,\text{Cov}\,[Y_1(f), \tau_1] + r^2(f)\,\text{Var}\,[\tau_1].$$

We always assume that $0 < \sigma^2(f) < \infty$. The case $\sigma^2(f) = 0$ is degenerate, and $\sigma^2(f) < \infty$ for most processes that have a finite state space. However, in some queueing systems, finite higher moments of service times and interarrival times are needed to ensure that $\sigma^2(f) < \infty$.

Now fix $n$, and let $\bar{Y}(n)$, $\bar{\tau}(n)$, $s_{11}(n)$, $s_{22}(n)$, and $s_{12}(n)$ be the usual unbiased estimators for $E[Y_1(f)]$, $E[\tau_1]$, Var $[Y_1(f)]$, Var $[\tau_1]$, and Cov $[Y_1(f), \tau_1]$, respectively; that is,

$$\bar{Y}(n) = \frac{1}{n}\sum_{k=1}^{n} Y_k(f),$$

$$\bar{\tau}(n) = \frac{1}{n}\sum_{k=1}^{n} \tau_k,$$

$$s_{11}(n) = \frac{1}{n-1}\sum_{k=1}^{n}\big(Y_k(f) - \bar{Y}(n)\big)^2,$$

$$s_{22}(n) = \frac{1}{n-1}\sum_{k=1}^{n}\big(\tau_k - \bar{\tau}(n)\big)^2,$$

and

$$s_{12}(n) = \frac{1}{n-1}\sum_{k=1}^{n}\big(Y_k(f) - \bar{Y}(n)\big)\big(\tau_k - \bar{\tau}(n)\big).$$

As a consequence of the strong law of large numbers for sequences of i.i.d. random variables, the estimator

$$\hat{r}(n) \stackrel{\text{def}}{=} \frac{\bar{Y}(n)}{\bar{\tau}(n)} \tag{1.2}$$

is strongly consistent for $r(f)$; that is, $\hat{r}(n) \to r(f)$ a.s. as $n \to \infty$. Moreover, the estimator

$$s(n) \stackrel{\text{def}}{=} \big\{s_{11}(n) - 2\hat{r}(n)s_{12}(n) + \hat{r}^2(n)s_{22}(n)\big\}^{1/2} \tag{1.3}$$

is strongly consistent for $\sigma(f)$ ($s(n) \to \sigma(f)$ a.s. as $n \to \infty$).

Theorem 1.4 establishes a CLT that leads to asymptotic confidence intervals for $r(f)$.

**Theorem 1.4.** *Assume that $0 < \sigma^2(f) < \infty$. Then*

$$\frac{\sqrt{n}\{\hat{r}(n) - r(f)\}}{\sigma(f)/E[\tau_1]} \Rightarrow N(0,1) \tag{1.5}$$

*as $n \to \infty$.*

*Proof.* The CLT for i.i.d. random variables with mean 0 and finite variance implies that

$$\frac{1}{\sigma(f)\sqrt{n}}\sum_{k=1}^{n} Z_k(f) \Rightarrow N(0,1) \tag{1.6}$$

as $n \to \infty$. This CLT can be rewritten as

$$\frac{1}{\sigma(f)\sqrt{n}}\left[\sum_{k=1}^{n} Y_k(f) - r(f)\sum_{k=1}^{n} \tau_k\right] \Rightarrow N(0,1)$$

as $n \to \infty$, and it follows that

$$\frac{\sqrt{n}\{\hat{r}(n) - r(f)\}}{(\sigma(f)/E[\tau_1])(E[\tau_1]/\bar{\tau}(n))} \Rightarrow N(0,1) \tag{1.7}$$

as $n \to \infty$. The strong law of large numbers ensures that $E[\tau_1]/\bar{\tau}(n) \Rightarrow 1$ as $n \to \infty$. Using Lemma 3.3 of Appendix A it follows that

$$\left(X_n, \frac{E[\tau_1]}{\bar{\tau}(n)}\right) \Rightarrow (N(0,1), 1)$$

as $n \to \infty$, where $X_n$ denotes the left side of (1.7). Now let $h$ be the mapping given by

$$h(x,y) = xy$$

and apply the continuous mapping theorem to conclude that

$$X_n \cdot \frac{E[\tau_1]}{\bar{\tau}(n)} \Rightarrow N(0,1) \cdot 1 \tag{1.8}$$

as $n \to \infty$. Since $N(0,1) \cdot 1$ has the same distribution as $N$(0,1), (1.8) is equivalent to (1.5). □

To obtain an asymptotic $100(1-2\gamma)\%$ confidence interval for $r(f)$, denote the distribution function of $N(0,1)$ by $\Phi$ and set

$$z_{1-\gamma} = \Phi^{-1}(1-\gamma)$$

for $0 < \gamma < 1/2$. Equation (1.5) implies that

$$\lim_{n\to\infty} P\left\{\frac{\sqrt{n}\,\{\hat{r}(n) - r(f)\}}{\sigma(f)/E[\tau_1]} \le x\right\} = \Phi(x)$$

for all $x$, so that

$$\lim_{n\to\infty} P\left\{\hat{r}(n) \le r(f) + \frac{\sigma(f)z_{1-\gamma}}{E[\tau_1]\sqrt{n}}\right\} = \Phi(z_{1-\gamma})$$

and

$$\lim_{n\to\infty} P\left\{\hat{r}(n) \ge r(f) - \frac{\sigma(f)z_{1-\gamma}}{E[\tau_1]\sqrt{n}}\right\} = 1 - \Phi(-z_{1-\gamma}).$$

Therefore,

$$\begin{aligned}
&\lim_{n\to\infty} P\Big\{\hat{r}(n) \ge r(f) - \frac{\sigma(f)z_{1-\gamma}}{E[\tau_1]\sqrt{n}}, \hat{r}(n) \le r(f) + \frac{\sigma(f)z_{1-\gamma}}{E[\tau_1]\sqrt{n}}\Big\} \\
&= \lim_{n\to\infty} P\Big\{\hat{r}(n) - \frac{\sigma(f)z_{1-\gamma}}{E[\tau_1]\sqrt{n}} \le r(f) \le \hat{r}(n) + \frac{\sigma(f)z_{1-\gamma}}{E[\tau_1]\sqrt{n}}\Big\} \\
&= \Phi(z_{1-\gamma}) - \Phi(-z_{1-\gamma}) \\
&= \Phi(z_{1-\gamma}) - (1 - \Phi(z_{1-\gamma})) \\
&= 1 - 2\gamma.
\end{aligned}$$

Thus, for large $n$ the interval with random end points $\hat{r}(n) - \sigma(f)z_{1-\gamma}/(E[\tau_1]\sqrt{n})$ and $\hat{r}(n) + \sigma(f)z_{1-\gamma}/(E[\tau_1]\sqrt{n})$ contains the unknown constant $r(f)$ approximately $100(1 - 2\gamma)\%$ of the time.

In general, the *standard deviation constant* $\sigma(f)/E[\tau_1]$ is not known. The most straightforward estimator is $s(n)/\bar{\tau}(n)$, and the strong law of large numbers ensures that $\sigma(f)/s(n) \Rightarrow 1$ as $n \to \infty$. The same argument that leads to (1.5) yields

$$\frac{\sqrt{n}\{\hat{r}(n) - r(f)\}}{s(n)/\bar{\tau}(n)} \Rightarrow N(0,1) \tag{1.9}$$

as $n \to \infty$, and it follows from this CLT that the interval

$$\left[\hat{r}(n) - \frac{s(n)z_{1-\gamma}}{\bar{\tau}(n)\sqrt{n}}, \hat{r}(n) + \frac{s(n)z_{1-\gamma}}{\bar{\tau}(n)\sqrt{n}}\right] \tag{1.10}$$

is an *asymptotic* $100(1-2\gamma)\%$ *confidence interval* for $r(f)$. Thus,

$$\lim_{n\to\infty} P\{r(f) \in \hat{I}(n)\} = 1 - 2\gamma,$$

where $\hat{I}(n)$ is the interval in (1.10). The asymptotic confidence interval $\hat{I}(n)$ has random end points and is symmetric about the point estimate $\hat{r}(n)$. Observe that the half-length of the confidence interval is $1/\sqrt{n}$ times a multiple $(z_{1-\gamma})$ of the estimate of the standard deviation constant $\sigma(f)/E[\tau_1]$, so that as $n$ increases, the length of the confidence interval converges to 0, and the midpoint converges to the true value. For 90% $(\gamma = .05)$ confidence intervals, $z_{1-\gamma} = 1.645$; for 95% $(\gamma = .025)$ confidence intervals, $z_{1-\gamma} = 1.96$.

The CLT in (1.5) (and thus (1.10)) holds provided that $\sigma(f) < \infty$. Observe that if

$$\sup_{s\in S} |f(s)| = b < \infty$$

and $E[\tau_1^2] < \infty$, then $\sigma(f) < \infty$. Alternative conditions which ensure that $\sigma(f) < \infty$ are available when the regenerative process is a GSMP with state space $S$ and event set $E$ satisfying the hypotheses of Theorem 3.13 of Chapter 2. Fix $e^* \in E$ and $S^* \subseteq S$, and let $\{T_n\colon n \geq 0\}$ be a finite-valued increasing sequence of stopping times for the underlying GSSMC $\{(S_n, C_n)\colon n \geq 0\}$ that are state transition times as in (3.7) of Chapter 2: $T_0 = 0$ and

$$T_n = \inf\{t > T_{n-1}\colon \text{at time } t \text{ event } e^* \text{ triggers} \\ \text{a state transition in some } s^* \in S^*\}$$

for $n \geq 1$. Assume that there exist $s, s_0' \in S$ and $\delta > 0$ such that $P\{X(T_n) = s_0' \mid X(T_{n-1}), \ldots, X(T_0)\} \geq \delta$ a.s. and for $s^* \in S^*$,

(i) $O(s_0'; s^*, e^*) = \emptyset$;

(ii) $N(s_0'; s^*, e^*) = N(s_0'; s, e^*)$; and
(iii) $F(\cdot\,; s_0', e', s^*, e^*) = F(\cdot\,; s_0', e', s, e^*)$ for all events $e' \in N(s_0'; s^*, e^*)$.

Then the successive random times at which $\{X(T_n) \colon n \geq 0\}$ hits state $s_0'$ form a sequence of regeneration points for the process $\{X(t) \colon t \geq 0\}$. The expected time between regeneration points is finite provided that

$$\sup_{n \geq 0} E[T_{n+1} - T_n] \leq b < \infty.$$

If $S$ is finite, or if $S$ is countably infinite and $f$ is bounded, $\sigma(f) < \infty$ provided that

$$\sup_{n \geq 0} E[(T_{n+1} - T_n)^{2+\epsilon}] \leq b < \infty$$

for some $\epsilon > 0$.

We now consider the bias of the estimator $\hat{r}(n)$. Write

$$\begin{aligned}\hat{r}(n) - r(f) &= \frac{\bar{Y}(n)}{\bar{\tau}(n)} - r(f)\\ &= \frac{\bar{Z}(n)}{\bar{\tau}(n)}\\ &= \bar{Z}(n)\Big(\frac{1}{E[\tau_1]} - \frac{(\bar{\tau}(n) - E[\tau_1])}{E^2[\tau_1]}\\ &\qquad + \frac{(\bar{\tau}(n) - E[\tau_1])^2}{E^3[\tau_1]} + O\big((\bar{\tau}(n) - E[\tau_1])^3\big)\Big).\end{aligned} \tag{1.11}$$

Taking expectations on both sides of (1.11), we obtain the formal expression

$$E[\hat{r}(n)] = r(f) - \frac{1}{n}\frac{E[Z_1\tau_1]}{E^2[\tau_1]} + \frac{1}{n^2}\frac{E[Z_1(\tau_1 - E[\tau_1])^2]}{E^3[\tau_1]} + O(n^{-3}) \tag{1.12}$$

and conclude that the estimator $\hat{r}(n)$ is biased. The bias expansion in (1.12) is not rigorous; in general, $E[\hat{r}(n)]$ does not even exist.

Several alternative estimators that attempt to reduce the bias have been proposed for $r(f)$. The first alternative estimator is known as the *Beale estimator*. If $\bar{\tau}(n) \neq 0$ and $\bar{Y}(n) \neq 0$, set

$$\hat{r}_1(n) = \hat{r}(n) \frac{\left(\frac{n-1}{n} + \frac{1}{n^2} \sum_{i=1}^n \frac{Y_i \tau_i}{\bar{Y}(n)\bar{\tau}(n)}\right)}{\left(\frac{n-1}{n} + \frac{1}{n^2} \sum_{i=1}^n \frac{\tau_i^2}{\bar{\tau}^2(n)}\right)};$$

if either $\bar{\tau}(n)$ or $\bar{Y}(n)$ is 0, set $\hat{r}_1(n) = \hat{r}(n)$.

The second alternative estimator is known as the *Tin estimator*. If $\bar{\tau}(n) \neq 0$, set

$$\hat{r}_2(n) = \hat{r}(n) + \frac{1}{n^2} \sum_{i=1}^n \frac{(Y_i - \hat{r}(n)\tau_i)\tau_i}{\bar{\tau}^2(n)};$$

if $\bar{\tau}(n) = 0$, set $\hat{r}_2(n) = \hat{r}(n)$.

The *jackknife estimator* is defined as follows. Let

$$\hat{r}_3(n) = \hat{r}(n) - \frac{n-1}{n} \sum_{i=1}^n \theta_i$$

where

$$\theta_i = \begin{cases} \frac{\sum_{k \neq i} Y_k}{\sum_{k \neq i} \tau_k} & \text{if } \sum_{k \neq i} \tau_k \neq 0 \\ 0 & \text{if } \sum_{k \neq i} \tau_k = 0 \,. \end{cases}$$

Formally expanding the expectations of these estimators in powers of $1/n$, we obtain

$$E[\hat{r}_i(n)] = r(f) + O(n^{-2})$$

for $i = 1, 2$, and 3. Thus, to order $1/n^2$, the three modified estimators are unbiased. Theorem 1.13 gives properties of these estimators. Portions of the proof require methods of advanced probability.

**Theorem 1.13.**

*(i) Assume that* $E[\| Y_1 \|] < \infty$, $E[\| \tau_1 \|] < \infty$, *and* $E[\tau_1] \neq 0$. *Then* $\hat{r}_1(n)$, $\hat{r}_2(n)$, *and* $\hat{r}_3(n)$ *are strongly consistent for* $r(f)$.

*(ii) Assume that* $E[\| Y_1 \|] < \infty$, $E[\| \tau_1 \|] < \infty$, $E[\tau_1] \neq 0$, *and* Var $[Z_1] < \infty$. *Then*

$$\sqrt{n}\,\{\hat{r}_i(n) - r\} \Rightarrow \sigma N(0,1)$$

*as* $n \to \infty$, *where* $\sigma^2 = \text{Var}\,[Z_1]/E^2[\tau_1]$.

*(iii) Under the same assumptions as (ii), set*

$$s_i^2(n) = \begin{cases} \frac{1}{n-1} \sum_{k=1}^{n} \frac{(Y_k - \hat{r}_i(n)\tau_k)^2}{\bar{\tau}^2(n)} & \text{if } \bar{\tau}(n) \neq 0 \\ 0 & \text{if } \bar{\tau}(n) = 0\ . \end{cases}$$

*Then* $s_i^2(n)$ *is weakly consistent for* $\sigma^2$ *in the sense that* $s_i^2(n)$ *converges in probability to* $\sigma^2$.

*(iv) In addition to the assumptions of (ii), assume that* Var $[Z_1] > 0$. *Set* $\hat{r}_0(n) = \hat{r}(n)$ *and* $s_0^2(n) = s^2(n)$. *Then (for* $i, j = 0, 1, 2, 3$*)*

$$\left[\hat{r}_i(n) - \frac{s_j(n) z_{1-\gamma}}{\sqrt{n}}, \hat{r}_i(n) + \frac{s_j(n) z_{1-\gamma}}{\sqrt{n}}\right]$$

*is an asymptotic* $100(1-2\gamma)\%$ *confidence interval for* $r(f)$.

According to part (iv) of Theorem 1.13, asymptotic confidence intervals may be based on any of the estimators $\hat{r}_0(n)$, $\hat{r}_1(n)$, $\hat{r}_2(n)$, and $\hat{r}_3(n)$.

**Algorithm 1.14.** *Regenerative Method*

1. *Select a sequence* $\{T_k\colon\ k \geq 0\}$ *of regeneration points for the process* $\{X(t)\colon\ t \geq 0\}$, *and observe a fixed number* $n$ *of cycles.*
2. *Measure* $(Y_1(f), \tau_1), (Y_2(f), \tau_2), \ldots, (Y_n(f), \tau_n)$.
3. *Form the point estimate*

$$\hat{r}(n) = \frac{\bar{Y}(n)}{\bar{\tau}(n)}$$

*for* $r(f)$.

4. *Form the asymptotic* $100(1-2\gamma)\%$ *confidence interval*

$$\left[\hat{r}(n) - \frac{s(n)z_{1-\gamma}}{\bar{\tau}(n)\sqrt{n}},\ \hat{r}(n) + \frac{s(n)z_{1-\gamma}}{\bar{\tau}(n)\sqrt{n}}\right].$$

*for* $r(f)$.

As a consequence of (1.5), the number $(n)$ of cycles required to obtain an asymptotic $100(1-2\gamma)\%$ confidence interval with a half-length that is $100\delta\%$ of $r(f)$ is approximately

$$n = \left(\frac{z_{1-\gamma}}{\delta}\right)^2 \left(\frac{\sigma(f)}{r(f)E[\tau_1]}\right)^2. \tag{1.15}$$

A two-stage procedure can be used. From a (short) preliminary *pilot run*, obtain point estimates for $\sigma(f)$, $r(f)$, and $E[\tau_1]$, and determine from (1.15) a value $(n)$ for the number of cycles required. Then simulate the regenerative process for this fixed number of cycles and use Algorithm 1.14 to obtain point and interval estimates for $r(f)$.

Alternatively, a sequential procedure can be used to determine the number of cycles. The procedure employs a sequential stopping rule that terminates the simulation when the relative width of the confidence intervals falls below a specified value. For $0 < \gamma < 1/2$ and $\delta > 0$, define a random variable

$$n(\delta, \gamma) = \min \left\{ n\colon\ n \geq 2, s(n) > 0, \frac{2z_{1-\gamma} s(n)}{\hat{r}(n)\bar{\tau}(n)\sqrt{n}} < \delta \right\},$$

and terminate the simulation when $n(\delta, \gamma)$ cycles have been simulated. Theorem 1.16 shows that for $\delta$ sufficiently small, the interval

$$\left[ \hat{r}(n(\delta,\gamma)) - \frac{s(n(\delta,\gamma))z_{1-\gamma}}{(\bar{\tau}(n(\delta,\gamma))\sqrt{n(\delta,\gamma)})}, \hat{r}(n(\delta,\gamma)) + \frac{s(n(\delta,\gamma))z_{1-\gamma}}{(\bar{\tau}(n(\delta,\gamma))\sqrt{n(\delta,\gamma)})} \right]$$

contains the unknown constant $r(f)$ approximately $100(1 - 2\gamma)\%$ of the time.

**Theorem 1.16.** *Assume that* $0 < E[\tau_1^2] < \infty$, $0 < E[Y_1^2(f)] < \infty$, *and* $P\{Y_1(f) - r(f)\tau_1 = 0\} < 1$. *Then*

$$\frac{\sqrt{n(\delta,\gamma)}\{\hat{r}(n(\delta,\gamma)) - r(f)\}}{s(n(\delta,\gamma))/\bar{\tau}(n(\delta,\gamma))} \Rightarrow N(0,1)$$

*as* $\delta \to 0$.

There is an analogous procedure for obtaining strongly consistent point estimates and asymptotic confidence intervals from a fixed length ($t$) simulation of a regenerative process. The procedure prescribes observation of a single sample path of $\{X(u)\colon 0 \leq u \leq t\}$. The idea is to follow Algorithm 1.14, except that we compute statistics for the random number $n(t)$ of cycles completed by time $t$; for example,

$$\hat{r}(n(t)) = \frac{\bar{Y}(n(t))}{\bar{\tau}(n(t))}.$$

The CLT for random partial sums implies that (1.5) holds with $n$ replaced by $n(t)$:

$$\frac{\sqrt{n(t)}\{\hat{r}(n(t)) - r(f)\}}{\sigma(f)/E[\tau_1]} \Rightarrow N(0,1) \tag{1.17}$$

as $t \to \infty$. The estimator $\hat{r}(n(t))$ is strongly consistent for $r(f)$ and has bias of order $1/t$.

An alternative procedure is to continue the simulation until the first regeneration point $T_{n(t)+1}$ after time $t$. The resulting estimator

$$\hat{r}(n(t)+1) = \frac{\bar{Y}(n(t)+1)}{\bar{\tau}(n(t)+1)}$$

has bias of order $1/t^2$, and a bias reduction is obtained.

**Example 1.18.** (Machine availability) Assume that service time at center 1 is exponentially distributed (as $V_1$) and service time at center 2 is distributed as a positive random variable $V_2$ with finite mean but otherwise arbitrary distribution. Let $X(t)$ be the number of customers waiting or in service at center 2 at time $t$. Denote by $L(t)$ the last state occupied by the process $\{X(t)\colon t \geq 0\}$ before jumping to $X(t)$, and set $V(t) = (L(t), X(t))$. Let $\zeta_n$ be the time of the $n$th state transition, $n \geq 0$. The process $\{V(\zeta_n)\colon n \geq 0\}$ is a regenerative process in discrete time. To see this, set $G = \{(s+1, s)\colon 0 \leq s < N\}$ and select $v_0 = (s_0 + 1, s_0) \in G$. Then the successive random times at which $\{V(\zeta_n)\colon n \geq 0\}$ hits state $v_0$ (that is, the times at which the process $\{X(t)\colon t \geq 0\}$ makes a state transition from $s_0 + 1$ to $s_0$) form a sequence of regeneration points, and the expected time between regeneration points is finite. Moreover, $V(\zeta_n) \Rightarrow V$ as $n \to \infty$.

Define the limiting probability that there is no spare available when a component fails as

$$P\{V = (N-1, N) | V \in G\} = \frac{P\{V = (N-1, N)\}}{P\{V \in G\}}.$$

Let $\gamma_m$ be the length (number of state transitions) of the $m$th cycle of $\{V(\zeta_n): n \geq 0\}$ defined by the successive random times at which the process hits the fixed state $v_0$. Set $\delta_m = \gamma_1 + \cdots + \gamma_m$ for $m \geq 1$. Now let $M_1$ be the total number of service completions at center 1 and $Y_1$ be the number of service completions at center 1 that leave center 1 idle in the first cycle of the process $\{V(\zeta_n): n \geq 0\}$; that is,

$$M_1 = \sum_{n=0}^{\delta_1 - 1} 1_G(V(\zeta_n))$$

and

$$Y_1 = \sum_{n=0}^{\delta_1 - 1} 1_{\{(N-1,N)\}}(V(\zeta_n)).$$

Denote the analogous quantities in the $k$th cycle by $M_k$ and $Y_k$. Since $\{V(\zeta_n): n \geq 0\}$ is a regenerative process, the sequence $\{(Y_k, M_k, \delta_k): k \geq 1\}$ consists of i.i.d. random vectors,

$$P\{V = (N-1, N)\} = \frac{E[Y_1]}{E[\delta_1]},$$

and

$$P\{V \in G\} = \frac{E[M_1]}{E[\delta_1]}.$$

Then

$$\frac{P\{V = (N-1, N)\}}{P\{V \in G\}} = \frac{E[Y_1]}{E[M_1]} \stackrel{\text{def}}{=} r.$$

From a fixed number $n$ of cycles, the regenerative method provides an estimator

$$\hat{r}(n) = \frac{\bar{Y}(n)}{\bar{M}(n)} = \frac{\sum_{k=1}^{n} Y_k}{\sum_{k=1}^{n} M_k}$$

that is strongly consistent for $r$. Asymptotic confidence intervals for $r$ are based on the CLT

$$\frac{\sqrt{n}\{\hat{r}(n) - r\}}{\sigma/E[M_1]} \Rightarrow N(0,1)$$

as $n \to \infty$, where $\sigma^2 \stackrel{\text{def}}{=} \text{Var}\,[Y_1 - rM_1]$.

**Example 1.19.** (Finite-capacity queue with preemptive-resume priority) Consider an open queueing system with one single-server center and two job types, numbered 1 and 2. Jobs of type $i$ arrive from an external source, queue and receive service at the center, and depart to an external sink. A job of type $i$ that arrives when the system already contains $B_i - 1$ ($\geq 0$) jobs of type $i$ causes the arrival process for jobs of type $i$ to shut down; the arrival process remains shut down until the first subsequent departure of a job of type $i$. Jobs of the same type receive service in order of their arrival at the center, and jobs of type 1 have preemptive-resume priority over jobs of type 2. A job of type 2 that has been preempted receives additional service at the center before any other job of type 2 receives service. Observe that there can be at most one preempted job of type 2 in the queue.

Assume that all interarrival times and service times are mutually independent and the interarrival times for job $i$ are identically distributed as a positive random variable $A_i$ with finite mean and a density function that is continuous and positive on $(0, +\infty)$. Also assume that service times for jobs of type $i$ are identically distributed as a positive random variable $V_i$ with finite mean and a density function that is continuous and positive on $(0, +\infty)$.

Set

$$X(t) = (B_2(t), B_1(t), M(t), L(t)), \tag{1.20}$$

where $B_i(t)$ is the number of jobs of type $i$ waiting in queue at time $t$,

$$L(t) = \begin{cases} 1 & \text{if the job at the head of the type 2 queue} \\ & \text{at time } t \text{ has been preempted} \\ 0 & \text{otherwise,} \end{cases}$$

and

$$M(t) = \begin{cases} 2 & \text{if there is a job of type 2 in service at time } t \\ 1 & \text{if there is a job of type 1 in service at time } t \\ 0 & \text{if there is no job in service at time } t. \end{cases}$$

Take event $e_i$ = "service completion for job of type $i$" and $e_{2+i}$ = "arrival of job of type $i$." Formal specification of the process $\{X(t)\colon t \geq 0\}$ is as a GSMP with finite state space

$$S = \big\{(b_2, b_1, m, l) \in \{0, 1, \ldots, B_2\} \times \{0, 1, \ldots, B_1\} \times \{0, 1, 2\} \times \{0, 1\} : \ m = 2 \text{ only if } b_1 = 0;\ m = 0 \text{ only if } b_2 = b_1 = 0;\ l = 1 \text{ only if } b_2 > 0 \text{ and } m = 1\big\}$$

and event set $E = \{e_1, e_2, e_3, e_4\}$.

For $s = (b_2, b_1, m, l) \in S$, event $e_1$ = "service completion for job of type 1" $\in E(s)$ if and only if $m = 1$. Event $e_2$ = "service completion for job of type 2" $\in E(s)$ if and only if $m = 2$ or $l = 1$. Event $e_{2+i}$ = "arrival of job of type $i$" $\in E(s)$ if and only if $b_i + 1_{\{i\}}(m) < B_i$. The speed $r(s, e_2)$ at which the clock associated with event $e_2$ runs in state $s = (b_2, b_1, 1, 1) \in S$ is equal to 0. All other speeds for active events are equal to 1.

If $e = e_1$ = "service completion for job of type 1," then the state transition probability $p(s'; s, e) = 1$ when

$$s = (b_2, b_1, 1, l) \text{ with } b_1 > 0 \text{ and } s' = (b_2, b_1 - 1, 1, l),$$

when

$$s = (b_2, 0, 1, l) \text{ with } b_2 > 0 \text{ and } s' = (b_2 - 1, 0, 2, l),$$

and when

$$s = (0, 0, 1, 0) \text{ and } s' = (0, 0, 0, 0).$$

If $e = e_2 =$ "service completion for job of type 2," then $p(s'; s, e) = 1$ when

$$s = (b_2, 0, 2, 0) \text{ with } b_2 > 0 \text{ and } s' = (b_2 - 1, 0, 2, 0)$$

and when

$$s = (0, 0, 2, 0) \text{ and } s' = (0, 0, 0, 0).$$

If $e = e_3 =$ "arrival of job of type 1," then $p(s'; s, e) = 1$ when

$$s = (0, 0, 0, 0) \text{ and } s' = (0, 0, 1, 0),$$

when

$$s = (b_2, 0, 2, 0) \text{ and } s' = (b_2 + 1, 0, 1, 1),$$

and when

$$s = (b_2, b_1, 1, l) \text{ and } s' = (b_2, b_1 + 1, 1, l).$$

If $e = e_4 =$ "arrival of job of type 2," then $p(s'; s, e) = 1$ when

$$s = (0, 0, 0, 0) \text{ and } s' = (0, 0, 2, 0)$$

and when

$$s = (b_2, b_1, m, l) \text{ with } m > 0 \text{ and } s' = (b_2 + 1, b_1, m, l).$$

All other state transition probabilities are equal to 0.

The set $N(s'; s, e_1)$ of new events when event $e_1$ triggers a state transition from $s = (b_2, b_1, m, l)$ to $s' = (b_2', b_1', m', m')$ equals $\{e_1\}$ if $m' = 1$, equals $\{e_2\}$ if $m' = 2$, and equals $\emptyset$

otherwise. The set $N(s'; s, e_2)$ equals $\{e_1\}$ if $m' = 1$, equals $\{e_2\}$ if $m' = 2$, and equals $\emptyset$ otherwise. The set $N(s'; s, e_3)$ equals $\{e_1, e_3\}$ if $s = (0,0,0,0)$ and equals $\{e_3\}$ otherwise. The set $N(s'; s, e_4)$ equals $\{e_2, e_4\}$ if $s = (0,0,0,0)$ and equals $\{e_4\}$ otherwise. The clock-setting distribution functions are as follows. If the new event $e' = e_i$ = "service completion for job of type $i$" and $s' = (b'_2, b'_1, i, l')$, then $F(x; s', e', s, e^*) = P\{V_i \leq x\}$. If $e' = e_{2+i}$ = "arrival of job of type $i$", then $F(x; s', e', s, e^*) = P\{A_i \leq x\}$ for $x \geq 0$.

Take $s_0 = (B_2, B_1 - 1, 1, 0)$ and $s'_0 = (B_2, B_1 - 2, 1, 0)$. A job of type 1 is in service, there are $B_1 - 1$ jobs of type 1 and $B_2$ jobs of type 2 waiting in queue, and none of the type 2 jobs have been preempted when the process $\{X(t)\colon t \geq 0\}$ is in state $s$. When event $e_1$ occurs and there is a state transition to $s'_0$, a job of type 1 goes into service. Observe that there is a state transition from $s_0$ to $s'_0$ only when event $e^* = e_1$ is the trigger event. Also observe that $\{X(t)\colon t \geq 0\}$ can be specified as an irreducible GSMP with a finite state space and each clock-setting distribution has finite mean and density function that is continuous and positive on $(0, +\infty)$. The successive random times $\{T_k\colon k \geq 0\}$ at which there is a state transition from $s_0$ to $s'_0$ form a sequence of regeneration points for the process $\{X(t)\colon t \geq 0\}$, and the expected time between regeneration points is finite by the argument in the proof of Theorem 3.2 of Chapter 2. Since the time between regeneration points is aperiodic, $X(t) \Rightarrow X$ as $t \to \infty$.

The regenerative method can be used to obtain point and interval estimates for the limiting expected number of jobs of type $i$ in the system. Define real-valued functions $f_i$ with domain $S$ as follows. For $s = (b_2, b_1, m, l) \in S$, set $f_i(s)$ equal to $b_i + 1_{\{i\}}(m)$. Then define the limiting expected number of jobs of type $i$ in the system as $E\big[f_i(X)\big]$. Set

$$Y_k(f_i) = \int_{T_{k-1}}^{T_k} f_i(X(u))du$$

and $\tau_k = T_k - T_{k-1}$, $k \geq 1$. The sequence $\{(Y_k(f_i), \tau_k)\colon\ k \geq 1\}$ consists of i.i.d. random vectors and

$$E\big[f_i(X)\big] = \frac{E\big[Y_1(f_i)\big]}{E\big[\tau_1\big]} \stackrel{\text{def}}{=} r(f_i).$$

Fix $n$ and measure the quantities $(Y_1(f_i), \tau_1)$, $(Y_2(f_i), \tau_2) \ldots$, $(Y_n(f_i), \tau_n)$. From $n$ cycles, the estimator

$$\hat{r}(n) = \frac{\bar{Y}(n)}{\bar{\tau}(n)} = \frac{\sum_{k=1}^{n} Y_k(f_i)}{\sum_{k=1}^{n} \tau_k}$$

is strongly consistent for $r(f_i)$. An asymptotic $100(1-2\gamma)\%$ confidence interval is

$$\left[\hat{r}(n) - \frac{s(n) z_{1-\gamma}}{\bar{\tau}(n)\sqrt{n}}, \hat{r}(n) + \frac{s(n) z_{1-\gamma}}{\bar{\tau}(n)\sqrt{n}}\right],$$

where the estimator $s^2(n)$ is strongly consistent for $\sigma^2(f_i) \stackrel{\text{def}}{=}$ Var $[Y_1(f_i) - r(f_i)\tau_1]$. Asymptotic confidence intervals for $r(f)$ are based on the CLT

$$\frac{\sqrt{n}\,\{\hat{r}(n) - r(f_i)\}}{\sigma(f_i)/E[\tau_1]} \Rightarrow N(0,1)$$

as $n \to \infty$.

**Example 1.21.** (Finite-capacity queue with preemptive-resume priority) Set $X(t) = \big(B_2(t), B_1(t), M(t), L(t)\big)$ as in (1.20). A *sojourn time* (total time in system) for a job of type $i$ starts when the job arrives and terminates when the job completes service. Observe that sojourn times are positive. Let $\zeta_n$ be the time of the $n$th state transition, $n \geq 0$. Set

$$\begin{aligned} \mathcal{S}_j(i) = \inf\{\zeta_n > \mathcal{T}_{j-1}(i)\colon\ & B_i(\zeta_n) = B_i(\zeta_{n-1}) + 1 \\ & \text{or } M(\zeta_n) = i, M(\zeta_{n-1}) = 0\} \end{aligned}$$

and

$$\mathcal{T}_j(i) = \inf\{\zeta_n > \mathcal{S}_j(i)\colon\ B_i(\zeta_n) = B_i(\zeta_{n-1}) - 1, M(\zeta_n) = M(\zeta_{n-1}) = i \text{ or } M(\zeta_n) \neq i, M(\zeta_{n-1}) = i\}$$

for $j \geq 1$ $(\mathcal{T}_0(i) = 0)$. Then the $j$th sojourn time for jobs of type $i$ is $D_j(i) = \mathcal{T}_j(i) - \mathcal{S}_j(i)$.

The regenerative method can be used to obtain point and interval estimates for the limiting expected sojourn time for jobs of type 1. Take $S^* = \{(b_2, b_1, m, l) \in S\colon\ b_1 = 0,\ m \neq 1\}$ and $e^* = e_3 =$ "arrival of job of type 1." Set $T_{-1} = -1$ and

$$T_n = \inf\{t > T_{n-1}\colon\ \text{at time } t \text{ event } e^* \text{ triggers a state transition in some state } s^* \in S^*\}$$

for $n \geq 0$, so that $T_n$ is the time of the $n$th arrival of a job of type 1 with no other job of type 1 in the system. Let $Y_k$ be the sum of the sojourn times for jobs of type 1 and $N_k$ be the total number of sojourn times for jobs of type 1 in the $k$th block defined by the random times $\{T_n\colon\ n \geq 0\}$. Then the sequence $\{(Y_k, N_k)\colon\ k \geq 1\}$ consists of i.i.d. random vectors and

$$E[D(1)] = \frac{E[Y_1]}{E[N_1]},$$

where $D(1)$ is the limiting sojourn time for jobs of type 1.

In general, the regenerative method cannot be used to obtain point and interval estimates for the limiting expected sojourn time for jobs of type 2. Observe that there are two arrival processes and jobs of type 1 have preemptive-resume priority over jobs of type 2. It is not possible to decompose the sequence of sojourn times for jobs of type $i$ so as to obtain i.i.d. random vectors and an appropriate ratio formula unless, for example, the interarrival times for jobs of type 1 are exponentially distributed. In particular, sojourn times for jobs of type 2 are not completely contained within cycles defined by the successive random times at which there is a state transition from $s_0 = (B_2, B_1 - 1, 1, 0)$ to $s_0' = (B_2, B_1 - 2, 1, 0)$.

**Example 1.22.** (Patrolling repairman) Assume that the walking time from machine $j$ to the next machine is a positive constant $W_j$ and the time to repair and restart machine $j$ is distributed as a positive random variable $R_j$ with finite mean. Also assume that the lifetime for machine $j$ is exponentially distributed as a random variable $L_j$. Set

$$X(t) = (Z_1(t), \ldots, Z_N(t), M(t), N(t)),$$

where

$$Z_j(t) = \begin{cases} 1 & \text{if machine } j \text{ is waiting for repair at time } t \\ 0 & \text{otherwise,} \end{cases}$$

$$M(t) = \begin{cases} j & \text{if machine } j \text{ is under repair at time } t \\ 0 & \text{if no machine is under repair at time } t, \end{cases}$$

$N(t) = 1$ if machine $N$ is under repair or the repairman is walking to machine 1 at time $t$, and $N(t) = j$ if machine $j - 1$ is under repair or the repairman is walking to machine $j$ at time $t$ $(j = 2, 3, \ldots, N)$. The successive random times at which there is a state transition from $(0, \ldots, 0, 0, 1)$ to $(0, \ldots, 0, 1, 2)$ form a sequence of regeneration points for the process $\{X(t)\colon t \geq 0\}$, and the expected time between regeneration points is finite. These random times correspond to the arrivals of the repairman at machine 1 with all machines running.

The regenerative method can be used to obtain point and interval estimates for repairman utilization and the number of machines that are stopped. Denote the state space of the process $\{X(t)\colon t \geq 0\}$ by $S$, and define real-valued functions $f$, $f_0$, $f_1$, ..., $f_N$ with domain $S$ as follows. For $s = (z_1, \ldots, z_N, m, n) \in S$, set

$$f(s) = \begin{cases} 1 & \text{if } m > 0 \\ 0 & \text{otherwise} \end{cases}$$

and

$$f_i(s) = \begin{cases} 1 & \text{if } z_1 + \cdots + z_N + 1_{\{1,2,\ldots,N\}}(m) = i \\ 0 & \text{otherwise} \end{cases}$$

$(i = 0, 1, \ldots, N)$. Define repairman utilization as $E[f(X)]$ and the limiting probability that $i$ machines are stopped as $E[f_i(X)]$. Point and interval estimates for these quantities are given in Table 3.1. The theoretical value for repairman utilization is 0.4462.

**Exercises**

1.1. (Shuttle train with finite waiting room) Consider a train that shuttles back and forth between two stations and has seats for $K$ $(\geq 1)$ passengers. Passengers that wish to board the train and ride to the other station arrive at station $i$ according to a Poisson process of rate $\lambda_i$, queue, and subsequently board the train in the order in which they arrive at the station. Assume that passengers board the train instantaneously and all passengers that board at station $i$ disembark instantaneously when the train arrives at the other station. At each station passengers disembark before any waiting passengers board the train. The waiting room at station $i$ has finite capacity $B_i$ $(\geq 1)$. A passenger that wishes to board the train at station $i$ and arrives when there are already $B_i$ passengers waiting to board is turned away. The time required for the train to travel from one station to the other is distributed as a positive random variable $L$ with finite mean. Travel times are independent of interarrival times, and the arrival process of passengers for boarding at station 1 is independent of the arrival process of passengers for boarding at station 2.

Describe how you would use regenerative simulation to obtain strongly consistent point estimates and asymptotic confidence intervals for the limiting expected number of passengers waiting to board at station 1 and the limiting probability that all seats on the train are occupied when the train leaves station 1. State the ratio formulas and central limit theorems that lead to the point and interval estimates.

1.2. (Airport shuttle) Consider an airport shuttle that provides transportation service to $N$ stations numbered $1, 2, \ldots, N$. The shuttle has seats for $K$ ($\geq 1$) passengers and moves from station to station in a strictly defined order: $1, 2, \ldots, N, N-1, \ldots, 1, 2, \ldots$. Passengers that wish to board the shuttle at station $i$ in the "northbound" (from station 1 to station $N$) direction arrive at station $i$ according to a Poisson process of rate $\lambda_i$ and disembark at station $j$ ($> i$) with probability $p_{i,j}$ ($i = 1, 2, \ldots, N-1$). Passengers that wish to board the shuttle at station $i+1$ in the "southbound" (from station $N$ to station 1) direction arrive at station $i+1$ according to a Poisson process of rate $\lambda_{2N-(i+1)}$ and disembark at station $j$ ($< i+1$) with probability $p_{2N-(i+1),j}$ ($i = 1, 2, \ldots, N-1$). Passengers that arrive for boarding at station $i$ in each direction queue and subsequently board the shuttle in the order in which they arrive at the station. At each station passengers disembark before any waiting passengers board the shuttle. Assume that passengers board the shuttle instantaneously and passengers disembark instantaneously. The time required for the shuttle to travel (in the northbound direction) from station $i$ to station $i+1$ is distributed as a positive random variable $L_i$ with finite first and second moments, and the time required for the shuttle to travel (in the southbound direction) from station $i+1$ to station $i$ is distributed as a positive random variable $L_{2N-(i+1)}$ with finite first and second moment. Travel times, interarrival times, and destinations are mutually independent. Assume that $\sum_{k=1}^{2(N-1)} \lambda_k \sum_{k=1}^{2(N-1)} E[L_k] < K$.

Describe how you would use regenerative simulation to obtain a strongly consistent point estimate and an asymptotic confidence interval for the limiting expected number of passengers waiting to board the shuttle at station $i$ and the limiting probability that all seats on the shuttle are occupied when it leaves station $i$ in the northbound direction.

1.3. (Telephone system) Consider a telephone system with $N$ ($\geq 2$) telephones connected to a switchboard by lines numbered $1, 2, \ldots, N$. The switchboard has $K$ ($< N$) links numbered $1, 2, \ldots, K$, each of which can connect any two lines subject to the restriction that only one connection at a time can be made to each line. If more than one link is available and the called line is not in use, a placed call is connected (instantaneously) on the lowest numbered available link. Assume that the system is a lost-call system in the sense that any call is immediately lost if no connection can be made when it is placed. A call is lost if there is at least one link available but the called line is in use (a *busy call*) and a call is lost if no link is available (a *blocked call*).

Assume that the duration of a call placed at line $i$ is distributed as a positive random variable $V_i$ with finite mean and a density function that is continuous and positive on $(0, +\infty)$. Also assume that calls are placed randomly in the sense that the time from the end of a call (placed or received) at line $i$ to the next call placed at line $i$ is exponentially distributed as a random variable $A_i$. When a call is placed, with independent probability $p_{ij}$ the called line is line $j$ ($\neq i$). After a lost call placed at line $i$, the time to the next call placed at line $i$ is distributed as $A_i$.

Describe how you would use regenerative simulation to obtain strongly consistent point estimates and asymptotic confidence intervals for the limiting probabilities that a call is successful, busy, and blocked. State the ratio formulas and central limit theorems that lead to the point and interval estimates.

**Table 3.1**

Simulation results for patrolling repairman

$$N = 4,\ E[L_1] = E[L_2] = E[L_3] = E[L_4] = 1$$

$$W_1 = W_2 = W_3 = W_4 = 0.05$$

$$E[R_1] = E[R_2] = E[R_3] = E[R_4] = 0.15$$

Exponentially distributed lifetimes for all machines
Constant repair times for all machines
Cycles defined by state transitions from
(0,0,0,0,0,1) to (0,0,0,0,0,2)

Number of cycles

| | 1000 | 2000 | 3000 | 4000 | 5000 |
|---|---|---|---|---|---|
| $E[f(X)]$ | 0.4473 | 0.4403 | 0.4400 | 0.4417 | 0.4432 |
| | ±0.0142 | ±0.0092 | ±0.0081 | ±0.0067 | ±0.0056 |
| $E[f_0(X)]$ | 0.3065 | 0.3138 | 0.3130 | 0.3101 | 0.3090 |
| | ±0.0105 | ±0.0078 | ±0.0063 | ±0.0054 | ±0.0048 |
| $E[f_1(X)]$ | 0.4233 | 0.4250 | 0.4224 | 0.4230 | 0.4219 |
| | ±0.0152 | ±0.0113 | ±0.0092 | ±0.0079 | ±0.0071 |
| $E[f_2(X)]$ | 0.2168 | 0.2102 | 0.2147 | 0.2158 | 0.2176 |
| | ±0.0111 | ±0.0081 | ±0.0065 | ±0.0057 | ±0.0050 |
| $E[f_3(X)]$ | 0.0491 | 0.0474 | 0.0466 | 0.0473 | 0.0478 |
| | ±0.0064 | ±0.0047 | ±0.0038 | ±0.0037 | ±0.0030 |
| $E[f_4(X)]$ | 0.0043 | 0.0036 | 0.0034 | 0.0038 | 0.0037 |
| | ±0.0015 | ±0.00096 | ±0.00075 | ±0.00069 | ±0.00062 |

## 3.2. Implementation Considerations

We first consider the computation of the quantities $s_{11}(n)$, $s_{22}(n)$, and $s_{12}(n)$ that appear in (1.3) and lead to an estimator $s(n)$ that is strongly consistent for $\sigma(f)$. A numerically stable, one-pass updating algorithm can be used to compute $s_{11}(n)$. Set $s_{11}^{(1)} = 0$ and compute

$$s_{11}^{(j+1)} = s_{11}^{(j)} + \frac{\left(\sum_{k=1}^{j} Y_k(f) - jY_{j+1}(f)\right)}{j}\frac{\left(\sum_{k=1}^{j} Y_k(f) - jY_{j+1}(f)\right)}{j+1}$$

for $j = 1, 2, \ldots, n-1$. Then $s_{11}(n) = s_{11}^{(n)}/(n-1)$. The sample variance $s_{22}(n)$ can be computed from the observations $\tau_1, \tau_2, \ldots, \tau_n$ in the same manner. Set $s_{22}^{(1)} = 0$ and compute

$$s_{22}^{(j+1)} = s_{22}^{(j)} + \frac{\left(\sum_{k=1}^{j} \tau_k(f) - j\tau_{j+1}(f)\right)}{j}\frac{\left(\sum_{k=1}^{j} \tau_k(f) - j\tau_{j+1}(f)\right)}{j+1}$$

for $j = 1, 2, \ldots, n-1$. Then $s_{22}(n) = s_{22}^{(n)}/(n-1)$. The analogous one-pass method to compute the sample covariance $s_{12}(n)$ is as follows. Set $s_{12}^{(1)} = 0$ and compute

$$s_{12}^{(j+1)} = s_{12}^{(j)} + \frac{\left(\sum_{k=1}^{j} Y_k(f) - jY_{j+1}(f)\right)}{j}\frac{\left(\sum_{k=1}^{j} \tau_k - j\tau_{j+1}\right)}{j+1}$$

for $j = 1, 2, \ldots, n-1$. Then $s_{12}(n) = s_{12}^{(n)}/(n-1)$. Given $s_{11}(n)$, $s_{22}(n)$, and $s_{12}(n)$, compute the quantity $s(n)$ as

$$s(n) = \left\{s_{11}(n) - 2\hat{r}(n)s_{12}(n) + \hat{r}^2(n)s_{22}(n)\right\}^{1/2}.$$

The main deficiency of this method arises from possible instability in the calculation of $s_{12}(n)$ with a resulting round-off and cancellation error.

An alternative stable two-pass method to compute $s^2(n)$ is

$$s^2(n) = \frac{1}{n-1}\sum_{k=1}^{n}(Y_k(f) - \hat{r}(n)\tau_k)^2,$$

where $\hat{r}(n)$ is computed on the first pass.

A regenerative stochastic process can have more than one sequence of regeneration points. However, with high probability, confidence intervals obtained from any two sequences of regeneration points are of the same length provided that the length of the simulation is large. To see this, let $\{X(t)\colon\ t \geq 0\}$ be a regenerative process in continuous time with state space $S$, and let $\{T_k(1)\colon\ k \geq 0\}$ and $\{T_k(2)\colon\ k \geq 0\}$ be two sequences of regeneration points. For a real-valued measurable function with domain $S$, set $T_{-1}(i) = 0$ and

$$Y_{k,i}(f) = \int_{T_{k-1}(i)}^{T_k(i)} f(X(u))du$$

for $k \geq 0$. Also set $\tau_k(i) = T_k(i) - T_{k-1}(i)$ for $k \geq 1$. Finally, set $Z_{k,i}(f) = Y_{k,i}(f) - r(f)\tau_k(i)$ and

$$\sigma_i^2(f) = \text{ Var } [Z_{1,i}(f)].$$

Ovserve that each sequence $\{Z_{k,i}(f)\colon\ k \geq 1\}$ consists of i.i.d. random vectors and $E[Z_{k,i}(f)] = 0$.

Theorem 2.5 depends crucially on an equality ((2.3) below) between certain moments associated with the two sequences of regeneration points. The equality is a consequence of the CLT's:

$$\frac{\int_0^t f(X(u))du - r(f)t}{\sqrt{t}\,\sigma_i(f)/\sqrt{E[\tau_1(i)]}} \Rightarrow N(0,1) \tag{2.1}$$

as $t \to \infty$.

**Lemma 2.2.** *Assume that $E[\tau_1(i)] < \infty$ for $i = 1, 2$. Also assume that $E[Y_{1,1}(|f|)] < \infty$ and $E[Y_{1,1}^2(|f - r(f)|)] < \infty$. Then*

$$\frac{\sigma_1^2(f)}{E[\tau_1(1)]} = \frac{\sigma_2^2(f)}{E[\tau_1(2)]}. \tag{2.3}$$

Now consider a simulation of fixed length $t$, and let $n_i(t)$ be the number of complete cycles defined by the sequence $\{T_k(i)\colon\ k \geq 0\}$. Recall from (1.15) that

$$\frac{\sqrt{n_i(t)}\{\hat{r}(n_i(t)) - r(f)\}}{\sigma_i(f)/E[\tau_1(i)]} \Rightarrow N(0,1) \tag{2.4}$$

as $t \to \infty$. Fix $\gamma$, and let $L_i(t)$ be the length of the $100(1-2\gamma)\%$ confidence interval for $r(f)$ derived from this CLT

**Theorem 2.5.** *Under the conditions of Lemma 2.2,*

$$\lim_{t\to\infty} \frac{L_1(t)}{L_2(t)} = 1$$

*with probability 1.*

*Proof.* By the strong laws for renewal processes and for partial sums,

$$\frac{n_i(t)}{t} \to \frac{1}{E[\tau_1(i)]}$$

and

$$s_i^2(n_i(t)) \to \sigma_i^2(f)$$

with probability 1 as $t \to \infty$. The CLT in (2.4) implies that

$$\sqrt{n_i(t)}L_i(t) = \frac{2s_i(n_i(t))z_{1-\gamma}}{\bar{\tau}_i(n_i(t))}.$$

Now let $t \to \infty$, and observe that

$$\sqrt{t}L_i(t) \to \frac{2\sigma_i(f)z_{1-\gamma}}{\sqrt{E[\tau_1(i)]}}$$

with probability 1 as $t \to \infty$. The desired result follows from Lemma 2.2. □

## 3.3. Discrete Time Conversion

There is an easily implemented variant of the regenerative method that is applicable to irreducible positive recurrent CTMC's. This "discrete time method" is computationally more efficient than the standard regenerative method in that fewer random numbers are required. The discrete time method is also statistically more efficient than the standard regenerative method in that it leads to shorter confidence intervals for a fixed number of regenerative cycles simulated.

Let $\{X(t)\colon t \geq 0\}$ be an irreducible positive recurrent CTMC with state space $S \subseteq \{0, 1, 2, \ldots\}$ and infinitesimal generator $\boldsymbol{Q} = \{q_{ij}\colon i, j \in S\}$. Recall that the holding time in state $i \in S$ has an exponential distribution with rate parameter $q_i = -q_{ii}$ and the elements of the jump matrix $\boldsymbol{R} = \{r_{ij}\colon i, j \in S\}$ of the CTMC are

$$r_{ij} = \begin{cases} q_{ij}/q_i & \text{if } j \neq i \\ 0 & \text{if } j = i. \end{cases} \tag{3.1}$$

The element $r_{ij}$ of the matrix $\boldsymbol{R}$ represents the probability that the CTMC makes a state transition from $i$ to $j$. Select $s_0 \in S$. The successive random times at which the process $\{X(t)\colon t \geq 0\}$ hits state $s_0$ form a sequence of regeneration points for the process, and the expected time between regeneration points is finite. Moreover, $X(t) \Rightarrow X$ as $t \to \infty$.

Let $\zeta_n$ be the time of the $n$th state transition, $n \geq 0$. Set $X(0) = s_0$, and let $T_n$ be the $n$th time at which $\{X(t)\colon t \geq 0\}$

hits state $s_0$: $T_0 = 0$ and

$$T_k = \inf\{\zeta_n > T_{k-1}\colon\ X(\zeta_n) = s_0\}$$

for $k \geq 1$. As usual, set $\tau_k = T_k - T_{k-1}$. Let $f$ be a real-valued measurable function with domain $S$, and assume that the ratio formula

$$E[f(X)] = \frac{E[Y_1(f)]}{E[\tau_1]} \stackrel{\text{def}}{=} r(f)$$

holds. Then from a fixed number of cycles, the regenerative method provides a strongly consistent point estimate for $r(f)$ and an asymptotic confidence interval.

Alternatively, let $\{X_n\colon\ n \geq 0\}$ be a DTMC with state space $S$ and transition matrix $\boldsymbol{R}$ given by (3.1). Take $X_0 = s_0$, and let $\gamma_k$ be the $k$th time at which the process $\{X_n\colon\ n \geq 0\}$ hits state $s_0$: $\gamma_0 = 0$ and

$$\gamma_k = \inf\{n > \gamma_{k-1}\colon\ X_n = s_0\}$$

for $k \geq 1$. In addition, set

$$S_k(f) = \sum_{n=\gamma_{k-1}}^{\gamma_k - 1} f(X_n) q_{X_n}^{-1}$$

and

$$\alpha_k = \sum_{n=\gamma_{k-1}}^{\gamma_k - 1} q_{X_n}^{-1}$$

The random indices $\{\gamma_k\colon\ k \geq 0\}$ form a sequence of regeneration points for the chain $\{X_n\colon\ n \geq 0\}$, and the expected time between regeneration points is finite. The regenerative property implies that the sequence $\{(S_k(f), \alpha_k)\colon\ k \geq 1\}$ consists of i.i.d. random vectors. The key observations are that

$$E[Y_k(f)] = E[S_k(f)]$$

and

$$E[\tau_k] = E[\alpha_k],$$

so that

$$r(f) = \frac{E[S_1(f)]}{E[\alpha_1]}.$$

These results imply that a strongly consistent point estimate for $r(f)$ and an asymptotic confidence interval can be obtained without generating exponential random numbers for the holding times in the CTMC $\{X(t)\colon t \geq 0\}$. The idea is to simulate the DTMC $\{X_n\colon n \geq 0\}$ and observe a fixed number of cycles defined by the random indices $\{\gamma_k\colon k \geq 0\}$. From $n$ cycles, the regenerative method provides an estimator

$$\hat{r}(n) = \frac{\bar{S}(n)}{\bar{\alpha}(n)}$$

that is strongly consistent for $r(f)$. An asymptotic $100(1 - 2\gamma)\%$ confidence interval is

$$\left[\hat{r}(n) - \frac{s(n)z_{1-\gamma}}{\bar{\alpha}(n)\sqrt{n}}, \hat{r}(n) + \frac{s(n)z_{1-\gamma}}{\bar{\alpha}(n)\sqrt{n}}\right],$$

where the estimator $s^2(n)$ is strongly consistent for $\mathrm{Var}\,[S_1(f) - r(f)\alpha_1]$. Computational methods (Theorems 4.1 and 4.10 below) for Markov chains can be used to show that

$$\mathrm{Var}\,[S_k(f) - r(f)\alpha_k] < \mathrm{Var}\,[Y_k(f) - r(f)\tau_k]. \tag{3.2}$$

The increased statistical efficiency of the discrete time method is a consequence of (3.2).

## 3.4. Theoretical Values for Markov Chains

Computational methods for discrete time and continuous time Markov chains can be used to assess the statistical efficiency of regenerative simulation. These computational methods provide theoretical values for variance constants that appear in the CLT's used to form confidence intervals.

Let $\{X_k\colon\ k \geq 0\}$ be an irreducible DTMC with finite state space $S = \{0, 1, \ldots, N\}$ and transition matrix

$$\boldsymbol{P} = \{p_{ij}\colon\ i, j \in S\}\,.$$

Denote the $n$-step transition probability from $i$ to $j$ by $p^n_{ij}$, and recall that

$$\boldsymbol{P}^n = \{p^n_{ij}\colon\ i, j \in S\}.$$

Throughout this section we use the following notation. For a fixed state $i \in S$, $P_i$ denotes the conditional probability associated with starting the chain in state $i$, and $E_i$ denotes the corresponding conditional expectation. Let $T_n(j)$ be the $n$th time at which $\{X_k\colon\ k \geq 0\}$ hits state $j$, $n \geq 1$; for example,

$$T_1(j) = \min\{k \geq 1\colon\ X_k = j\}\,.$$

Set $\tau_1(j) = T_1(j)$ and $\tau_n(j) = T_n(j) - T_{n-1}(j)$ for $n > 1$. This notation is consistent with the notation introduced in Section 3.2 for regenerative processes. Since the chain hits every fixed state $j \in S$ infinitely often with probability 1, $\{T_n(j)\colon\ n \geq 1\}$ is a (possibly delayed) renewal process. If $X_0 = j$, the process $\{T_n(j)\colon\ n \geq 1\}$ is an ordinary renewal process.

It is convenient to consider vectors such as $(v(0), v(1), \ldots, v(N))$ to be column vectors. We view real-valued functions such as $f$ and $g$ with domain $S$ in this way and denote these functions by $\boldsymbol{f}$ and $\boldsymbol{g}$. In this context, the symbol $E[\,\cdot\,]$ denotes the vector $(E_0[\,\cdot\,], E_1[\,\cdot\,], \ldots, E_N[\,\cdot\,])$. In addition, for vectors $\boldsymbol{u}$ and $\boldsymbol{v}$ the symbol $\boldsymbol{u} \circ \boldsymbol{v}$ denotes the Hadamard product of vectors; that is,

$$\boldsymbol{u} \circ \boldsymbol{v} = (u(0)v(0), u(1)v(1), \ldots, u(N)v(N)).$$

Denote the component of the vector $\boldsymbol{v}$ corresponding to state $j$ by $[\boldsymbol{v}]_j$. For a matrix $\boldsymbol{A} = (a_0, a_1, \ldots, a_m)$, set

$$\boldsymbol{u} \circ \boldsymbol{A} = \boldsymbol{A} \circ \boldsymbol{u} = (\boldsymbol{u} \circ a_0, \boldsymbol{u} \circ a_1, \ldots, \boldsymbol{u} \circ a_m).$$

Finally, for a matrix $\boldsymbol{B} = (b_0, b_1, \ldots, b_m)$, set

$$\boldsymbol{A} \circ \boldsymbol{B} = (a_0 \circ b_0, a_1 \circ b_1, \ldots, a_m \circ b_m).$$

Now we show how to compute the quantities $E_i[Y_1(f)]$ and $E_i[Y_1(f)Y_1(g)]$ for general real-valued functions $f$ and $g$ with domain $S$ and $i \in S$. We only consider cycles of the regenerative process $\{X_k\colon\ k \geq 0\}$ formed by the successive random times at which the process hits state 0, and henceforth we suppress the 0 in the notation $T_n(0), \tau_n(0)$, etc. Observe that we have placed no real restriction and equally well we could choose any other state $i \in S$. For $i, j \in S$ and $n = 0, 1, \ldots$, set

$$_0p_{ij}^n = P_i\left\{\tau_1 > n, X_n = j\right\}$$

and

$$_0\boldsymbol{P}^n = \left\{_0p_{ij}^n\colon\ i, j \in S\right\}.$$

Obtain $_0\boldsymbol{P}^1 = {_0\boldsymbol{P}}$ from $\boldsymbol{P}$ by setting the 0-column of $\boldsymbol{P}$ equal to 0. It is easy to see that $_0\boldsymbol{P}^n$ is the product of $n$ copies of $_0\boldsymbol{P}$ and $_0p_{i0}^n = 0$ for all $n \geq 1$. Set

$$Y_1(f) = \sum_{k=0}^{T_1 - 1} f(X_k)$$

and

$$Z_1(f) = Y_1(f) - r(f)\tau_1.$$

**Theorem 4.1.** *For an irreducible discrete time Markov chain with finite state space and transition matrix* $\boldsymbol{P}$*,*

$$E[Y_1(f)] = (\boldsymbol{I} - {_0\boldsymbol{P}})^{-1}\boldsymbol{f} \tag{4.2}$$

*and*

$$E[Y_1(f)Y_1(g)] = (\boldsymbol{I} - {_0\boldsymbol{P}})^{-1}\boldsymbol{h}, \tag{4.3}$$

where $\boldsymbol{h} = \boldsymbol{f} \circ E[Y_1(g)] + \boldsymbol{g} \circ E[Y_1(f)] - \boldsymbol{f} \circ \boldsymbol{g}$.

*Proof.* First write

$$Y_1(f) = \sum_{n=0}^{\infty} f(X_n) 1_{\{T_1 > n\}} = \sum_{n=0}^{\infty} 1_{\{T_1 > n\}} \sum_{j \in S} f(j) 1_{\{X_n = j\}}. \tag{4.4}$$

Since $E_i[1_{\{T_1>n\}} 1_{\{X_n=j\}}] = {}_0p_{ij}^n$,

$$E_i[Y_1(f)] = \sum_{n=0}^{\infty} \sum_{j \in S} {}_0p_{ij}^n f(j) = \sum_{j \in S} (\boldsymbol{I} - {}_0\boldsymbol{P})_{ij}^{-1} f(j).$$

This equation is equivalent to (4.2). The various interchanges of $\Sigma$ and $E_i$ are justified since

$$\sum_{n=0}^{\infty} \sum_{j \in S} {}_0p_{ij}^n |f(j)| < \infty.$$

To establish (4.3), write $Y_1(g)$ in the form given by (4.4). The product $Y_1(f)Y_1(g)$ can be written as

$$\begin{aligned} Y_1(f)Y_1(g) &= \sum_{n=0}^{\infty} a_n \sum_{m=0}^{\infty} b_m \\ &= \sum_{n=0}^{\infty} a_n \sum_{m=n}^{\infty} b_m + \sum_{n=0}^{\infty} a_n \sum_{m=0}^{n} b_m - \sum_{n=0}^{\infty} a_n b_n, \end{aligned} \tag{4.5}$$

where the sequences $\{a_n\colon\ n \geq 0\}$ and $\{b_m\colon\ m \geq 0\}$ are defined by (4.4). Next observe that

$$E_i\left[1_{\{T_1>n\}} 1_{\{X_n=j\}} 1_{\{T_1>m+n\}} 1_{\{X_{n+m}=k\}}\right] = {}_0p_{ij}^n {}_0p_{jk}^m.$$

Then taking the first term on the right side of (4.5),

$$
\begin{aligned}
E_i\left[\sum_{n=0}^{\infty} a_n \sum_{m=0}^{\infty} b_{n+m}\right] &= \sum_{n=0}^{\infty}\sum_{m=0}^{\infty}\sum_{j\in S}\sum_{k\in S} f(j)g(k)\,{}_0p_{ij}^n\,{}_0p_{jk}^m \\
&= \sum_{n=0}^{\infty}\sum_{j\in S} f(j)\,{}_0p_{ij}^n \sum_{m=0}^{\infty}\sum_{k\in S} g(k)\,{}_0p_{jk}^m \\
&= \sum_{n=0}^{\infty}\sum_{j\in S} (\boldsymbol{f}\circ E[Y_1(g)])(j)\,{}_0p_{ij}^n \\
&= \sum_{j\in S} (\boldsymbol{I}-{}_0\boldsymbol{P})_{ij}^{-1}(\boldsymbol{f}\circ E[Y_1(g)])(j).
\end{aligned}
$$

The second term on the right side of (4.5) can be handled in exactly the same way by simply interchanging $\boldsymbol{f}$ and $\boldsymbol{g}$ since

$$
\sum_{n=0}^{\infty}\sum_{m=0}^{\infty} a_n b_m = \sum_{m=0}^{\infty}\sum_{n=m}^{\infty} b_m a_n.
$$

Finally, the last term on the right side of (4.5) can be obtained from (4.2) with $\boldsymbol{f}$ replaced by $\boldsymbol{f}\circ\boldsymbol{g}$. These three terms together yield (4.3). □

**Corollary 4.6.** *For an irreducible discrete time Markov chain with finite state space and transition matrix* $\boldsymbol{P}$,

$$
E[\tau_1] = (\boldsymbol{I}-{}_0\boldsymbol{P})^{-1}\boldsymbol{e} \tag{4.7}
$$

*and*

$$
E[Z_1^2(f)] = (\boldsymbol{I}-{}_0\boldsymbol{P})^{-1}\boldsymbol{h}, \tag{4.8}
$$

*where* $\boldsymbol{e} = (1,\ldots,1)$ *and* $\boldsymbol{h} = 2(\boldsymbol{f}-r(f)\boldsymbol{e})\circ E[Y_1(f-r(f)e)] - (\boldsymbol{f}-r(f)\boldsymbol{e})\circ(\boldsymbol{f}-r(f)\boldsymbol{e})$.

*Proof.* Using the fact that $\tau_1 = Y_1(e)$, (4.7) follows from (4.2). Alternatively,

$$E_i[T_1] = \sum_{n=0}^{\infty} P_i\{T_1 > n\} = \sum_{n=0}^{\infty}\sum_{j=1}^{N} {}_0p_{ij}^n = \left[(\boldsymbol{I} - {}_0\boldsymbol{P})^{-1}\boldsymbol{e}\right]_i.$$

To establish (4.8), observe that

$$Z_1(f) = \sum_{n=0}^{T_1-1}\big(f(X_n) - r(f)\big) = Y_1(f - r(f)e)$$

and apply (4.3). □

**Example 4.9.** (Machine replacement) Consider $N$ identical machines that are subject to failure according to a given probability distribution. New machines are ordered at the beginning of each week to replace machines that have failed. Machines are delivered one week after being ordered. Let $X_n$ be the number of available machines at the beginning of the $n$th week, and let $d(X_n)$ be the number of machines that fail during the $n$th week, $n \geq 0$. Then

$$X_{n+1} = N - d(X_n),$$

and the process $\{X_n \colon n \geq 0\}$ is a DTMC with state space $S = \{0, 1, \ldots, N\}$. Assume that the machines fail according to a binomial distribution; that is, for some $p$ $(0 < p < 1)$

$$P\{d(X_n) = j | X_n = i\} = \binom{i}{j} p^j (1-p)^{i-j}$$

for $j = 0, 1, \ldots, i$. Then the elements of the transition matrix $\boldsymbol{P} = \{p_{ij} \colon i, j \in S\}$ are

$$p_{ij} = \begin{cases} \binom{i}{N-j} p^{N-j}(1-p)^{i-(N-j)} & \text{if } N - i \leq j \leq N \\ 0 & \text{if } 0 \leq j < N - i, \end{cases}$$

and the limiting distribution $\{\pi_j\colon\ j \in S\}$ of $\{X_n\colon\ n \geq 0\}$ is

$$\pi_j = \binom{N}{j}\left(\frac{1}{1+p}\right)^j\left(\frac{p}{1+p}\right)^{N-j}$$

for $0 \leq j \leq N$. Since the limiting distribution is binomial, it follows that

$$E[X] = \frac{N}{1+p}.$$

Numerical results obtained from Theorem 4.1 and Corollary 4.6 are given in Table 3.2. The function $f$ is the identity function.

Now let $\{X(t)\colon\ t \geq 0\}$ be a CTMC with finite state space $S = \{0, 1, \ldots, N\}$, infinitesimal generator $\boldsymbol{Q} = \{q_{ij}\colon\ i, j \in S\}$, and transition functions $\boldsymbol{P}^t = \{P^t(i,j)\colon\ i, j \in S\}$ for $t \geq 0$. The holding time in state $i \in S$ is exponentially distributed with rate parameter $q_i = -q_{ii}$, and the elements of the jump matrix $\boldsymbol{R} = \{r_{ij}\colon\ i, j \in S\}$ of the CTMC are

$$r_{ij} = \begin{cases} q_{ij}/q_i & \text{if } j \neq i \\ 0 & \text{if } j = i\ . \end{cases}$$

Assume that $0 < q_i < \infty$ for all $i \in S$, so that all states are stable and nonabsorbing. In addition, assume that $\sum_{j=0}^{N} q_{ij} = 0$. This last assumption guarantees that, starting from any state $i \in S$, the CTMC makes a state transition to a next state $j \in S$. Also assume that $\boldsymbol{R}$ is irreducible so that the CTMC is irreducible and therefore positive recurrent. As before, denote by $P_i$ and $E_i$ the conditional probability and conditional expectation, respectively, associated with starting in state $i \in S$. Denote by $T_n(j)$ the $n$th time at which $\{X(t)\colon\ t \geq 0\}$ hits state $j$, $n \geq 1$; for example,

$$T_1(j) = \inf\{u > 0\colon\ X(u-) \neq j, X(u) = j\}\,.$$

Then set $\tau_1(j) = T_1(j)$ and $\tau_n(j) = T_n(j) - T_{n-1}(j)$ for $n > 1$.

We compute the quantities $E_i[Y_1(f)]$ and $E_i[Y_1(f)Y_1(g)]$ for real-valued functions $\boldsymbol{f}$ and $\boldsymbol{g}$ with domain $S$ and $i \in S$. As in the case of DTMC's, we restrict attention to regenerative cycles formed by the successive random times at which the process hits state 0 and suppress the 0 in the notation. Set

$$_0P^t(i,j) = P_i\{\tau_1 > t, X(t) = j\}$$

and

$$_0\boldsymbol{P}^t = \{_0P^t(i,j):\ i,j \in S\}.$$

Next construct the matrix $_0\boldsymbol{R}^n$ from $\boldsymbol{R}$ in the same manner as $_0\boldsymbol{P}^n$ was constructed from $\boldsymbol{P}$. For a real-valued function $f$ with domain $S$, set

$$Y_1(f) = \int_0^{T_1} f(X(u))du$$

and

$$Z_1(f) = Y_1(f) - r(f)\tau_1.$$

**Theorem 4.10.** *For an irreducible continuous time Markov chain with finite state space, jump matrix* $\boldsymbol{R}$, *and vector* $\boldsymbol{q}$ *of rate parameters for holding times,*

$$E[Y_1(f)] = \int_0^\infty {}_0\boldsymbol{P}^t \boldsymbol{f} dt = (\boldsymbol{I} - {}_0\boldsymbol{R})^{-1}(\boldsymbol{f} \circ \boldsymbol{q}^{-1}) \qquad (4.11)$$

*and*

$$E[Y_1(f)Y_1(g)] = \int_0^\infty {}_0\boldsymbol{P}^t \boldsymbol{h} dt = (\boldsymbol{I} - {}_0\boldsymbol{R})^{-1}(\boldsymbol{h} \circ \boldsymbol{q}^{-1}), \quad (4.12)$$

where $\boldsymbol{h} = \boldsymbol{f} \circ E[Y_1(g)] + \boldsymbol{g} \circ E[Y_1(f)]$ and the column vector $\boldsymbol{q}^{-1} = (q_0^{-1}, q_1^{-1}, \ldots, q_N^{-1})$.

*Proof.* As in case of DTMC's, first write

$$\begin{aligned} Y_1(f) &= \int_0^\infty f(X(t)) 1_{\{T_1 > t\}} dt \\ &= \int_0^\infty 1_{\{T_1 > t\}} \sum_{j \in S} f(j) 1_{\{X(t)=j\}} dt. \end{aligned}$$

Taking the expectation with respect to $E_i$ on both sides and using

$$E_i\left[1_{\{T_1 > t\}} 1_{\{X(t)=j\}}\right] = {}_0P^t(i,j),$$

we obtain

$$E_i[Y_1(f)] = \int_0^\infty \sum_{j \in S} f(j)\, {}_0P^t(i,j) dt,$$

the first expression in (4.11). Again the interchanges of $E_i$, $\int$, and $\sum$ are justified since $E_i[T_j] < \infty$ and

$$\int_0^\infty \sum_{j \in S} |f(j)|\, {}_0P^t(i,j) dt < \infty.$$

To obtain the second expression in (4.11), set $\zeta_0 = 0$ and let $\zeta_n$ be the time of the $n$th state transition, $n \geq 1$. Since $\{X(t)\colon\ t \geq 0\}$ is irreducible and $S$ is finite, $\zeta_n \to \infty$ with probability 1 for each initial state $i \in S$. Since

$$[0, +\infty) = \bigcup_{n=0}^{\infty} [\zeta_n, \zeta_{n+1}),$$

we can write

$$E_i\Big[\int_0^\infty f(X(t))1_{\{T_1>t\}}dt\Big] = \sum_{n=0}^{\infty} E_i\Big[\int_{\zeta_n}^{\zeta_{n+1}} f(X(t))1_{\{T_1>t\}}dt\Big]. \tag{4.13}$$

Let $\{X_n\colon n \geq 0\}$ be the embedded jump chain of the CTMC, and denote by $\delta_n(j)$ the length of the $n$th cycle formed by the successive random times at which the chain hits state $j$. Then

$$\begin{aligned} E_i\Big[\int_{\zeta_n}^{\zeta_{n+1}} f(X(t))1_{\{T_1>t\}}dt\Big] &= E_i\big[f(X_n)1_{\{\delta_1>n\}}(\zeta_{n+1}-\zeta_n)\big] \\ &= E_i\big[f(X_n)1_{\{\delta_1>n\}}E_i[\zeta_{n+1}-\zeta_n|X_n]\big] \\ &= \sum_{j\in S} {}_0r_{ij}^n\,(f(j)q_j^{-1}) \\ &= \big[{}_0\boldsymbol{R}^n(\boldsymbol{f}\circ\boldsymbol{q}^{-1})\big]_i. \end{aligned} \tag{4.14}$$

Combining (4.13) and (4.14), we obtain

$$E[Y_1(f)] = \sum_{n=0}^{\infty} {}_0\boldsymbol{R}^n(\boldsymbol{f}\circ\boldsymbol{q}^{-1}) = (\boldsymbol{I}-{}_0\boldsymbol{R})^{-1}(\boldsymbol{f}\circ\boldsymbol{q}^{-1}),$$

the second part of (4.11).

To establish (4.12), write

$$\begin{aligned} Y_1(f)Y_1(g) &= \Big(\int_{t=0}^{\infty}\Big)\Big(\int_{u=0}^{\infty}\Big) \\ &= \Big(\int_{t=0}^{\infty}\Big)\Big(\int_{u=t}^{\infty}\Big) + \Big(\int_{t=0}^{\infty}\Big)\Big(\int_{u=0}^{t}\Big). \end{aligned} \tag{4.15}$$

Take the first term on the right side of (4.15) and write this term as

$$\begin{aligned}
&\Big(\int_{t=0}^{\infty}\Big)\Big(\int_{u=t}^{\infty}\Big)\\
&\quad= \int_{t=0}^{\infty} dt \int_{u=0}^{\infty} f(X(t))1_{\{T_1>t\}}g(X(t+u))1_{\{T_1>t+u\}}du\\
&\quad= \int_{t=0}^{\infty} dt \int_{u=0}^{\infty} 1_{\{T_1>t\}} \sum_{j\in S} f(j)1_{\{X(t)=j\}}1_{\{T_1>t+u\}}\\
&\qquad\qquad\qquad \sum_{k\in S} g(k)1_{\{X(t+u)=k\}}du.
\end{aligned}$$

Now take expectations on both sides and obtain

$$\begin{aligned}
&E_i\Big[\int_{t=0}^{\infty}\int_{u=t}^{\infty}\Big]\\
&\quad= \int_0^{\infty} \sum_{j\in S} f(j)_0P^t(i,j)dt \int_0^{\infty} \sum_{k\in S} g(k)_0P^u(j,k)du\\
&\quad= \int_0^{\infty} \sum_{j\in S} {}_0P^t(i,j)\left[f(j)E_j[Y_1(g)]\right]dt.
\end{aligned}$$

This term is exactly the contribution to (4.12) from the first term of $\boldsymbol{h}$. The second term on the right side of (4.15) can be handled in exactly the same way. The second expression in (4.12) follows from (4.11). □

**Corollary 4.16.** *For an irreducible continuous time Markov chain with finite state space, jump matrix* $\boldsymbol{R}$, *and vector* $\boldsymbol{q}$ *of rate parameters for holding times,*

$$E[\tau_1] = (\boldsymbol{I} - {}_0\boldsymbol{R})^{-1}\boldsymbol{q}^{-1} \tag{4.17}$$

*and*

$$E[Z_1^2(f)] = (\boldsymbol{I} - {}_0\boldsymbol{R})^{-1}(\boldsymbol{h} \circ \boldsymbol{q}^{-1}), \tag{4.18}$$

where $\boldsymbol{h} = 2(\boldsymbol{f} - r(f)\boldsymbol{e}) \circ E[Y_1(f - r(f)e)]$.

*Proof.* Using the fact that $T_1 = Y_1(e)$, (4.17) follows from (4.11). Similarly, (4.18) follows from the fact that

$$Z_1(f) = \int_0^{T_1} \big(f(X(u)) - r(f)\big)du$$

and (4.12). □

## 3.5. Statistical Efficiency

The computational results for Markov chains can be used to assess the statistical efficiency of the regenerative method when the regenerative process is an irreducible positive recurrent CTMC. Let $\{X(t)\colon t \geq 0\}$ be an irreducible CTMC with finite state space $S = \{0, 1, \ldots, N\}$, and let $f$ be a real-valued measurable function with domain $S$. Suppose that we wish to obtain point and interval estimates for the quantity

$$r(f) \stackrel{\text{def}}{=} \frac{E_0[Y_1(f)]}{E_0[\tau_1(0)]}$$

based on regenerative cycles defined by the successive random times at which the CTMC hits state 0. As before, denote by $T_n(0)$ the $n$th time at which $\{X(t)\colon t \geq 0\}$ hits state 0, $n \geq 1$; for example,

$$T_1(0) = \inf\{u > 0\colon\ X(u-) \neq 0, X(u) = 0\}.$$

Set $T_0(0) = 0$ and $\tau_n(0) = T_n(0) - T_{n-1}(0)$ for $n \geq 1$. It is convenient to have a CLT in terms of the fixed length $t$ of the simulation.

**Theorem 5.1.** *Assume that* $E_0[\tau_1(0)] < \infty$. *Also assume that* $E_0[Y_1(|f|)] < \infty$ *and* $E_0\left[Y_1^2(|f - r(f)|)\right] < \infty$. *Then*

$$\frac{\int_0^t f(X(u))du - r(f)t}{\sqrt{t}\,\sigma_0(f)/\sqrt{E_0[\tau_1(0)]}} \Rightarrow N(0,1) \tag{5.2}$$

*as* $t \to \infty$, *where* $\sigma_0^2(f) = E_0[Z_1^2(f)]$.

The half-length of a confidence interval for $r(f)$ obtained from a simulation of fixed length $t$ is proportional to the quantity appearing in the denominator of (5.2). Since the numerator in this CLT is independent of the state (0) selected to form regenerative cycles, so is the denominator by the convergence of types theorem. Thus, the quantity $\sigma_0(f)/\sqrt{E_0[\tau_1(0)]}$ is an appropriate measure of the statistical efficiency of the simulation. This quantity can be computed using Corollary 4.16.

**Example 5.3.** (Machine availability) Suppose that service time at center $i$ is exponentially distributed as a random variable $V_i$. Let $X(t)$ be the number of jobs waiting or in service at center 2 at time $t$. Under these assumptions, the process $\{X(t)\colon t \geq 0\}$ is an irreducible CTMC with finite state space $S = \{0, 1, \ldots, N\}$. The successive random times at which the process $\{X(t)\colon t \geq 0\}$ hits any fixed state form a sequence of regeneration points, and the expected time between regeneration points is finite.

Take $f = 1_{\{0,1,\ldots,N-1\}}$, and set $r(f) = E_0[Y_1(f)]/E_0[\tau_1(0)]$. Tables 3.3 and 3.4 give theoretical values associated with estimation of machine availability $r(f)$ by the regenerative method when there are two spare components ($N = 3$).

**Table 3.2**

Theoretical values for machine replacement

$N = 4,\ p = 0.25$

Binomially distributed failure times

Return state $i$

| | 0 | 1 | 2 | 3 | 4 |
|---|---|---|---|---|---|
| $E_i[\tau_1(i)]$ | 625.0000 | 39.0625 | 6.5104 | 2.4414 | 2.4414 |
| $E_i[Y_1(f)]$ | 2000.0000 | 125.0000 | 20.8333 | 7.8125 | 7.8123 |
| $E_i[Y_1(1_{\{0\}})]$ | 1.0000 | 0.0625 | 0.0104 | 0.0039 | 0.0039 |
| $E_i[Y_1(1_{\{1\}})]$ | 16.0000 | 1.0000 | 0.1667 | 0.0625 | 0.0625 |
| $E_i[Y_1(1_{\{2\}})]$ | 96.0000 | 6.0000 | 1.0000 | 0.3750 | 0.3750 |
| $E_i[Y_1(1_{\{3\}})]$ | 256.0000 | 16.0000 | 2.6667 | 1.0000 | 1.0000 |
| $E_i[Y_1(1_{\{4\}})]$ | 256.0000 | 16.0000 | 2.6667 | 1.0000 | 1.0000 |
| $E[X]$ | 3.2000 | 3.2000 | 3.2000 | 3.2000 | 3.2000 |
| $\pi_0$ | 0.0016 | 0.0016 | 0.0016 | 0.0016 | 0.0016 |
| $\pi_1$ | 0.0256 | 0.0256 | 0.0256 | 0.0256 | 0.0256 |
| $\pi_2$ | 0.1536 | 0.1536 | 0.1536 | 0.1536 | 0.1536 |
| $\pi_3$ | 0.4096 | 0.4096 | 0.4096 | 0.4096 | 0.4096 |
| $\pi_4$ | 0.4096 | 0.4096 | 0.4096 | 0.4096 | 0.4096 |
| $\sigma_i^2(f)$ | 0.00001 | 5.0000 | 2.5000 | 0.9375 | 0.9375 |
| $\sigma_i^2(f)/E_i[\tau_1(i)]$ | 0.3840 | 0.3840 | 0.3840 | 0.3840 | 0.3840 |

**Table 3.3**

Theoretical values for machine availability

$N = 3,\ E[V_1] = 0.25,\ E[V_2] = 1.0$

Exponentially distributed lifetimes
Exponentially distributed repair times

Return state $i$

| | 0 | 1 | 2 | 3 |
|---|---|---|---|---|
| $E_i[Y_1(f)]$ | 5.2500 | 1.0500 | 0.2625 | 0.3281 |
| $E_i[\tau_1(i)]$ | 21.2500 | 4.2500 | 1.0625 | 1.3281 |
| $E_i[Y_1(f)]/E_i[\tau_1(i)]$ | 0.2471 | 0.2471 | 0.2471 | 0.2471 |
| $\sigma_i(f)$ | 1.5707 | 0.7024 | 0.3512 | 0.3927 |
| $\sigma_i^2(f)/E_i[\tau_1(i)]$ | 0.1161 | 0.1161 | 0.1161 | 0.1161 |

**Table 3.4**

Theoretical values for machine availability

$N = 3$, $E[V_1] = 0.80$, $E[V_2] = 1.0$

Exponentially distributed lifetimes
Exponentially distributed repair times

Return state $i$

| | 0 | 1 | 2 | 3 |
|---|---|---|---|---|
| $E_i[Y_1(f)]$ | 3.0500 | 1.3556 | 1.0844 | 1.9520 |
| $E_i[\tau_1(i)]$ | 4.6125 | 2.0500 | 1.6400 | 2.9520 |
| $E_i[Y_1(f)]/E_i[\tau_1(i)]$ | 0.6612 | 0.6612 | 0.6612 | 0.6612 |
| $\sigma_i(f)$ | 1.4520 | 0.9680 | 0.8658 | 1.1616 |
| $\sigma_i^2(f)/E_i[\tau_1(i)]$ | 0.4571 | 0.4571 | 0.4571 | 0.4571 |

## Chapter 4

# Networks of Queues

Networks of queues with preemptive or nonpreemptive priorities among job classes are used frequently as models for complex congestion phenomena, and simulation is often the only available means for analyzing the behavior of such models. An appropriate state vector for simulation of a closed network of queues with priorities among job classes is a linear "job stack," an enumeration of the jobs by center and job class. Regenerative methods can be used to simulate networks of queues with general service times and a special "single-state" in which all jobs are at the same single-server center. The structure of the network of queues ensures that the "job stack process" probabilistically restarts whenever there is a state transition to some fixed state from some fixed single-state. However, it is nontrivial to determine conditions on the building blocks of a network of queues under which the job stack process hits a fixed single-state infinitely often with probability 1.

The arguments given to establish the applicability of the regenerative method use the assumptions that all interruptions of service are of the "preemptive-repeat" type and all distributions of service time have finite first and second moments and a density function that is continuous and positive on $(0, +\infty)$. Using "new better than used" assumptions on the distributions of service time along with sample path behavior of the underlying job stack process, criteria for recurrence and regeneration that avoid these restrictions can be obtained. Estimation procedures for networks of queues with "preemptive-resume" priorities follow from these results.

## 4.1. The Job Stack Process

We consider closed networks of queues with a finite number $N$ of stochastically identical jobs (customers), a finite number $L$ of single-server or multiple-server centers numbered $1, 2, \ldots, L$, and a finite number $K$ of mutually exclusive job classes numbered $1, 2, \ldots, K$. At every point of continuous time each job is of exactly one job class, but jobs may change job class as they traverse the network. Upon completion of service at center $i$, a job of job class $j$ moves to center $k$ and becomes a job of job class $l$ with probability $p_{ij,kl}$. The routing matrix

$$\boldsymbol{P} = \{p_{ij,kl}\colon\ (i,j),(k,l) \in C\}$$

is a given irreducible stochastic matrix, and

$$C \subseteq \{1,2,\ldots,L\} \times \{1,2,\ldots,K\}$$

is the set of (center, class) pairs in the network. In accordance with the routing matrix $\boldsymbol{P}$, some centers may never see jobs of certain job classes. At each center, jobs queue and receive service according to a fixed priority scheme among job classes; the priority scheme may differ from center to center. Within a job class at a center, jobs receive service according to a fixed queueing discipline that depends only on the order of arrival of the jobs. According to a fixed procedure for each center, a job in service may or may not be preempted when a job of higher priority joins the queue at the center. A job may not be preempted by another job of the same job class. If more than one job subject to preemption is in service at a center when a job of higher priority joins the queue at the center, the preempted job is selected according to a fixed scheme that depends only on the job classes in service at the center. Any interruption of service is of the "preemptive-repeat" type; that is, service starts afresh when the preempted job goes back into service. A preempted job goes into service at a center before any other job of the same job class that is waiting in queue at the center. We assume that service times are mutually independent and the service time for a job of job class $j$ at center $i$, distributed as a positive random variable $V(i,j)$, has finite first and second moments and a density function that is continuous and positive on $(0,+\infty)$.

Let $S_i(t)$ be the job class of the job in service at center $i$ at time $t$, $i=1,2,\ldots,L$; by convention $S_i(t)=0$ if there are no jobs at center $i$ at time $t$. If center $i$ has more than one server, we take $S_i(t)$ to be a vector that records the job class of the job in service from each server at center $i$. Specifically, we enumerate the servers at center $i$ as $1,2,\ldots,m(i)$ and set

$$S_i(t)=(S_{i,1}(t),S_{i,2}(t),\ldots,S_{i,m(i)}(t)),$$

where $S_{i,m}(t)$ is the job class of the job in service from server $m$ at center $i$ at time $t$. If there is more than one available server, a job that goes into service is served by the lowest numbered available server. We denote the job classes served at center $i$ and ordered by decreasing priority by $j_1(i),j_2(i),\ldots,j_{k(i)}(i)$, and we let $C_{j_1}^{(i)}(t),\ldots,C_{j_{k(i)}}^{(i)}(t)$ be the number of jobs waiting in queue at time $t$ of the various job classes served at center $i$, $i=1,2,\ldots,L$.

We order the $N$ jobs in a linear stack and define the vector $Z(t)$ by

$$\begin{aligned}Z(t)=\big(&C_{j_{k(1)}}^{(1)}(t),\ldots,C_{j_1}^{(1)}(t),S_1(t);\ldots;\\&C_{j_{k(L)}}^{(L)}(t),\ldots,C_{j_1}^{(L)}(t),S_L(t)\big).\end{aligned}$$

The *job stack at time t* corresponds to the nonzero components in the vector $Z(t)$ and is an ordering of the jobs by job class at the individual centers. Within a job class at a particular center, jobs appear in the job stack in order of their arrival at the center, the latest to arrive being closest to the top of the job stack. The job at the top of the job stack is said to be in position 1 of the job stack, the next job in position 2, etc. A job that has been preempted appears at the head of the queue of jobs of its job class. The process $\{Z(t)\colon\ t \geq 0\}$ is called the *job stack process*.

Formal specification of the job stack process $\{Z(t)\colon\ t \geq 0\}$ is as a GSMP with a finite state space $D^*$ that is the set of all vectors $z = (c^{(1)}_{j_{k(1)}}, \ldots, c^{(1)}_{j_1}, s_{1,1}, \ldots, s_{1,m(1)}; \ldots; c^{(L)}_{j_{k(L)}}, \ldots, c^{(L)}_{j_1}, s_{L,1}, \ldots, s_{L,m(L)}) \in \prod_{i=1}^{L} \{0, 1, \ldots, N-1\}^{k(i)} \times \{0, j_1(i), j_2(i), \ldots, j_{k(i)}(i)\}^{m(i)}$ satisfying

(i) $N = c^{(1)}_{j_{k(1)}} + \cdots + c^{(1)}_{j_1} + 1_{\{j_1(1),\ldots,j_{k(1)}(1)\}}(s_{1,1}) + \cdots + 1_{\{j_1(1),\ldots,j_{k(1)}(1)\}}(s_{1,m(1)}) + \cdots + c^{(L)}_{j_{k(L)}} + \cdots + c^{(L)}_{j_1} + 1_{\{j_1(L),\ldots,j_{k(1)}(L)\}}(s_{L,1}) + \cdots + 1_{\{j_1(L),\ldots,j_{k(1)}(1)\}}(s_{L,m(L)})$;

(ii) for all $i$ and $m$, $s_{i,m} = 0$ only if $c^{(i)}_{j_{k(i)}} + \ldots + c^{(i)}_{j_1} = 0$; and

(iii) for all $i$ and $k$ such that jobs of job class $j_k(i)$ are subject to preemption at center $i$,

$$c^{(i)}_{j_{k-1}} = 0, \ldots, c^{(i)}_{j_1} = 0 \text{ whenever}$$
$$1_{j_k(i)}(s_{i,1}) + \cdots + 1_{j_k(i)}(s_{i,m(i)}) > 0.$$

The event set

$$E = \{e_{1,1}, \ldots, e_{1,m(1)}, \ldots, e_{L,1}, \ldots, e_{L,m(L)}\}, \tag{1.1}$$

where event $e_{i,m}$ = "service completion by server $m$ at center $i$." For $z = (c^{(1)}_{j_{k(1)}}, \ldots, c^{(1)}_{j_1}, s_{1,1}, \ldots, s_{1,m(1)}; \ldots; c^{(L)}_{j_{k(L)}}, \ldots, c^{(L)}_{j_1}, s_{L,1}, \ldots, s_{L,m(L)}) \in D^*$, event $e_{i,m}$ = "service completion by server $m$ at center $i$" is active in state $z$ ($e_{i,m} \in E(z)$) if and only if $s_{i,m} > 0$. According to this definition, $e_{i,m} \in E(z)$ if and only if a job receives service from server $m$ at center $i$ when the job stack process is in state $z$. All speeds $r(z, e)$ for active events are equal to 1. Recall that by assumption any interruption of service is of the preemptive-repeat type.

For $z \in D^*$, $(i, j), (k, l) \in C$, and $m \in \{1, 2, \ldots, m(i)\}$, let $q(z; i, j, k, l; m)$ be the next state of the job stack process when in state $z$ there is a completion of service by server $m$ to a job of job class $j$ at center $i$, and the job moves to center $k$ and becomes a job of job class $l$. If $e = e_{i,m}$, then the state transition probability $p(z'; z, e) = p_{ij,kl}$ when

$$z = \left(c^{(1)}_{j_{k(1)}}, \ldots, s_{1,m(1)}; \ldots; c^{(L)}_{j_{k(L)}}, \ldots, s_{L,m(L)}\right)$$
$$\text{with } s_{i,m} = j \text{ and } z' = q(z; i, j, k, l; m).$$

All other state transition probabilities $p(z'; z, e)$ are equal to 0. The clock-setting distribution functions are as follows. If the new event $e' = e_{i,m}$, then $F(x; z', e', z, e^*) = P\{V(i, j) \leq x\}$ for $x \geq 0$, where $s'_{i,m} = j$.

We assume that the initial distribution $\mu$ of the underlying GSSMC satisfies

$$\mu(A) = q(z_0) \prod_{e_i \in E(z_0)} F_0(a_i; e_i, z_0)$$

for all sets

$$A = \{z_0\} \times \{(c_{0,1}, \ldots, c_{0,M}) \in C(z_0)\colon\ 0 \leq c_{0,i} \leq a_i \text{ for } 1 \leq i \leq M\},$$

where $M = m(1) + m(2) + \cdots + m(L)$, $q$ is a probability distribution on $D^*$, and $F_0(\,\cdot\,; e, z)$ is a proper distribution function for $z \in D^*$ and $e \in E(z)$.

For any single-server center $i$ that serves only one job class (that is, $k(i) = 1$), the state vector $Z(t)$ can be simplified by replacing $C^{(i)}_{j_{k(i)}}(t), S_i(t)$ by the number $Q_i(t)$ of jobs waiting or in service at center $i$ at time $t$. The definition of the job stack process does not take into account explicitly that the total number of jobs in the network is fixed. For a complex network, the use of the resulting somewhat larger state space facilitates generation of the process. For relatively simple networks, it may be desirable to remove the redundancy.

**Example 1.2.** (Cyclic queues with preemptive priority) Consider a network of queues with two single-server centers ($L = 2, m(1) = m(2) = 1$) and two job classes ($K = 2$) that provides service to $N$ ($\geq 2$) jobs. Take $C = \{(1,1),(2,1),(2,2)\}$ to be the set of (center, class) pairs so that center 1 serves jobs of job class 1 and center 2 serves jobs of job class 1 and job class 2. All queueing disciplines are FCFS. Jobs of job class 2 have preemptive priority at center 2 over jobs of job class 1. A job of job class 1 that completes service at center 2 moves to center 1 with fixed probability $p$ ($0 < p < 1$) and with probability $1 - p$ moves to center 2 and becomes a job of job class 2. A job of job class 2 that completes service at center 2 moves to center 1. Upon completion of service at center 1, a job moves to center 2 and becomes a job of job class 1. Thus, the nonzero elements of the routing matrix $\boldsymbol{P}$ are $p_{11,21} = 1$, $p_{21,11} = p$, $p_{21,22} = 1 - p$, and $p_{22,11} = 1$.

Set

$$Z(t) = \left(Q_1(t), C_1^{(2)}(t), C_2^{(2)}(t), S_2(t)\right),$$

where $Q_1(t)$ is the number of jobs waiting or in service at center 1 at time $t$, $S_2(t)$ is the job class of the job receiving service at center 2 at time $t$, and $C_j^{(2)}(t)$ is the number of jobs of job class $j$ waiting in queue at center 2 at time $t$. Formal specification of the job stack process $\{Z(t)\colon\ t \geq 0\}$ is as a GSMP with a finite state space $D^*$ and event set $E = \{e_{1,1}, e_{2,1}\}$.

When there are $N = 2$ jobs in the network, the state space $D^*$ of the job stack process is $D^* = \big\{(2,0,0,0),(1,0,0,1),$

$(0,1,0,1),(0,1,0,2),(1,0,0,2),(0,0,1,2)\}$. Observe that since jobs of job class 2 have preemptive priority at center 2 over jobs of job class 1, (0,0,1,1) is not a stable state of the job stack process. When $N = 5$, $z = (3,1,0,2)$ is an element of the state space of the job stack process. The job in position 1 of the job stack associated with state $z$ is at the tail of the queue at center 1; the job in position 4 of the job stack is waiting in queue at center 2 as a job of job class 1.

**Example 1.3.** (Multiprogrammed computer system) Consider a network of queues that consists of two sequential stages, the $\alpha-$stage and the $\beta-$stage, in a loop. Two servers, numbered 1 and 2, provide service to $N$ $(\geq 2)$ stochastically identical jobs. Upon completion of service at the $\alpha-$stage, a job moves to the $\beta-$stage, and upon completion of service at the $\beta-$stage the job returns to the $\alpha-$stage. Within the $\alpha-$stage, a job receives an $\alpha_1-$, an $\alpha_2-$, and an $\alpha_3-$service in this order, and within the $\beta-$stage, a job receives a $\beta_1-$, a $\beta_2-$, and a $\beta_3-$service in this order. Only server 2 can provide a $\beta_2-$service and only server 1 can provide an $\alpha_1-$, $\alpha_2-$, $\alpha_3-$, $\beta_1-$, or $\beta_3-$service. Servers 1 and 2 provide service concurrently, subject to the restriction that server 1 cannot provide a $\beta_1-$ or a $\beta_3-$service when server 2 provides a $\beta_2-$service. After having received an $\alpha_3-$service, a job moves instantaneously from the $\alpha-$stage to the tail of the queue in the $\beta-$stage and after having received a $\beta_3-$ service, moves instantaneously from the $\beta-$stage to the tail of the queue in the $\alpha-$stage. Interruption of an $\alpha_2-$service occurs at the completion of a concurrent $\beta_2-$service. The "$\beta_2-$ complete" interruption of an $\alpha_2-$service is of the preemptive-repeat type. All other services in the system complete without interruption.

At the completion of an $\alpha_1-$, $\alpha_2-$, $\alpha_3-$, $\beta_1-$, or $\beta_3-$service or at an interruption of an $\alpha_2-$service, server 1 chooses the next service according to a "rule of priority" as follows:

(i) if there is a job waiting for $\beta_3$−service, begin this service;
(ii) if there is a job waiting for $\beta_1$−service, begin this service unless there is a $\beta_2$−service in progress;
(iii) if the last service provided in the $\alpha$−stage was a completed $\alpha_2$−service, begin an $\alpha_3$−service;
(iv) if the last service provided in the $\alpha$−stage was an interrupted $\alpha_2$−service, resume the $\alpha_2$−service;
(v) if the last service provided in the $\alpha$−stage was an $\alpha_1$−service, begin an $\alpha_2$ service; and
(vi) if the last service provided in the $\alpha$−stage was an $\alpha_3$−service and the queue in the $\alpha$−stage is not empty, begin an $\alpha_1$−service.

If no claim is made on server 1 according to the rule of priority, server 1 remains idle until the completion of the next $\beta_2$−service, at which time the rule of priority is invoked again. The queueing discipline in the $\alpha$−stage is FCFS as is the queueing discipline in the $\beta$−stage. Assume that service times are mutually independent and $\alpha_j$−service time is distributed as a positive random variable $A_j$ with finite mean but otherwise arbitrary distribution. Also assume that $\beta_j$−service time is distributed as a positive random variable $B_j$ with finite mean but otherwise arbitrary distribution.

This system can be represented as a network of queues with two single-server centers ($L = 2, m(1) = m(2) = 1$) and six job classes ($K = 6$). At every point of continuous time each job is of exactly one job class, but jobs may change job class as they traverse the network. Upon completion of service at center $i$, a job of job class $j$ moves to center $k$ and becomes a job of job class $l$ with probability $p_{ij,kl}$. Take

$$C = \{(1,1),(1,2),(1,3),(1,4),(1,5),(2,6)\}$$

to be the set of (center, class) pairs. Service to jobs of job class 1 corresponds to $\beta_3$-service, and service to jobs of job class 2 corresponds to $\beta_1$−service. Service to jobs of job class 3 corresponds to $\alpha_3$−service, service to jobs of job class 4 corresponds to $\alpha_2$−service, and service to jobs of job class 5 corresponds to $\alpha_1$−service. Service to jobs of job class 6 corresponds to $\beta_2$−service. Jobs queue and receive service at center 1 according to a fixed priority scheme among job classes. The job classes served at center 1 and ordered by decreasing priority are $j_1(1) = 1$, $j_2(1) = 2$, $j_3(1) = 3$, $j_4(1) = 4$, $j_5(1) = 5$. This priority ordering is subject to the constraint that there can be no service at center 1 to a job of job class 1 or job class 2 while service (to a job of job class 6) is in progress at center 2.

A job of job class 4 at center 1 is subject to preemption when a job of higher priority joins the queue at center 1, and the interruption of service is of the preemptive-repeat type. All other jobs in the network are not subject to preemption. The nonzero elements of the routing matrix $\boldsymbol{P}$ are $p_{11,15} = p_{12,26} = p_{13,12} = p_{14,13} = p_{15,14} = p_{26,14} = 1$.

Set

$$Z(t) = \big(C_5(t), C_4(t), C_3(t), C_2(t), C_1(t), S_1(t), Q_2(t)\big),$$

where $Q_2(t)$ is the number of jobs waiting or in service at center 2, $S_1(t)$ is the job class of the job in service at center 1, and $C_j(t)$ is the number of jobs of job class $j$ waiting in queue at center 1 at time $t$. The job stack process $\{Z(t)\colon t \geq 0\}$ has a finite state space $D^*$ that is the set of all vectors $(c_5, c_4, c_3, c_2, c_1, s_1, q_2, m) \in \{0, 1, \ldots, N\} \times \{0, 1\} \times \{0, 1\} \times \{0, 1, \ldots, N\} \times \{0, 1\} \times \big\{0, 1, \ldots, 5\big\} \times \{0, 1\} \times \{0, 1\}$ satisfying

(i) $c_1 + c_2 + \cdots + c_5 + 1_{\{s_1 > 0\}} + q_2 = N$;
(ii) $s_1 \neq 1, 2$ and $c_1 = 0$ when $q_2 = 1$;
(iii) $c_4 c_3 = 0$; and
(iv) $c_4 = c_3 = 0$ when $s_1 = 3, 4$, or $5$.

Without further restrictions on the building blocks of a network of queues with priorities among job classes, the job stack process defined by (1.1) need not have a limiting distribution independent of the initial conditions.

**Example 1.4.** (Cyclic queues with three job classes) Consider a network of queues with two single-server centers ($L = 2, m(1) = m(2) = 1$) and three job classes ($K = 3$) that provides service to $N$ ($\geq 2$) jobs. Take $C = \{(1,1),(1,2),(1,3),(2,1),(2,2),(2,3)\}$ to be the set of (center, class) pairs so that each center serves jobs of job classes 1, 2, and 3. All queueing disciplines are FCFS. At each center jobs of job class 1 have nonpreemptive priority over jobs of job class 2, and jobs of job class 2 have nonpreemptive priority over jobs of job class 3. Upon completion of service at center 1, a job of job class 1 moves to center 2 and becomes a job of job class 2. A job of job class 2 that completes service at center 1 moves to center 2 and becomes a job of job class 3, and a job of job class 3 that completes service at center 1 moves to center 2 and becomes a job of job class 1. Upon completion of service at center 2, a job moves to center 1, but does not change job class. Thus, the nonzero elements of the routing matrix $\boldsymbol{P}$ are $p_{11,22} = p_{12,23} = p_{13,21} = p_{21,11} = p_{22,12} = p_{23,13} = 1$.

Set

$$\begin{aligned} Z(t) = \big(&C_3^{(1)}(t), C_2^{(1)}(t), C_1^{(1)}(t), S_1(t),\\ &C_3^{(2)}(t), C_2^{(2)}(t), C_1^{(2)}(t), S_2(t)\big), \end{aligned}$$

where $S_i(t)$ is the job class of the job in service at center $i$ at time $t$ and $C_j^{(i)}(t)$ is the number of jobs of job class $j$ waiting in queue at center $i$. Assume that service time for jobs of job class $j$ at center $i$ is exponentially distributed with rate parameter $\lambda(i,j)$ as a random variable $V(i,j)$. Under these assumptions the job stack process is a CTMC with a finite state space $D^*$. When there are $N = 2$ jobs in the network, the set $D^*$ has twenty-seven elements. All states of the job stack process are recurrent, but there are two irreducible closed sets of recurrent states. One of the sets consists of the nine states (0,1,0,3,0,0,0,0), (0,0,0,0,0,1,0,3), (0,0,1,2,0,0,0,0), (0,0,0,0,0,0,1,2), (1,0,0,1,0,0,0,0), (0,0,0,0,1,0,0,1), (0,0,0,1,0,0, 0,3), (0,0,0,3,0,0,0,2), and (0,0,0,2,0,0,0,1).

For $z, z' \in D^*$, we say that $z'$ is a *successor state* of $z$ and write $z \to z'$ if $p(z'; z, e_{i,m}) > 0$ for some event $e_{i,m} \in E(z)$. State $z'$ is said to be *accessible* from $z$, and we write $z \rightsquigarrow z'$, if there exist states $z_0 = z, z_1, z_2, \ldots, z_{n-1}, z_n = z'$ and (center, server) pairs $(i_0, m_0)$, $(i_1, m_1), \ldots$, $(i_{n-1}, m_{n-1})$ such that

$$p(z_1; z_0, e_{i_0,m_0})p(z_2; z_1; e_{i_1,m_1}) \cdots p(z_n; z_{n-1}, e_{i_{n-1},m_{n-1}}) > 0.$$

When $z \rightsquigarrow z'$ and $z' \rightsquigarrow z$ states $z$ and $z'$ are said to *communicate*. We assume throughout that there exists a *target state* $z^*$ of the job stack process such that $z \rightsquigarrow z^*$ for all $z \in D^*$. This assumption ensures that the set

$$D = \{z \in D^*\colon\ z^* \rightsquigarrow z\}$$

of states that are accessible from $z^*$ is *irreducible* in the sense that $z$ and $z'$ communicate for all $z, z' \in D$. (To see this, let $z_1, z_2 \in D$. The definition of $z^*$ ensures that $z_1 \rightsquigarrow z^*$ and $z^* \rightsquigarrow z_2$ because $z_2 \in D$. Therefore, $z_1 \rightsquigarrow z_2$.) We also assume that the initial distribution $q$ is concentrated on the set $D$ so that $Z(0) \in D$. Thus, we consider the job stack process restricted to the set $D$.

The set $D^* - D$ may be nonempty even if the routing matrix $\boldsymbol{P}$ is irreducible.

**Example 1.5.** (Cyclic queues with preemptive priority) Set $Z(t) = (Q_1(t), C_1^{(2)}(t), C_2^{(2)}(t), S_2(t))$ as in Example 1.2. When there are $N = 2$ jobs in the network, $(0,0,1,2) \to (1,0,0,2)$, $(1,0,0,2) \to (2,0,0,0)$, $(1,0,0,2) \to (0,1,0,2)$, $(2,0,0,0) \to (1,0,0,1)$, $(0,1,0,2) \to (1,0,0,1)$, $(1,0,0,1) \to (0,1,0,1)$,

$(1,0,0,1) \to (2,0,0,0)$, $(1,0,0,1) \to (1,0,0,2)$, $(0,1,0,1) \to (1,0,0,1)$, and $(0,1,0,1) \to (0,1,0,2)$. Since $k(1) = 1$, $z_1^* = (2,0,0,0)$ is a target state of the job stack process, and the set $D$ of all states that are accessible from $z_1^*$ is

$$D = \{(2,0,0,0), (1,0,0,1), (0,1,0,1), (0,1,0,2), (1,0,0,2)\}.$$

Since (0,0,1,2) is not accessible from any state of the job stack process, $(0,0,1,2) \in D^* - D$. The remaining five elements of the set $D^*$ comprise the set $D$, and all pairs of states in $D$ communicate.

**Example 1.6.** (Data base management system) As a model of resource contention in a data base management system, consider a network of queues with two single-server centers ($L = 2, m(1) = m(2) = 1$) and seven job classes ($K = 7$) that provides service to $N$ ($\geq 2$) jobs. At every point of continuous time each job is of exactly one job class, but jobs may change job class as they traverse the network. Take $C = \{(1,2), (1,3), \ldots, (1,7), (2,1)\}$ to be the set of (center, class) pairs. The job classes served at center 1 and ordered by decreasing priority are $j_1(1) = 2$, $j_2(1) = 3, \ldots, j_6(1) = 7$. Service to a job of job class 7 (at center 1) is preempted when any other job of higher priority joins the queue at center 1. No job of any other job class is subject to preemption. Assume that for fixed $0 < p_1, p_2 < 1$ the nonzero elements of the routing matrix $\boldsymbol{P}$ are $p_{12,13} = p_{14,13} = 1 - p_1$, $p_{13,12} = p_{15,12} = 1 - p_2$, $p_{12,21} = p_{14,21} = p_1$, $p_{13,17} = p_{15,17} = p_2$, and $p_{16,14} = p_{17,16} = p_{21,15} = 1$.

Set

$$Z(t) = (C_7^{(1)}(t), \ldots, C_2^{(1)}(t), S_1(t), Q_2(t)),$$

where $C_j^{(1)}(t)$ is the number of jobs of job class $j$ waiting in queue at center 1 at time $t$, $S_1(t)$ is the job class in service at center 1, and $Q_2(t)$ is the number of jobs waiting or in service at center 2. Both $z_1^* = (N-1,0,0,0,0,0,7,0)$ and $z_2^* = (0,0,0,0,0,0,0,N)$ are target states of the job stack process $\{Z(t)\colon t \geq 0\}$. Take $i_0 = 1$ or $i_0 = 2$, and set $D = \{z \in D^*\colon z_{i_0}^* \rightsquigarrow z\}$. Then the set $D^* - D$ is nonempty, and $z = (N-1,0,0,0,0,0,5,0)$ is an element of the set $D$.

Theorem 1.7 postulates the existence of a distinguished center $i_0$: either center $i_0$ serves only one job class or service at center $i_0$ to a job of the lowest priority job class $j_{k(i_0)}(i_0)$ served at center $i_0$ is preempted when any other job of higher priority joins the queue. Let $z_{i_0}^*$ be the state in which all $N$ jobs are of the lowest priority job class $j_{k(i_0)}(i_0)$ at center $i_0$. Then $z_{i_0}^*$ is a target state of the job stack process, and all pairs of states that are accessible from $z_{i_0}^*$ communicate. The proof is constructive and rests on the existence of a finite length "path" from $(i_0, j_{k(i_0)}(i_0))$ to $(i_0, j_{k(i_0)}(i_0))$ that includes all (center, class) pairs. Recall that the job at the top of the job stack is said to be in position 1 of the job stack, the next job in position 2, etc.

**Theorem 1.7.** *Assume that the routing matrix* $\boldsymbol{P}$ *is irreducible and for some center* $i_0$*, either* $k(i_0) = 1$ *or service to a job of job class* $j_{k(i_0)}(i_0)$ *at center* $i_0$ *is preempted when any other job of higher priority joins the queue. Let* $z_{i_0}^*$ *be the state of the job stack process in which all* $N$ *jobs are of job class* $j_{k(i_0)}(i_0)$ *at center* $i_0$*. Then* $z$ *and* $z'$ *communicate for all* $z, z' \in D = \{z \in D^*\colon z_{i_0}^* \rightsquigarrow z\}$.

*Proof.* It is sufficient to show that there exists $z^* \in D^*$ such that $z \rightsquigarrow z^*$ for all $z \in D^*$. Without loss of generality, assume that either $k(1)$=1 or that service to jobs of job class $j_{k(1)}(1)$ at center 1 is prempted when jobs of higher priority join the queue. Let $z^* = z_1^*$ be the state of the job stack process in which all $N$ jobs are of job class $j_{k(1)}(1)$ at center 1. Since the routing matrix $\boldsymbol{P}$ is irreducible, there exists a finite sequence $(i_1, j_1), (i_2, j_2), \ldots, (i_M, j_M) \in C$ such that

(i) $(i_1, j_1) = (i_M, j_M) = (1, j_{k(1)}(1))$;
(ii) for any (center, class) pair $(i, j) \in C$, there exists $m$ $(1 \le m \le M)$ such that $(i_m, j_m) = (i, j)$; and
(iii) $p_{i_m j_m, i_{m+1} j_{m+1}} > 0$ for $m = 1, 2, \ldots, M-1$.

Let $l_m$ be the index of the first occurrence of $(1, j_{k(1)}(1))$ following $(i_m, j_m)$ in $(i_1, j_1), (i_2, j_2), \ldots, (i_M, j_M)$:

$$l_m = \min\{l \ge m\colon\ (i_l, j_l) = (1, j_{k(1)}(1))\}$$

for $m = 1, 2, \ldots, M-1$. Fix $(i, j) \in C$. Then there exists a subsequence $(i_m, j_m), (i_{m+1}, j_{m+1}), \ldots, (i_{l_m}, j_{l_m})$ of $(i_1, j_1), (i_2, j_2), \ldots, (i_M, j_M)$ such that $(i_m, j_m) = (i, j)$. Select the shortest such subsequence. If there are several of equal length, shorter than all other, select the one for which the index of the first element is smallest. Let $m(i, j)$ be the index of the first element of this subsequence:

$$m(i,j) = \min\{m\colon\ (i_m, j_m) = (i, j) \text{ and } l_m - m \le l_n - n \text{ for all } n \text{ such that } (i_n, j_n) = (i, j)\}.$$

In terms of this index, define the *successor (center, class) pair* $g(i, j)$ of $(i, j)$ as

$$g(i, j) = (i_{m(i,j)+1}, j_{m(i,j)+1}).$$

Recall that $q(z; i, j, k, l; m)$ denotes the next state of the job stack process when in state $z$ there is a completion of service by server $m$ to a job of job class $j$ at center $i$, and the job moves to center $k$ and becomes a job of job class $l$. For $z, z' \in D^*$, $z'$ is said to be a neighbor of $z$ if $z' = q(z : i, j, k, l, m)$ for some $(i, j), (k, l) \in C$ and $1 \le m \le m(i)$. Denote by $U(z)$ the set of all (center, class) pairs $(i,j) \in C - \{(1, j_{k(1)}(1))\}$ such that when the job stack process is in state $z$ there is at least one job of job class $j$ in service at center $i$. Let $h$ be a function taking values in $C$ and with domain $D^* \times \{1, 2, \ldots, N\}$ such that $h(z,n)$ is $(i,j)$ when the job in position $n$ of the job stack associated with state $z \in D^*$ is of job class $j$ at center $i$. Define a nonnegative "distance" from $z$ to $z^*$ as follows. For the job in position $n$ of the job stack associated with state $z$, set

$$d(z, n; z^*) = \min\{l_m - m \colon\ (i_m, j_m) = h(z,n)\}$$

for $n = 1, 2, \ldots, N$. Define the *distance* $d(z; z^*)$ from $z$ to $z^*$ according to

$$d(z, z^*) = \sum_{n=1}^{N} d(z, n; z^*).$$

First suppose that $z \neq z^*$. Then $U(z)$ is nonempty since either $k(1)$=1 or service to jobs of job class $j_{k(1)}(1)$ at center 1 is preempted. Select $(k,l) \in U(z)$. Let $z_1$ be a neighbor of $z$ such thst $z_1 = q(z; k, l, i_{m(k,l)+1}, j_{m(k,l)+1}, m)$ for some $1 \leq m < m(k)$. As $(k,l) \in U(z)$, it follows from the definition of the successor (center, class) pair $g(k,l)$ of $(k,l)$ that $z \to z_1$. Moreover, $d(z_1; z^*) < d(z; z^*)$. Next, select $(k_1, l_1) \in U(z_1)$. Let $z_2$ be a neighbor of $z_1$ such that $z_1 = q(z; k_1, l_1, k_2, l_2, m)$ for some $1 \leq m < m(k_1)$, where $(k_2, l_2) = g(k_1, l_1)$. Clearly, $z_1 \to z_2$ and necessarily $d(z_2; z^*) < d(z_1; z^*)$. Continuing in this way for at most a finite number of steps, the distance to $z^*$ decreases to 0 and it follows that $z \rightsquigarrow z^*$. Now suppose that $z = z^*$. Then there exists $z' \neq z^*$ such that $z^* \to z'$, and by the previous argument $z \rightsquigarrow z^*$. □

**Example 1.8.** (Cyclic queues with preemptive priority) Set $Z(t) = (Q_1(t), C_1^{(2)}(t), C_2^{(2)}(t), S_2(t))$ as in Example 1.2. As center 1 serves only one job class, the argument in the proof of Theorem 1.7 shows that $z^* = (N, 0, 0, 0)$ is a target state of the job stack process in the sense that $z \rightsquigarrow z^*$ for all $z \in D^*$. To illustrate, suppose that $N = 4$ and let $z = (1, 2, 0, 2)$. Take $(i_1, j_1) = (1, 1)$, $(i_2, j_2) = (2, 1)$, $(i_3, j_3) = (2, 2)$, and $(i_4, j_4) = (1, 1)$, and observe that since $p_{11,21}, p_{21,11}, p_{22,11} > 0$, $(i_1, j_1)$, $(i_2, j_2)$, $(i_3, j_3)$, $(i_4, j_4)$ is a path from (1,1) to (1,1) that includes all (center, class) pairs. Set $z_1 = (2, 1, 0, 1)$, $z_2 = (2, 1, 0, 2)$, $z_3 = (3, 0, 0, 1)$, and $z_4 = (3, 0, 0, 2)$. Then $z \to z_1$, $z_1 \to z_2$, $z_2 \to z_3$, $z_3 \to z_4$, and $z_4 \to z^*$. Since jobs of job class 2 have preemptive priority over jobs of job class 1 at center 2, $z_2^* = (0, N-1, 0, 1)$ is also a target state of the job stack process.

As in the proof of Theorem 1.7 let $h$ be a function taking values in $C$ and with domain $D^* \times \{1, 2, \ldots, N\}$ such that $h(z, n) = (i, j)$ when the job in position $n$ of the job stack associated with state $z$ is of job class $j$ at center $i$. Recall that $j_1(i_0), j_2(i_0), \ldots, j_{k(i_0)}(i_0)$ are the job classes served at center $i_0$ and ordered by decreasing priority.

**Definition 1.9.** *An element $z_0$ of the set $D$ is said to be a single-state of the job stack process if*

$$h(z_0, n) \in \{(i_0, j_1(i_0)), (i_0, j_2(i_0)), \ldots, (i_0, j_{k(i_0)}(i_0))\}$$

*for some single-server center $i_0$ $(n = 1, 2, \ldots, N)$.*

According to this definition, $z_0$ is a single-state of the job stack process if all jobs are at the same center ($i_0$) with exactly one job in service when the job stack process is in state $z_0$. We assume that a single-state of the job stack process exists. Observe that if center $i_0$ is a single-server center satisfying the hypotheses of Theorem 1.7, then $z_{i_0}^*$ is a single-state of the job stack process.

Additional conditions on the distributions of service time are needed to ensure that the job stack process hits a fixed single-state infinitely often with probability 1.

**Example 1.10.** (Cyclic queues) Consider a network of queues with two single-server centers ($L = 2, m(1) = m(2) = 1$) and one job class ($K = 1$) that provides service to $N$ ($\geq 2$) jobs. Then $C = \{(1,1),(2,1)\}$ is the set of (center, class) pairs in the network. Suppose that both queueing disciplines are FCFS. Also suppose that a job that completes service at center 1 moves to center 2 and a job that completes service at center 2 moves to center 1. Denote the distribution function of service time at center $i$ by $F_i$.

Set

$$Z(t) = (Q_1(t), Q_2(t)),$$

where $Q_i(t)$ is the number of jobs waiting or in service at center $i$ at time $t$. Formal specification of the job stack process $\{Z(t)\colon t \geq 0\}$ is as a GSMP with finite state space $S = \{(q_1, N - q_1)\colon 0 \leq q_1 \leq N\}$ and event set $E = \{e_{1,1}, e_{2,1}\}$.

States $(0, N)$ and $(N, 0)$ are single-states of the job stack process in the sense of Definition 1.9. The process $\{Z(t)\colon t \geq 0\}$ probabilistically restarts whenever there is a state transition from $(N, 0)$ to $(N-1, 1)$. The job stack process also probabilistically restarts whenever there is a state transition from $(0, N)$ to $(1, N-1)$. Let $V_1$ and $V_2^{N-1}$ be independent random variables distributed as $F_1$ and the $(N-1)$-fold convolution of $F_2$, respectively. If $F_1$ and $F_2$ are such that $P\left\{V_2^{N-1} < V_1\right\} = 1$, then beginning in any state with any clock-reading vector the process $\{Z(t)\colon t \geq 0\}$ makes a state transition from $(N, 0)$ to $(N-1, 1)$ infinitely often with probability 1. However, the process makes a state transition from $(0, N)$ to $(1, N-1)$ only a finite number of times.

## 4.2. Regenerative Job Stack Processes

Conditions on the building blocks of a network of queues are needed to ensure that the job stack process is a regenerative process in continuous time and the expected time between regeneration points is finite. The idea is that the job stack process restricted to the set of states that are accessible from a fixed target state can be represented as an irreducible GSMP with state space $D$ and event set $E$ given by (1.1).

Select a single-state $z_0$ of the job stack process and a successor state $z_0'$ ($z_0 \rightarrow z_0'$). Recall that by definition, $z_0$ is an element of the set $D$. Set $Z(0) = z_0'$, and let $L(t)$ be the last state occupied by the job stack process $\{Z(t) \colon t \geq 0\}$ before jumping to the state occupied at time $t$. Also set

$$Y(t) = (L(t), Z(t)). \tag{2.1}$$

Let $\zeta_n$ be the time of the $n$th state transition, $n \geq 0$.

**Theorem 2.2.** *Assume that the routing matrix $\boldsymbol{P}$ is irreducible and for some center $i_0$, either $k(i_0) = 1$ or service to a job of job class $j_{k(i_0)}(i_0)$ at center $i_0$ is preempted when any other job of higher priority joins the queue. Then $P\{Y(\zeta_n) = (z, z') \text{ i.o.}\} = 1$ for all $z, z' \in D$ such that $z \rightarrow z'$.*

*Proof.* The process $\{Y(t) \colon t \geq 0\}$ can be represented as a GSMP with a finite state space $B = \{(u, w) \in D \times D \colon u \rightarrow w\}$ and event set $E$ given by (1.1). For $y = (u, w) \in B$ with $u, w \in D$, event $e_{i,m} =$ "service completion by server $m$ at center $i$" is active in state $y$ if and only if $e_{i,m} \in E(w)$. Observe that for $y = (u, w), y' = (w, v) \in B$ and $e \in E$, $p(v; w, e)$ is the probability that the new state is $y'$ given that event $e$ triggers a state transition in state $y$. Also observe that for $y = (u, w), y' = (w, v) \in B$, $e \in E$, and $e' \in N(y'; y, e^*)$, the clock-setting distribution function $F(\,\cdot\,; y', e', y, e^*) = F(\,\cdot\,; v, e', w, e^*)$. Proposition 3.1 of Chapter 2 ensures that $P\{Y(\zeta_n) = (z, z') \text{ i.o.}\} = 1$ for all $(z, z')$ provided that

(i) the process $\{Y(t) \colon t \geq 0\}$ is irreducible in the sense that $y$ and $y'$ communicate for all states $y, y'$ of the process; and

(ii) each clock-setting distribution has finite mean and a density function that is continuous and positive on $(0, +\infty)$.

Condition (i) holds as a consequence of Theorem 1.7; $z_2 \rightsquigarrow z_1'$ implies $y \rightsquigarrow y'$ for $y = (z_1, z_2)$ and $y' = (z_1', z_2')$. Condition (ii) holds by assumption. □

Theorem 2.2 implies that $P\{Y(\zeta_n) = (z_0, z_0') \text{ i.o.}\} = 1$, and, when $Y(\zeta_n) = (z_0, z_0')$, the only (at most two) clocks that are active have just been set. The jumps of the job stack process from state $z_0$ are governed by some probability mass function $p(\,\cdot\,; z_0, e)$. Thus, the future evolution of the process $\{Z(t)\colon\ t \geq 0\}$ is independent of the history of the process before time $\zeta_n$ and has the same distribution as it does when $Z(0) = z_0'$. Thus, the successive random times $\{\zeta_{\beta_n}\colon\ n \geq 0\}$ at which there is a state transition from $z_0$ to $z_0'$ form a sequence of regeneration points for the job stack process.

Using properties of the GSSMC associated with state transitions of the job stack process, it can be shown that the expected value of the number $L_n$ of service completions in the time interval $[\zeta_{\beta_{n-1}}, \zeta_{\beta_n})$ is finite. For $(k, l) \in C$, let $L_n(k, l)$ be the number of service completions by a job of job class $l$ at center $k$ in the time interval $[\zeta_{\beta_{n-1}}, \zeta_{\beta_n})$. Also let $V_{m,n}(k, l)$ be the service time of the $m$th such job, $m = 1, 2, \ldots, L_n(k, l)$. Observe that

$$\zeta_{\beta_n} - \zeta_{\beta_{n-1}} \leq \sum_{(k,l)\in C} \sum_{m} V_{m,n}(k, l).$$

Then, applying Wald's identity to the inner sum,

$$\begin{aligned} E[\zeta_{\beta_n} - \zeta_{\beta_{n-1}}] &\leq \sum_{(k,l)\in C} E\Big[\sum_{m} V_{m,n}(k, l)\Big] \\ &= \sum_{(k,l)\in C} E[L_n(k, l)]\, E[V(k, l)] \end{aligned}$$

$$\leq E\Big[\sum_{(k,l)\in C} L_n(k,l)\Big] \max_{(k,l)\in C} E[V(k,l)]$$
$$= E[L_n] \max_{(k,l)\in C} E[V(k,l)].$$

Since the set $C$ is finite and each random variable $V(k,l)$ has finite mean, it follows that $E[\zeta_{\beta_n} - \zeta_{\beta_{n-1}}] < \infty$.

**Theorem 2.3.** *Assume that the conditions of Theorem 2.2 hold. Then the job stack process $\{Z(t)\colon t \geq 0\}$ is a regenerative process in continuous time, and the expected time between regeneration points is finite.*

Theorem 2.3 holds even if each distribution of service time does not have a density function that is continuous and positive on $(0, +\infty)$, provided that $P\{Y(\zeta_n) = (z_0, z_0') \text{ i.o.}\} = 1$ and the expected value of the number $L_n$ of service completions in the time interval $[\zeta_{\beta_{n-1}}, \zeta_{\beta_n})$ is finite. A similar argument shows that the successive random times at which the job stack process leaves state $z_0$ form a sequence of regeneration points and the expected time between regeneration points is finite.

For a real-valued measurable function $f$ with domain $D$, set $\zeta_{\beta_{-1}} = 0$ and

$$Y_k(f) = \int_{\zeta_{\beta_{k-1}}}^{\zeta_{\beta_k}} f(Z(u))\, du$$

for $k \geq 0$. Also set $\tau_k = \zeta_{\beta_k} - \zeta_{\beta_{k-1}}$ for $k \geq 1$. It follows directly from Theorem 2.3 that the sequence $\{(Y_k(f), \tau_k)\colon k \geq 1\}$ consists of i.i.d. random vectors.

**Proposition 2.4.** *For any real-valued measurable function $f$ with domain $D$ such that $Y_0(|f|) < \infty$ a.s. and $E[Y_1(|f|)] < \infty$,*

$$\lim_{t\to\infty} \frac{1}{t} \int_0^t f\big(Z(u)\big)\, du = \frac{E\big[Y_1(f)\big]}{E\big[\tau_1\big]} \stackrel{\text{def}}{=} r(f) \text{ a.s..}$$

Observe that the distribution of the time between regeneration points is aperiodic since each clock-setting distribution has a density function that is continuous and positive on $(0, +\infty)$. As the expected time between regeneration points is finite, it follows that $Z(t) \Rightarrow Z$ as $t \to \infty$.

**Proposition 2.5.** *For any real-valued measurable function $f$ with domain $D$ such that $E[|f(Z)|] < \infty$ or $E[Y_1(|f|)] < \infty$,*

$$E[f(Z)] = \frac{E[Y_1(f)]}{E[\tau_1]}.$$

*Moreover,*

$$\lim_{t \to \infty} E\big[f(Z(t))\big] = E[f(Z)]$$

*for all bounded functions $f$ such that $P\{Z \in D(f)\} = 0$.*

Using Wald's second moment identity and the fact that the distributions of service time have finite second moment, it can be shown that $E[\tau_1^2] < \infty$. Since the state space of the job stack process is finite, it follows that the variance constant $\sigma^2(f) = \text{Var}\,[Y_1(f) - r(f)\tau_1] < \infty$. From $n$ cycles, the regenerative method provides an estimator

$$\hat{r}(n) \stackrel{\text{def}}{=} \frac{\bar{Y}(n)}{\bar{\tau}(n)} = \frac{\sum_{k=1}^{n} Y_k(f)}{\sum_{k=1}^{n} \tau_k}$$

that is strongly consistent for $r(f)$. An asymptotic $100(1 - 2\gamma)\%$ confidence interval is

$$\left[\hat{r}(n) - \frac{s(n) z_{1-\gamma}}{\bar{\tau}(n)\sqrt{n}}, \hat{r}(n) + \frac{s(n) z_{1-\gamma}}{\bar{\tau}(n)\sqrt{n}}\right],$$

where the estimator $s^2(n)$ is strongly consistent for $\sigma^2(f)$. Asymptotic confidence intervals for $r(f)$ are based on the CLT

$$\frac{\sqrt{n}\,\{\hat{r}(n) - r(f)\}}{\sigma(f)/E[\tau_1]} \Rightarrow N(0,1)$$

as $n \to \infty$.

**Algorithm 2.6.** *Regenerative method for networks of queues*

1. *Select a single-state $z_0$ of the job stack process as in Definition 1.9 and a successor state $z_0'$ ($z_0 \to z_0'$).*
2. *Set $Z(0) = z_0'$, and observe a fixed number $n$ of cycles of the job stack process defined by the successive random times $\{\zeta_{\beta_k}\colon k \geq 0\}$ at which the job stack process makes a state transition from $z_0$ to $z_0'$.*
3. *Measure $(Y_1(f), \tau_1), (Y_2(f), \tau_2), \ldots, (Y_n(f), \tau_n)$.*
4. *Form the point estimate*

$$\hat{r}(n) = \frac{\bar{Y}(n)}{\bar{\tau}(n)}$$

   *for $r(f)$.*

5. *Form the asymptotic $100(1-2\gamma)\%$ confidence interval*

$$\left[\hat{r}(n) - \frac{s(n)z_{1-\gamma}}{\bar{\tau}(n)\sqrt{n}}, \hat{r}(n) + \frac{s(n)z_{1-\gamma}}{\bar{\tau}(n)\sqrt{n}}\right]$$

   *for $r(f)$.*

This algorithm prescribes selection of a single-state $z_0 \in D$ of the job stack process. For complex networks, it is nontrivial to determine the set $D$ by inspection. Let $z_i^*$ be the state of the job stack process in which all $N$ jobs are of job class $j_{k(i)}(i)$ at center $i$ ($i = 1, 2, \ldots, L$). It is easy to show that the set $D$ consists of those states that are accessible from some $z_i^*$.

**Proposition 2.7.** *Assume that the routing matrix $\boldsymbol{P}$ is irreducible and for some center $i_0$, either $k(i_0) = 1$ or service to a job of job class $j_{k(i_0)}(i_0)$ at center $i_0$ is preempted when any other job of higher priority joins the queue. Let $z \in D^*$. Then $z \in D$ if and only if $z_i^* \rightsquigarrow z$ for some $i = 1, 2, \ldots, L$.*

## Exercises

2.1. (Cyclic queues with feedback and three centers) Consider a network of queues with three single-server centers ($L = 3, m(1) = m(2) = m(3) = 1$) and a single job class ($K = 1$) that provides service to $N$ ($\geq 2$) jobs. Take $C = \{(1,1), (2,1), (3,1)\}$ to be the set of (center, class) pairs. A job that completes service at center 1 moves to center 2. With fixed probability $p$ ($0 < p < 1$), a job that completes service at center 2 moves to center 2 and with probability $1 - p$ moves to center 3. A job that completes service at center 3 moves to center 1. Assume that all queueing disciplines are FCFS. Also assume that service times are mutually independent and the service time at center $i$ has finite mean and a density function that is continuous and positive on $(0, +\infty)$.

Describe how you would use the regenerative method to obtain a strongly consistent point estimate and an asymptotic confidence interval for the limiting probability that there is at least one job at center 1. Be explicit about the sequence of regeneration points, the quantities to be measured, and the construction of the point and interval estimates. State the ratio formula and central limit theorem that lead to the estimates.

2.2. (Central server) Consider a network of queues with $L$ ($> 1$) single-server centers ($m(1) = m(2) = \ldots = m(L) = 1$) and a single job class ($K = 1$) that provides service to $N$ ($\geq 2$) jobs. Take $C = \{(1,1), (2,1), \ldots, (L,1)\}$ to be the set of (center, class) pairs. With fixed probability $p_i$ ($\sum_{i=2}^{L} p_i = 1$) a job that completes service at center 1 moves to center $i$ ($2 \leq i \leq L$). A job that completes service at center $i$ moves to center 1. Assume that all queueing disciplines are FCFS. Also assume that service times are mutually independent and the service time at center $i$ has finite mean and a density function that is continuous and positive on $(0, +\infty)$.

Describe how you would use the regenerative method to obtain a strongly consistent point estimate and an asymptotic confidence interval for the limiting expected number of jobs at center 1. Be explicit about the sequence of regeneration points, the quantities to be measured, and the construction of the point and interval estimates. State the ratio formula and central limit theorem that lead to the estimates.

2.3. (Finite-capacity queue with preemptive-repeat priority) Consider a queueing system with one single-server center and two job types numbered 1 and 2. Jobs of type $i$ arrive from an external source, queue and receive service at the center, and depart to an external sink. A job of type $i$ that arrives when the system already contains $B_i - 1$ $(\geq 0)$ jobs of type $i$ causes the arrival process for jobs of type $i$ to shut down. The arrival process remains shut down until the first subsequent departure of a job of type $i$. Jobs of the same type receive service in order of their arrival at the center and jobs of type 1 have preemptive-repeat priority over jobs of type 2. A job of type 2 that has been preempted receives additional service at the center before any other job of type 2 receives service. Assume that the interarrival times for jobs of type $i$ are i.i.d. as a positive random variable $A_i$ with finite mean and a density function that is continuous and positive on $(0, +\infty)$. Also assume that the service times for jobs of type $i$ are i.i.d. as a positive random variable $L_i$ with finite mean and a density function that is continuous and positive on $(0, +\infty)$.

Set

$$X(t) = (B_2(t), B_1(t), M(t)),$$

where

$$M(t) = \begin{cases} 2 & \text{if there is a job of type 2 in service at time } t \\ 1 & \text{if there is a job of type 1 in service at time } t \\ 0 & \text{if there is no job in service at time } t \end{cases}$$

and $B_i(t)$ is the number of jobs of type $i$ waiting in queue at time $t$. Show that $\{X(t): t \geq 0\}$ is a regenerative process in continuous time and the expected time between regeneration points is finite. Also describe how you would use the regenerative method to obtain a strongly consistent point estimate and an asymptotic confidence interval for the limiting probability that there are at most $k$ jobs in the system ($k = 0, 1, \ldots, B_1 + B_2$). Be explicit about the sequence of regeneration points, the quantities to be measured, and the construction of the point and interval estimates. State the ratio formula and central limit theorem that lead to the estimates.

2.4. (Network with stochastically nonidentical jobs) Consider a network of queues with $N+1$ single-server centers numbered $0, 1, \ldots, N$ that can provide service concurrently to $N$ stochastically nonidentical jobs labelled $1, 2, \ldots, N$. Center 0 provides nonpreemptive service distributed as a random variable $V_0$ to each of the $N$ jobs. Center $j$ provides service only to job $j$ ($j = 1, 2, \ldots, N$). Upon completion of service to job $j$ at center 0, with fixed probability $p_j$ ($0 \leq p_j < 1$) job $j$ joins the tail of the queue at center 0 and with probability $1 - p_j$ moves to center $j$ where, with no queueing delay, service begins. Service at center 0 is according to a fixed priority ordering of jobs, lower labelled jobs having higher priority. With respect to this priority scheme, a job that joins a nonempty queue at center 0 upon completion of service at center 0 does not contend for the next service. Assume that successive service times at center $i$ form a sequence of independent random variables, identically distributed as a positive random variable $V_i$ with finite first and second moment and a density function that is continuous and positive on $(0, +\infty)$. Also assume that the service times are mutually independent and independent of the routing of jobs from center 0.

Set

$$X(t) = (X_1(t), X_2(t), \ldots, X_N(t)),$$

where

$$X_j(t) = \begin{cases} 2 & \text{if job } j \text{ is in service at center 0 at time } t \\ 1 & \text{if job } j \text{ is in queue at center 0 at time } t \\ 0 & \text{if job } j \text{ is in service at center } j \text{ at time } t. \end{cases}$$

Show that $\{X(t)\colon\ t \geq 0\}$ is a regenerative process in continuous time and the expected time between regeneration points is finite. Let $W_n(j)$ be the $n$th waiting time (exclusive of service) at center 0 for job $j$, and suppose that $W_n(j) \Rightarrow W(j)$ as $n \to \infty$. Describe how you would use the regenerative method to obtain a strongly consistent point estimate and an asymptotic confidence interval for $P\{W(j) > x\}$ ($x$ fixed).

## 4.3. Geometric Trials Criteria for Recurrence

The geometric trials lemma (Lemma 3.8 of Chapter 2) can be used in conjunction with "new better than used" assumptions and a positivity condition to obtain criteria for recurrence results in networks of queues. Recall that the distribution of a positive random variable $A$ is said to be NBU if

$$P\{A > x + y\} \leq P\{A > x\}\ P\{A > y\}$$

for all $x, y \geq 0$.

**Example 3.1.** (Cyclic queues) Denote the distribution function of service time at center $i$ by $F_i$. Set $Z(t) = (Q_1(t), Q_2(t))$, where $Q_i(t)$ is the number of jobs waiting or in service at center 1 at time $t$. Assume that the distribution of service time at center 2 is NBU. Let $T_n$ be the time of the $n$th service completion at center 1, $n \geq 1$. Then

$$P\{Z(T_n-) = (N, 0) \text{ i.o.}\} = 1.$$

To see this, let $T_{n-1}^{+}$ be the first time greater than or equal to $T_{n-1}$ at which a job begins service at center 1 and suppose that $Z(T_{n-1}^{+}) = (q_1^{+}, N - q_1^{+})$ for some $1 \le q_1^{+} \le N$. If $q_1^{+} = N$, then clearly $Z(T_n-) = (N, 0)$. If $1 \le q_1^{+} < N$, then $Z(T_n-) = (N, 0)$ provided that all of the $N - q_1^{+}$ jobs at center 2 at time $T_{n-1}^{+}$ complete service before the job at center 1. Let $B_{\gamma_{n-1},1}$ be the service time for the job that starts service at center 1 at time $T_{n-1}^{+}$, $B_{\gamma_{n-1},2}$ be the remaining service time for the job in service at center 2 at time $T_{n-1}^{+}$ ($B_{\gamma_{n-1},2} = 0$ if no job is in service at center 2). Also let $\{V_{2,k}\colon k = 1, \ldots, N - q_1^{+} - 1\}$ be mutually independent random variables with common distribution function $F_2$. These random variables represent the service times of the other jobs, if any, at center 2. Then for any $z_0, z_1, \ldots, z_{n-1}$,

$$\begin{aligned} P\{Z(T_n-) &= (N,0) \mid Z(T_{n-1}^{+}) = (q_1^{+}, N - q_1^{+}), \\ &\quad Z(T_{n-1}-) = z_{n-1}, \ldots, Z(T_1-) = z_1\} \\ &\ge P\Big\{ B_{\gamma_{n-1},2} + \sum_{k=1}^{N-q_1^{+}-1} V_{2,k} \le B_{\gamma_{n-1},1} \,\Big|\, Z(T_{n-1}^{+}) = \\ &\quad (q_1^{+}, N - q_1^{+}), Z(T_{n-1}-) = z_{n-1}, \ldots, Z(T_1-) = z_1 \Big\} \\ &= P\{B_{\gamma_{n-1},2} \le R_{n-1}(q_1^{+}) \mid Z(T_{n-1}^{+}) = (q_1^{+}, N - q_1^{+}), \\ &\quad Z(T_{n-1}-) = z_{n-1}, \ldots, Z(T_1-) = z_1\}, \end{aligned}$$

where $R_{n-1}(q_1^{+})$ is distributed as $B_{\gamma_{n-1},1} - \sum_{k=1}^{N-q_1^{+}-1} V_{2,k}$.

Since the distribution of service time at center 2 is NBU, it is plausible that the residual service time $B_{\gamma_{n-1},2}$ is stochastically dominated by an independent sample from the distribution function $F_2$. Then

$$\begin{aligned} P\{B_{\gamma_{n-1},2} \le R_{n-1}(q_1^+) \mid Z(T_{n-1}^+) = (q_1^+, N - q_1^+), \\ Z(T_{n-1}-) = z_{n-1}, \ldots, Z(T_1-) = z_1\} \\ \ge P\Big\{V_2 \le R_{n-1}(q_1^+) \mid Z(T_{n-1}^+) = (q_1^+, N - q_1^+), \\ Z(T_{n-1}-) = z_{n-1}, \ldots, Z(T_1-) = z_1\Big\} \\ = P\left\{V_2 \le R_{n-1}(q_1^+)\right\}, \end{aligned}$$

where the random variable $V_2$ has distribution function $F_2$. The last equality is a consequence of the fact that $V_2$ and $R_{n-1}(q_1^+)$ are independent of $\{Z(T_{n-1}^+), Z(T_{n-1}-), \ldots, Z(T_1-)\}$. The difficulty is that although the distribution of service time at center 2 is NBU, the random variable $B_{\gamma_{n-1},2}$ under consideration is a residual service time observed at the random time $T_{n-1}^+$. Using the fact that the job stack process can be specified as a GSMP and $T_{n-1}^+$ is a stopping time, it can be shown that the above inequality holds.

Now let $V_1$ and $V_2^{N-1}$ be independent random variables with distribution function $F_1$ and the $(N-1)$-fold convolution of $F_2$, respectively, and set

$$\delta = P\left\{V_2^{N-1} \le V_1\right\}.$$

Since $P\left\{V_2 \le R_{n-1}(q_1^+)\right\} \ge \delta$, we can combine the preceding inequalities to obtain

$$\begin{aligned} P\{Z(T_n-) = (N,0) \mid Z(T_{n-1}^+) = (q_1^+, N - q_1^+), \\ Z(T_{n-1}-) = z_{n-1}, \ldots, Z(T_1-) = z_1\} \ge \delta \end{aligned}$$

and, using the law of total expectation,

$$P\left\{Z(T_n-) = (N,0) \mid Z(T_{n-1}-), \ldots, Z(T_1-)\right\} \ge \delta \text{ a.s.}$$

for all $n$. Thus, $P\{Z(T_n-) = (N,0) \text{ i.o.}\} = 1$ provided that $\delta > 0$. This positivity condition is an assumption about the distributions of service time that can be verified *a priori*. The proof of Theorem 4.5 uses these ideas.

To clarify the role of the NBU assumption, let $Y_{\gamma_{n-1},2}$ be the elapsed service time for the job in service at center 2 at time $T_{n-1}^+$; $Y_{\gamma_{n-1},2} = 0$ if no job is in service at center 2 at time $T_{n-1}^+$. It can be shown that for $1 \le q_1^+ < N$,

$$\begin{aligned}
P\Big\{ B_{\gamma_{n-1},2} &\le R_{n-1}(q_1^+) \,\Big|\, Z(T_{n-1}^+) = (q_1^+, N - q_1^+),\\
&Z(T_{n-1}-) = z_{n-1}, \ldots, Z(T_1-) = z_1 \Big\}\\
&= E\Big[1 - \Big(\overline{F}_2(Y_{\gamma_{n-1},2} + R_{n-1}(q_1^+)) \Big/ \overline{F}_2(Y_{\gamma_{n-1},2})\Big)\\
&\quad \Big|\; Z(T_{n-1}^+) = (q_1^+, N - q_1^+), Z(T_{n-1}-) = z_{n-1}, \ldots,\\
&\quad Z(T_1-) = z_1 \Big],
\end{aligned}$$

where $\overline{F}_2 = 1 - F_2$. The NBU assumption provides an a.s. upper bound (namely, $\overline{F}_2(R_{n-1}(q_1^+))$) on the quantity $\overline{F}_2(Y_{\gamma_{n-1},2} + R_{n-1}(q_1^+)) \big/ \overline{F}_2(Y_{\gamma_{n-1},2})$ that is independent of $Y_{\gamma_{n-1},2}$, so that

$$\begin{aligned}
P\Big\{ B_{\gamma_{n-1},2} &\le R_{n-1}(q_1^+) \mid Z(T_{n-1}^+) = (q_1^+, N - q_1^+),\\
&\quad Z(T_{n-1}-) = z_{n-1}, \ldots, Z(T_1-) = z_1 \Big\}\\
&\ge E\Big[1 - \overline{F}_2(R_{n-1}(q_1^+)) \mid Z(T_{n-1}^+) = (q_1^+, N - q_1^+),\\
&\quad Z(T_{n-1}-) = z_{n-1}, \ldots, Z(T_1-) = z_1 \Big]\\
&= E[F_2(R_{n-1}(q_1^+))] \ge \delta.
\end{aligned}$$

## 4.4. Preemptive-Resume Priorities

The geometric trials lemma can be used to obtain criteria for recurrence and regeneration that avoid the preemptive-repeat and positive density assumptions used previously. The principal result is that the successive random times at which the job stack process leaves a fixed single-state $z_{i_0}^*$ form a sequence of regeneration points for the job stack process. Using a "distance function" defined on the set of (center, class) pairs in the network along with Wald's moment identities and an assumption that the first and second moments of the distributions of service time are finite, we then show that the time between regeneration points has finite first and second moments. Under these conditions, the regenerative method can be used to obtain strongly consistent point estimates and asymptotic confidence intervals for steady-state means.

We consider closed networks of queues with a finite number $N$ of jobs (customers), a finite number $L$ of centers numbered $1, 2, \ldots, L$, and a finite number $K$ of mutually exclusive job classes numbered $1, 2, \ldots, K$. We assume throughout that $N \geq 2$ and $LK > 1$. At every point of continuous time each job is of exactly one job class, but jobs may change job class as they traverse the network. Upon completion of service at center $i$, a job of job class $j$ moves to center $k$ and becomes a job of job class $l$ with probability $p_{ij,kl}$. The routing matrix

$$\boldsymbol{P} = \{p_{ij,kl}\colon\ (i,j),(k,l) \in C\}$$

is a given irreducible stochastic matrix, and

$$C \subseteq \{1,2,\ldots,L\} \times \{1,2,\ldots,K\}$$

is the set of (center, class) pairs in the network. We assume that each center is a single-server center and jobs queue and receive service according to a fixed priority scheme among job classes; the priority scheme may differ from center to center. Within a job class at a center, jobs receive service according to a fixed queueing discipline that depends only on the order of arrival of the jobs. According to a fixed procedure for each center, a job in service may or may not be preempted by all jobs of a given higher priority job class. A job may not be preempted by another job of the same job class. Any interruption of service is of the "preempt-resume" type in that service continues from the point of interruption when the preempted job goes back into service. A preempted job goes into service at a center before any other job of the same job class that is waiting in queue at the center. We assume that all service times are a.s. finite and mutually independent and the service time for a job of job class $j$ at center $i$ has distribution function $F_{ij}$. We denote by $\mu_{ij}^m$ the $m$th moment of $F_{ij}$ and write $\overline{F}_{ij} = 1 - F_{ij}$.

Let $S_i(t)$ be the job class of the job in service at center $i$ at time $t$; by convention $S_i(t) = 0$ if at time $t$ there are no jobs at center $i$. We denote the job classes served at center $i$ and ordered by decreasing priority by $j_1(i), j_2(i), \ldots, j_{k(i)}(i)$, and we let $C_{j_1}^{(i)}(t), \ldots, C_{j_{k(i)}}^{(i)}(t)$ be the number of jobs waiting in queue at time $t$ of the various job classes served at center $i$.

We order the $N$ jobs in a linear stack and define the vector $Z(t)$ by

$$\begin{aligned} Z(t) = \bigl(&C_{j_{k(1)}}^{(1)}(t), \ldots, C_{j_1}^{(1)}(t), S_1(t); \ldots; \\ &C_{j_k(L)}^{(L)}(t), \ldots, C_{j_1}^{(L)}(t), S_L(t)\bigr). \end{aligned}$$

The job stack at time $t$ corresponds to the nonzero components in the vector $Z(t)$ and is an ordering of the jobs by job class at the individual centers. Within a job class at a particular center, jobs appear in the job stack in order of their arrival at the center, the latest to arrive being closest to the top of the job stack. A job that has been preempted appears at the head of the queue of jobs of its job class. The process $\{Z(t)\colon\ t \geq 0\}$ is called the *job stack process*.

The process $\{Z(t)\colon\ t \geq 0\}$ has a finite state space $D^*$ that is the set of all vectors $s = (c^{(1)}_{j_{k(1)}}, \ldots, c^{(1)}_{j_1}, s_{1,1}, \ldots, s_{1,m(1)}; \ldots; c^{(L)}_{j_{k(L)}}, \ldots, c^{(L)}_{j_1}, s_{L,1}, \ldots, s_{L,m(L)}) \in \prod_{i=1}^{L} \{0, 1, \ldots, N-1\}^{k(i)} \times \{0, j_1(i), j_2(i), \ldots, j_{k(i)}(i)\}^{m(i)}$ satisfying

(i) $N = c^{(1)}_{j_{k(1)}} + \cdots + c^{(1)}_{j_1} + 1_{\{j_1(1), \ldots, j_{k(1)}(1)\}}(s_{1,1}) + \cdots + 1_{\{j_1(L), \ldots, j_{k(1)}(1)\}}(s_{1,m(1)}) + \cdots + c^{(L)}_{j_{k(L)}} + \cdots + c^{(L)}_{j_1} + 1_{\{j_1(L), \ldots, j_{k(1)}(L)\}}(s_{L,1}) + \cdots + 1_{\{j_1(L), \ldots, j_{k(1)}(1)\}}(s_{L,m(L)})$;

(ii) for all $i$ and $m$, $s_{i,m} = 0$ only if $c^{(i)}_{j_{k(i)}} + \cdots + c^{(i)}_{j_1} = 0$; and

(iii) for all $i$ and $k$ such that jobs of job class $j_k(i)$ are subject to preemption at center $i$,

$$c^{(i)}_{j_{k-1}} = 0, \ldots, c^{(i)}_{j_1} = 0 \text{ whenever}$$
$$1_{j_k(i)}(s_{i,1}) + \cdots + 1_{j_k(i)}(s_{i,m(i)}) > 0.$$

Also set

$$M(t) = \big(M_{1,j_{k(1)}(1)}(t), \ldots, M_{1,j_1(1)}(t); \ldots; M_{L,j_{k(L)}(L)}(t), \ldots, M_{L,j_1(L)}(t)\big),$$

where

$$M_{i,j}(t) = \begin{cases} 1 & \text{if at time } t \text{ a preempted job is at the head of the job class } j \text{ queue at center } i \\ 0 & \text{otherwise.} \end{cases}$$

Take event $e_{i,j}$ = "service completion to a job of job class $j$ at center $i$." Formal specification of the process $\{(Z(t), M(t))\colon t \geq 0\}$ is as a GSMP with state space $D^* \times \{0,1\}^{|C|}$ and event set

$$E = \left\{e_{1,j_1(1)}, \ldots, e_{1,j_{k(1)}(1)}, \ldots, e_{L,j_1(L)}, \ldots, e_{L,j_{k(L)}(L)}\right\}.$$

For $s = (c^{(1)}_{j_{k(1)}}, \ldots, c^{(1)}_{j_1}, s_1; \ldots; c^{(L)}_{j_{k(L)}}, \ldots, c^{(L)}_{j_1}, s_L; m_{1,j_{k(1)}}, \ldots, m_{1,j_1}; \ldots; m_{L,j_{k(L)}}, \ldots, m_{L,j_1}) \in D^* \times \{0,1\}^{|C|}$, event $e_{i,j} =$

"service completion to a job of job class $j$ at center $i$" is active in state $s$ if and only if $s_i = j$ or $m_{i,j} = 1$. Thus, $e_{i,j} \in E(s)$ if and only if in state $s$ either a job of job class $j$ receives service at center $i$ or a preempted job is at the head of the queue of jobs of job class $j$ at center $i$. The speed at which the clock associated with event $e_{i,j}$ runs in state $s$ is 0 if $m_{i,j} = 1$. All other speeds for active events are equal to 1. Recall that by assumption there is only one server at each center and any interruption of service is of the preemptive-resume type.

For $s \in D^* \times \{0,1\}^{|C|}$ and $(i,j), (k,l) \in C$, denote by $q(s; i, j, k, l)$ the new state of the process $\{(Z(t), M(t))\colon\ t \geq 0\}$ when in state $s$ there is a completion of service to a job of job class $j$ at center $i$, and the job moves to center $k$ and becomes a job of job class $l$. If $e = e_{i,j}$, then the state transition probability $p(s'; s, e) = p_{ij,kl}$ when

$$s = (c^{(1)}_{j_{k(1)}}, \ldots, s_1; \ldots; c^{(L)}_{j_{k(L)}}, \ldots, s_L; m_{1,j_{k(1)}}, \ldots, m_{1,j_1}; \ldots;\\ m_{L,j_{k(L)}}, \ldots, m_{L,j_1}) \text{ with } s_i = j \text{ and } s' = q(s; i, j, k, l).$$

All other state transition probabilities $p(s'; s, e)$ are equal to 0. The clock-setting distribution functions are as follows. If the new event $e' = e_{i,j}$, then $F(x; s', e', s, e^*) = F_{ij}(x)$.

Theorem 4.5 asserts that under a mild restriction on the priorities among job classes, the job stack process leaves a fixed single-state $z^*_{i_0}$ infinitely often provided that certain distributions of service time are NBU and a positivity condition is satisfied. The proof uses a geometric trials criterion for recurrence. The successive random times at which there is a service completion for jobs of a fixed job class at a fixed center serve as the $\{T_n\colon\ n \geq 0\}$ in the geometric trials lemma. Lemma 4.1 shows that these random times are a.s. finite. Set $T_0(i,j) = 0$, and let $T_n(i,j)$ be the time of the $n$th service completion by a job of job class $j$ at center $i$, $n > 0$.

**Lemma 4.1.** *Assume that the routing matrix* $\boldsymbol{P}$ *is irreducible and service times for a job of job class* $j$ *at center* $i$ *are a.s. finite for all* $(i, j) \in C$. *Then* $P\{T_n(i, j) < \infty\} = 1$ *for all* $(i, j) \in C$ *and* $n \geq 0$.

*Proof.* Let $A_{ij}^n$ be the event "the $n$th service completion occurs at center $i$ for a job of job class $j$," and let $A_{ij}$ be the event "infinitely many service completions occur at center $i$ for a job of job class $j$": $A_{ij} = \{ A_{ij}^n \text{ i.o.} \}$. Similarly, for $(k, l) \in C$ let $B_{ij,kl}^n$ be the event "upon the $n$th service completion of a job of job class $j$ at center $i$, the job moves to center $k$ and becomes a job of job class $l$" and set $B_{ij,kl} = \{ B_{ij,kl}^n \text{ i.o.} \}$.

Now fix $(k, l) \in C$, and observe that since the routing matrix $\boldsymbol{P}$ is irreducible, there exists a (center, class) pair $(i, j) \in C$ such that $(i, j) \neq (k, l)$ and $p_{ij,kl} > 0$. Since $N$ is finite by assumption, it follows that

$$P\{A_{ij} \cap A_{kl}^c\} \leq P\{B_{ij,kl}^c\},$$

where $A^c$ denotes the complement of an event $A$. Next observe that since the events $B_{ij,kl}^n$ are independent and $P\{ B_{ij,kl}^n \} = p_{ij,kl} > 0$ for all $n \geq 1$, $P\{ B_{ij,kl}^c \} = 0$ by the Borel-Cantelli lemma. Thus,

$$P\{A_{kl}^c\} = P\{A_{kl}^c \cap A_{ij}\} + P\{A_{kl}^c \cap A_{ij}^c\} = P\{A_{kl}^c \cap A_{ij}^c\}.$$

Using the irreducibility of the routing matrix $\boldsymbol{P}$ and applying the above argument repeatedly, we obtain

$$P\{A_{kl}^c\} = P\Big\{ \bigcap_{(i,j) \in C} A_{ij}^c \Big\}. \tag{4.2}$$

Since all service times are a.s. finite, the left side of (4.2) is 0. Then $P\{A_{kl}\} = 1$ and $T_n(k, l)$ is a.s. finite for all $n \geq 0$. $\square$

In order to apply the geometric trials lemma, we exploit properties of distributions that are NBU. Observe that if $B_1$, $B_2, \ldots, B_n$ are mutually independent random variables with distributions that are NBU, then

$$
\begin{aligned}
&P\Big\{ \sum_{i=1}^{n}(B_i - y_i) > x \Big\} \\
&\qquad \le P\Big\{ \sum_{i=1}^{n} B_i > x \Big\} P\Big\{ B_i > y_i, \; i = 1, \ldots, n \Big\}
\end{aligned}
\tag{4.3}
$$

for all $x, y_1, y_2, \ldots, y_n \ge 0$.

The proof of Theorem 4.5 uses an integer-valued "distance function" on the set $C$ of (center, class) pairs, along with a "successor function" on $C$ and taking values in $C$. Definition of the distance function is in terms of an $n$-step successor function $g^n$ defined recursively by $g^0(i,j) = (i,j)$ and

$$g^n(i,j) = (g_1^n(i,j), g_2^n(i,j)) = g(g^{n-1}(i,j))$$

for $n > 0$. For each $(i,j) \in C$, there exists an integer $n(i,j)$ $(0 \le n(i,j) \le LK)$ such that

$$g^{n(i,j)}(i,j) = (1, j_{k(1)}(1)).$$

Then the *distance function* $r$ is given by

$$r(i,j) = \inf \big\{ n \ge 0\colon \; g^n(i,j) = (1, j_{k(1)}(1)) \big\} \tag{4.4}$$

for $(i,j) \in C$. Observe that the successor function can be chosen so that

$$g^m(i,j) \ne g^n(i,j)$$

for all $0 \le m < n \le r(i,j)$ and $(i,j) \in C$.

Formal definition of the successor function is as follows. Since the routing matrix $\boldsymbol{P}$ is irreducible, there exists a finite sequence $(i_1,j_1),(i_2,j_2),\ldots,(i_M,j_M)$ of (center, class) pairs such that (i) $(i_1,j_1) = (i_M,j_M) = (1,j_{k(1)}(1))$; (ii) for any $(k,l) \in C$ there exists $m$ $(1 \le m \le M)$ such that $(i_m,j_m) = (k,l)$; and (iii) $p_{i_m j_m, i_{m+1} j_{m+1}} > 0$ for $m = 1,2,\ldots,M-1$. Let $l_m$ be the index of the first occurrence of $(1,j_{k(1)}(1))$ following $(i_m,j_m)$ in $(i_1,j_1),(i_2,j_2),\ldots,(i_M,j_M)$:

$$l_m = \min\left\{l \ge m\colon\ (i_l,j_l) = (1,j_{k(1)}(1))\right\}$$

for $m = 1,2,\ldots,M-1$. Fix $(k,l) \in C$. Then there exists a subsequence $(i_m,j_m),(i_{m+1},j_{m+1}),\ldots,(i_{l_m},j_{l_m})$ of $(i_1,j_1),(i_2,j_2),\ldots,(i_M,j_M)$ such that $(i_m,j_m) = (k,l)$. Select the shortest such subsequence. If there are several of equal length, shorter than all others, select the one for which the index of the first element is smallest. Let $m(k,l)$ be the index of the first element of this subsequence:

$$m(k,l) = \min\bigl\{m\colon\ (i_m,j_m) = (k,l) \text{ and } l_m - m \le l_n - n \text{ for all } n \text{ such that } (i_n,j_n) = (k,l)\bigr\}.$$

In terms of this index, define the *successor function* $g$ by

$$g(k,l) = (g_1(k,l),g_2(k,l)) = (i_{m(k,l)+1},j_{m(k,l)+1}).$$

For a center $i_0$, let $z^*_{i_0}$ be the state of the job stack process in which all $N$ jobs are of job class $j_{k(i_0)}(i_0)$ at center $i_0$. Set $C_{i_0} = C - \{(i_0,j_{k(i_0)}(i_0))\}$, and observe that $z^*_{i_0}$ is a single-state of the job stack process in the sense that all jobs are at the same single-server center $i_0$. Also observe that in state $z^*_{i_0}$ no jobs have been preempted.

**Theorem 4.5.** *Assume that the routing matrix* $\boldsymbol{P}$ *is irreducible. Also assume that*

*(i) for some center* $i_0$*, either* $k(i_0) = 1$ *or service to a job of job class* $j_{k(i_0)}(i_0)$ *is preempted when any other job of higher priority joins the queue;*

*(ii) the distribution of service time for a job of job class $j$ at center $i$ is NBU for all $(i,j) \in C_{i_0}$; and*

*(iii) $\sum_{(i,j)\in C_{i_0}} (N-1)\, v(i,j) < u(i_0, j_{k(i_0)}(i_0))$, where $v(i,j) = \sup\{x\colon\ F_{ij}(x) = 0\}$ and $u(i,j) = \inf\{x\colon\ F_{ij}(x) = 1\}$.*

*Set*

$$T_n = T_n(i_0, j_{k(i_0)}(i_0)) \tag{4.6}$$

*for $n \geq 0$. Then there exists $\delta > 0$ such that*

$$P\left\{Z(T_n-) = z^*_{i_0} \mid Z(T_{n-1}-), \ldots, Z(T_1-)\right\} \geq \delta \text{ a.s.} \tag{4.7}$$

*so that $P\left\{Z(T_n-) = z^*_{i_0} \text{ i.o.}\right\} = 1$.*

*Proof.* Assume without loss of generality that $i_0 = 1$ so that $C_1$ is the set of all (center, class) pairs other than $(1, j_{k(1)}(1))$. Let $T^+_{n-1}$ be the first time greater than or equal to $T_{n-1}$ at which a job of job class $j_{k(1)}(1)$ starts service at center 1. Also let $T^+_{n,k}$ be the first time greater than or equal to $T^+_{n-1}$ at which the job in position $k$ of the job stack at time $T^+_{n-1}$ appears at center 1 as a job of job class $j_{k(1)}(1)$, and let $C_{n,k}$ be the total amount of time that the job is in service during $(T^+_{n-1}, T^+_{n,k}]$. Denote by $k^+_{n-1}$ the position in the job stack of the job in service at center 1 at time $T^+_{n-1}$, and let $H^+_{n-1}$ be the total service time of that job at center 1. Since interruptions of service are of the preempt-resume type, it follows from Condition (i) that $T_n \geq \max_{k\neq k^+_{n-1}} T^+_{n,k}$ whenever $\sum_{k\neq k^+_n} C_{n,k} < H^+_{n-1}$, so that $Z(T_n-) = z^*_1$. Thus,

$$\begin{aligned} P&\left\{Z(T_n-) = z^*_1 \mid Z(T_{n-1}-), \ldots, Z(T_1-)\right\} \\ &\geq P\Big\{ \sum_{k\neq k^+_n} C_{n,k} < H^+_{n-1} \mid Z(T_{n-1}-), \ldots, Z(T_1-) \Big\} \text{ a.s..} \end{aligned} \tag{4.8}$$

Denote by $A_n$ the event that all jobs are routed according to the successor function $g$ during the time interval $(T^+_{n-1}, T_n)$.

Consider the distribution of the quantity $\sum_{k \neq k_n^+} C_{n,k}$ when $A_n$ occurs. For $(i,j) \in C_1$, let $B(i,j)$ be a random variable distributed as $\sum_{r=0}^{r(i,j)-1} V(g_1^r(i,j), g_2^r(i,j))$, where $V(i,j)$ is a random variable with distribution $F_{ij}$. The total amount of service, for a job that moves to center $i$ as a job of job class $j$ and is routed according to the successor function $g$ until it moves to center 1 and becomes a job of job class $j_{k(1)}(1)$, is distributed as $B(i,j)$. Set

$$I_{n-1}^+ = \{\, (i,j) \in C_1\text{: a job of job class } j \text{ at center } i \text{ has not completed its most recent service at time } T_{n-1}^+ \,\}.$$

If the job in position $k \neq k_{n-1}^+$ of the job stack at time $T_{n-1}^+$ is a job of job class $j$ at center $i$ and $(i,j) \notin I_{n-1}^+$, then $C_{n,k}$ is distributed as $B(i,j)$. Otherwise, $C_{n,k}$ is distributed as $H_{n-1}^+(i,j) + B(g_1(i,j), g_2(i,j))$, where $H_{n-1}^+(i,j)$ is the remaining service time of the job. Denote by $n(i,j,z)$ the number of jobs of job class $j$ at center $i$ when the job stack process is in state $z$. It follows that $\sum_{k \neq k_n^+} C_{n,k}$ is distributed as $\sum_{(i,j) \in I_{n-1}^+} H_{n-1}^+(i,j) + B_{n-1}^+$ whenever event $A_n$ occurs, where $B_{n-1}^+$ is distributed as $\sum_{(i,j) \in C_1} G_{n-1}^+(i,j)$. The random variable $G_{n-1}^+(i,j)$ is distributed as the $n(i,j,Z(T_{n-1}^+))$-fold convolution of $B(i,j)$ for $(i,j) \notin I_{n-1}^+$ and of the sum of $B(g_1(i,j), g_2(i,j))$ and the $(n(i,j,Z(T_{n-1}^+)) - 1)$-fold convolution of $B(i,j)$ for $(i,j) \in I_{n-1}^+$.

Equation (4.8) then implies that

$$\begin{aligned} &P\{Z(T_n-) = z_1^* \mid Z(T_{n-1}-), \ldots, Z(T_1-)\} \\ &\quad \geq P\Big\{A_n,\ \sum_{(i,j) \in I_{n-1}^+} H_{n-1}^+(i,j) + B_{n-1}^+ < H_{n-1}^+ \\ &\qquad\qquad \Big|\ Z(T_{n-1}-), \ldots, Z(T_1-)\Big\} \text{ a.s..} \end{aligned}$$

The assumption that the routing of jobs is Markovian ensures that

$$P\Big\{A_n \,\Big|\, \sum_{(i,j)\in I^+_{n-1}} H^+_{n-1}(i,j) + B^+_{n-1} < H^+_{n-1},$$
$$Z(T_{n-1}-),\ldots,Z(T_1-)\Big\} \ge \rho^{(N-1)(LK-1)} \text{ a.s.},$$

where $\rho = \min\{p_{ij,kl}\colon\ p_{ij,kl} > 0\}$, so that by (4.7)

$$\begin{aligned} P\{Z(T_n-) = z_1^* \mid Z(T_{n-1}-),\ldots,Z(T_1-)\} \\ \ge \rho^{(N-1)(LK-1)} P\Big\{ \sum_{(i,j)\in I^+_{n-1}} H^+_{n-1}(i,j) < R^+_{n-1} \\ \Big|\, Z(T_{n-1}-),\ldots,Z(T_1-)\Big\} \text{ a.s.,} \end{aligned} \tag{4.9}$$

where $R^+_{n-1} = H^+_{n-1} - B^+_{n-1}$. For a job of job class $j$ at center $i$ that has not completed its most recent service at time $T^+_{n-1}$, let $Y^+_{n-1}(i,j)$ be the amount of service received so far. It then can be shown that the random variables $H^+_{n-1}(i,j)$ for $(i,j) \in I^+_{n-1}$ are conditionally independent (given $R^+_{n-1}$, $I^+_{n-1}$, $Y^+_{n-1}(i,j)$ for $(i,j) \in I^+_{n-1}$ and $Z(T_{n-1}-),\ldots,Z(T_1-)$) with marginal distributions defined by

$$P\Big\{H^+_{n-1}(i,j) \le x \,\Big|\, R^+_{n-1}; I^+_{n-1}; Y^+_{n-1}(i,j),\ (i,j)\in I^+_{n-1};$$
$$Z(T_{n-1}-),\ldots,Z(T_1-)\Big\} = 1 - \frac{\overline{F}_{ij}(x + Y^+_{n-1}(i,j))}{\overline{F}_{ij}(Y^+_{n-1}(i,j))} \text{ a.s.}$$

for $(i,j) \in I^+_{n-1}$. Let $V^k(i,j)$ be a random variable distributed as the $k$-fold convolution of $F_{ij}$. Using Condition (ii) and (4.3),

$$
\begin{aligned}
P\Big\{ \sum_{(i,j)\in I^+_{n-1}} & H^+_{n-1}(i,j) < R^+_{n-1} \;\Big|\; Z(T_{n-1}-),\ldots,Z(T_1-)\Big\} \\
&= E\Big[ P\Big\{ \sum_{(i,j)\in I^+_{n-1}} H^+_{n-1}(i,j) < R^+_{n-1} \;\Big|\; R^+_{n-1}; I^+_{n-1}; \\
&\quad Y^+_{n-1}(i,j),(i,j)\in I^+_{n-1}; Z(T_{n-1}-),\ldots,Z(T_1-)\Big\} \\
&\quad \Big|\; Z(T_{n-1}-),\ldots,Z(T_1-)\Big] \\
&= E\Big[ P\Big\{ \sum_{(i,j)\in I^+_{n-1}} (V(i,j) - Y^+_{n-1}(i,j)) < R^+_{n-1} \\
&\quad \Big|\; V(i,j) > Y^+_{n-1}(i,j),(i,j)\in I^+_{n-1}; R^+_{n-1}; I^+_{n-1}; \\
&\quad Y^+_{n-1}(i,j),(i,j)\in I^+_{n-1}\Big\} \;\Big|\; Z(T_{n-1}-),\ldots,Z(T_1-)\Big] \\
&\ge E\Big[ P\Big\{ \sum_{(i,j)\in I^+_{n-1}} V(i,j) < R^+_{n-1} \;\Big|\; R^+_{n-1}; I^+_{n-1}\Big\} \\
&\quad \Big|\; Z(T_{n-1}-),\ldots,Z(T_1-)\Big] \\
&\ge P\Big\{ \sum_{(i,j)\in C_1} V^{N-1}(i,j) \le V^1(i_0, j_{k(i_0)}(i_0))\Big\} \stackrel{\text{def}}{=} \eta \text{ a.s.,}
\end{aligned}
\tag{4.10}
$$

where $\eta$ is strictly positive by Condition (iii). The last inequality in (4.10) follows from the fact that the random variable $H^+_{n-1}$ is distributed as $V^1(1, j_{k(1)}(1))$ and $\sum_{(i,j)\in C_1} V^{N-1}(i,j)$ stochastically dominates $\sum_{(i,j)\in I^+_{n-1}} V(i,j) + B^+_{n-1}$. Set

$$
\delta = \eta\, \rho^{(N-1)(LK-1)}.
$$

It then follows from (4.9) and (4.10) that

$$P\{\, Z(T_n-) = z_1^* \mid Z(T_{n-1}-), \ldots, Z(T_1-) \,\} \geq \delta \text{ a.s.},$$

and the proof is complete. □

Observe that the positivity condition of Condition (iii) is satisfied if service times at center $i_0$ are unbounded (that is, $u(i_0, j_{k(i_0)}(i_0)) = +\infty$), or if service time for a job of job class $j$ at center $i$ has a density function that is positive on $(0, b)$ for some $b > 0$ and all $(i, j) \in C$.

**Example 4.11.** (Cyclic queues) Denote the distribution function of service time for a job of job class $j$ at center $i$ by $F_{ij}$. Also assume that the distribution of service time at center 2 is NBU and $(N-1)\,v(2,1) < u(1,1)$, where $v(i,j) = \sup\{x\colon\ F_{ij}(x) = 0\}$ and $u(i,j) = \inf\{x\colon\ F_{ij}(x) = 1\}$. Set

$$Z(t) = \bigl(C_1^{(1)}(t), S_1(t), C_1^{(2)}(t), S_2(t)\bigr),$$

where $S_i(t)$ equals 1 if there is a job in service at center $i$ at time $t$ (and equals 0 otherwise) and $C_1^{(i)}(t)$ is the number of jobs (of job class 1) waiting in queue at center $i$. Then the job stack process $\{Z(t)\colon\ t \geq 0\}$ leaves state $(N-1, 1, 0, 0)$ (and hits state $(N-2, 1, 0, 1)$) infinitely often with probability 1.

Theorem 4.16 below asserts that the job stack process $\{Z(t)\colon\ t \geq 0\}$ is a regenerative process in continuous time and the time between regeneration points has finite first and second moments. This result requires the conditions of Theorem 4.5 and the assumption that the distribution of service time for jobs of the lowest priority job class at the distinguished center $i_0$ has finite first and second moments. We first show that for all (center, class) pairs $(i, j)$ the expected value of the number $K_n(i, j)$ of service completions in the time interval $(T_{n-1}(i, j), T_n(i, j)]$ is uniformly bounded in $n$. Recall that

$T_n(i,j)$ is the time of the $n$th service completion by a job of job class $j$ at center $i$.

**Lemma 4.12.** *Set* $\kappa = 2N(LK-1)\rho^{-N(LK-1)}$ *and* $\rho = \min\{p_{ij,kl}\colon p_{ij,kl} > 0\}$. *Then*

$$E[K_n(i,j)] \le \kappa \tag{4.13}$$

*for all* $(i,j) \in C$ *and* $n \ge 1$.

*Proof.* Assume without loss of generality that (center, class) pair $(i,j) = (1, j_{k(1)}(1))$, and set $T_k = T_k(1, j_{k(1)}(1))$ for $k \ge 0$. Fix $n \ge 1$. As in the proof of Theorem 4.5, let $T^+_{n-1}$ be the first time greater than $T_{n-1}$ at which a job of job class $j_{k(1)}(1)$ starts service at center 1. We claim that

$$E[K^+_{n-1}] \le N(LK-1)\rho^{-N(LK-1)}, \tag{4.14}$$

where $K^+_{n-1}$ is the number of service completions in the time interval $(T_{n-1}, T^+_{n-1}]$. Define a distance $d^*(z)$ for $z \in D^*$ as follows. Let $n(i,j,z)$ be the number of jobs of job class $j$ at center $i$ when the job stack process is in state $z$. Using the distance function $r$ defined by (4.4), set

$$d(z) = \sum_{(i,j)\in C} n(i,j,z)\, r(i,j)$$

for $z \in D^*$. For $z = (c^{(1)}_{j_{k(1)}}, \ldots, c^{(1)}_{j_1}, s_1; \ldots; c^{(L)}_{j_k(L)}, \ldots, c^{(L)}_{j_1}, s_L) \in D^*$, set

$$d^*(z) = \begin{cases} d(z) & \text{if } s_1 \ne j_{k(1)}(1) \\ 0 & \text{if } s_1 = j_{k(1)}(1). \end{cases} \tag{4.15}$$

Let $\zeta_k$ be the time of the $k$th service completion, and observe that the distance $d^*(z)$ defined by (4.15) satisfies:

(i) $0 \leq d^*(z) \leq N(LK-1)$;
(ii) $T_{n-1}^+ \leq \zeta_k$ whenever $\zeta_k \geq T_{n-1}$ and $d^*(Z(\zeta_k)) = 0$; and
(iii) $P\bigl\{d^*(Z(\zeta_k+1)) < d^*(Z(\zeta_k)) \bigm| d^*(Z(\zeta_k)) > 0; Z(t),\, t \leq \zeta_k\bigr\} \geq \rho > 0$ a.s..

Next define the random variables $\{W_n\colon n \geq N(LK-1)\}$ as follows. Set $W_n = 1$ if $d^*(Z(\zeta_k)) = 0$ for some $n - N(LK-1) \leq k \leq n$; otherwise, set $W_n = 0$. Also define the random index $\beta_0$ by $\zeta_{\beta_0} = T_{n-1}$, and set

$$\beta_k = \beta_{k-1} + N(LK-1)$$

for $k > 0$. Properties (i)–(iii) of the distance $d^*(z)$ imply that

$$P\left\{W_{\beta_n} = 1 \bigm| W_{\beta_{n-1}}, \ldots, W_{\beta_0}\right\} \geq \rho^{N(LK-1)} \text{ a.s..}$$

By the same reasoning as in the proof of the geometric trials lemma,

$$E[\min\{k \geq 0\colon\ W_{\beta_k} = 1\}] \leq \rho^{-N(LK-1)},$$

and (4.14) holds. A similar argument shows that the expected number of service completions in $(T_{n-1}^+, T_n]$ is less than or equal to $N(LK-1)\,\rho^{-N(LK-1)}$ so that (4.13) holds. □

**Theorem 4.16.** *Assume that the conditions of Theorem 4.5 hold. Then the job stack process $\{Z(t)\colon t \geq 0\}$ is a regenerative process in continuous time. Moreover, if the distribution of service time for jobs of job class $j_{k(i_0)}(i_0)$ at center $i_0$ has finite first and second moments, then the time between regeneration points has finite first and second moments.*

*Proof.* For convenience, select an initial state $Z(0) \neq z_{i_0}^*$ according to $p((\,\cdot\,,0,\ldots,0);(z_{i_0}^*,0,\ldots,0),e_{i_0,j_{k(i_0)}(i_0)})$; thus, the job stack process $\{Z(t)\colon t \geq 0\}$ behaves at time $t = 0$ as if it has just left state $z^*_{i_0}$. Define $\{T_n: n \geq 0\}$ as in (4.6). Set $T_{\beta_0} = 0$ and

$$T_{\beta_k} = \inf\left\{T_n > T_{\beta_{k-1}}: Z(T_n-) = z^*_{i_0} \text{ and } Z(T_n) \neq z^*_{i_0}\right\} \tag{4.17}$$

for $k \geq 1$. According to this definition, $T_{\beta_k}$ is the $k$th time that the job stack process leaves state $z^*_{i_0}$. As $LK > 1$ and the routing matrix $\boldsymbol{P}$ is irreducible, $p_{ij,kl} > 0$ for $(i,j) = (i_0, j_{k(i_0)}(i_0))$ and some $(k,l) \neq (i,j)$. Since

$$P\left\{Z(T_n) \neq z^*_{i_0} \mid Z(T_n-) = z^*_{i_0}\right\} \geq p_{ij,kl}$$

for all $n \geq 0$, Theorem 4.5 and the Borel-Cantelli lemma together imply that each time $T_{\beta_k}$ is a.s. finite. By construction, each $T_{\beta_k}$ is a stopping time for the job stack process. Moreover, the only service times in progress have just begun, and the assumption that the routing of jobs is Markovian ensures that the joint distribution of these new service times is the same at each $T_{\beta_k}$. Thus, the successive random times $\{T_{\beta_k}: k \geq 0\}$ form a sequence of regeneration points for the process $\{Z(t): t \geq 0\}$.

To see that $E[T_{\beta_1} - T_{\beta_0}] = E[T_{\beta_1}] < \infty$, observe that all distributions of service time have finite first and second moments. This assertion follows from the finiteness assumptions on the moments of service times for jobs of job class $j_{k(i_0)}(i_0)$ at center $i_0$ and the fact that a random variable with a distribution that is NBU possesses moments of all orders. Lemma 4.12 shows that there exists a constant $\kappa < \infty$ such that $E[K_n] < \kappa$ for all $n \geq 1$, where $K_n$ is the number of service completions in the time interval $(T_{n-1}, T_n]$. It follows from (4.7) that $E[L_1] < \infty$, where $L_1$ is the number of service completions in the time interval $(0, T_{\beta_1}]$. For $(k,l) \in C$, let $L_1(k,l)$ be the number of service completions by a job of job class $l$ at center $k$ in the time interval $(0, T_{\beta_1}]$, and let $V_m(k,l)$ be the service time of the $m$th such job, $m = 1, 2, \ldots, L_1(k,l)$. Then

$$T_{\beta_1} \leq \sum_{(k,l)\in C} \sum_m V_m(k,l). \tag{4.18}$$

Applying Wald's identity to the inner sum in (4.18),

$$\begin{aligned} E[T_{\beta_1}] &\le \sum_{(k,l)\in C} E\Big[\sum_m V_m(k,l)\Big] \\ &= \sum_{(k,l)\in C} E[L_1(k,l)]\,\mu_{kl} \\ &\le E\Big[\sum_{(k,l)\in C} L_1(k,l)\Big] \max_{(k,l)\in C} \mu_{kl} \\ &= E[L_1] \max_{(k,l)\in C} \mu_{kl}. \end{aligned}$$

Since the set $C$ of (center, class) pairs is finite, (4.18) implies that $E[T_{\beta_1}] < \infty$. Using Wald's second moment identity, a similar argument shows that $E[T_{\beta_1}^2] < \infty$. □

For a real-valued measurable function $f$ with domain $D$, set $T_{\beta_{-1}} = 0$ and

$$Y_k(f) = \int_{T_{\beta_{k-1}}}^{T_{\beta_k}} f(Z(u))\,du$$

for $k \ge 0$. Also set $\tau_k = T_{\beta_k} - T_{\beta_{k-1}}$ for $k \ge 1$. It follows direclty from Theorem 4.16 that the sequence $\{(Y_k(f), \tau_k) \colon k \ge 1\}$ consists of i.i.d. random vectors.

**Proposition 4.19.** *For any real-valued measurable function $f$ with domain $D$ such that $Y_0(|f|) < \infty$ a.s. and $E[Y_1(|f|)] < \infty$,*

$$\lim_{t\to\infty} \frac{1}{t} \int_0^t f\big(Z(u)\big)\,du = \frac{E\big[Y_1(f)\big]}{E\big[\tau_1\big]} \stackrel{\text{def}}{=} r(f) \text{ a.s..}$$

Observe that Theorem 4.16 does not guarantee that the time $\tau_1$ between regeneration points is aperiodic.

**Proposition 4.20.** *Assume that $\tau_1$ is aperiodic. Then $Z(t) \Rightarrow Z$ as $t \to \infty$ and*

$$E[f(Z)] = \frac{E[Y_1(f)]}{E[\tau_1]}$$

*for any real-valued measurable function $f$ with domain $D$ such that $E[|f(Z)|] < \infty$ or $E[Y_1(|f|)] < \infty$. Moreover,*

$$\lim_{t\to\infty} E\big[f(Z(t))\big] = E[f(Z)]$$

*for all bounded functions $f$ such that $P\{Z \in D(f)\} = 0$.*

Observe that since the state space of the job stack process is finite and $E[\tau_1^2] < \infty$, the variance constant $\sigma^2(f) = \text{Var}\,[Y_1(f) - r(f)\,\tau_1]$ is finite. From $n$ cycles, the regenerative method provides an estimator

$$\hat{r}(n) = \frac{\bar{Y}(n)}{\bar{\tau}(n)} = \frac{\sum_{k=1}^{n} Y_k(f)}{\sum_{k=1}^{n} \tau_k}$$

that is strongly consistent for $r(f)$. Asymptotic confidence intervals for $r(f)$ are based on the CLT

$$\frac{\sqrt{n}\,\{\hat{r}(n) - r(f)\}}{\sigma(f)/E[\tau_1]} \Rightarrow N(0,1)$$

as $n \to \infty$.

**Algorithm 4.21.** *Regenerative method for networks of queues (preemptive-resume priorities)*

1. *Select a center $i_0$ satisfying Condition (i) of Theorem 4.5 and an initial state $z_0 \neq z_{i_0}^*$ for the job stack process according to $p((\,\cdot\,,0,\ldots,0);(z_{i_0}^*,0,\ldots,0),e_{i_0,j_{k(i_0)}(i_0)})$;*

2. *Set* $Z(0) = z_0$, *and observe a fixed number* $n$ *of cycles of the job stack process defined by the successive random times* $\{T_{\beta_k}\colon k \geq 0\}$ *at which the job stack process leaves state* $z^*_{i_0}$.
3. *Measure* $(Y_1(f), \tau_1), (Y_2(f), \tau_2), \ldots, (Y_n(f), \tau_n)$.
4. *Form the point estimate*

$$\hat{r}(n) = \frac{\bar{Y}(n)}{\bar{\tau}(n)}$$

*for* $r(f)$.

5. *Form the asymptotic* $100(1-2\gamma)\%$ *confidence interval*

$$\left[\hat{r}(n) - \frac{s(n)z_{1-\gamma}}{\bar{\tau}(n)\sqrt{n}}, \hat{r}(n) + \frac{s(n)z_{1-\gamma}}{\bar{\tau}(n)\sqrt{n}}\right]$$

*for* $r(f)$.

**Exercises**

4.1. (Multiprogrammed computer system) Consider a network of queues with two single-server centers ($L = 2, m(1) = m(2) = 1$), and six mutually exclusive job classes ($K = 6$) that provides service to $N$ ($\geq 2$) jobs. At every point of continuous time each job is in exactly one job class, but jobs may change job class as they traverse the network. Take $C = \{(1,1),(1,2),(1,3),(1,4),(1,5),(2,6)\}$ to be the set of (center, class) pairs. Jobs queue and receive service at center 1 according to a fixed priority scheme among job classes. The job classes served at center 1 and ordered by decreasing priority are $j_1(1) = 1$, $j_2(1) = 2$, $j_3(1) = 3$, $j_4(1) = 4$, and $j_5(1) = 5$. This priority ordering is subject to the constraint that there can be no service at center 1 to a job of job class 1 or job class 2 while service (to a job of job class 6) is in progress at center 2. Jobs of job class 4 at center 1 are preempted when a job of higher priority joins the queue at center

1. The interruption of service to a job of job class 4 is of the preemptive-resume type. All other services in the network are not subject to interruption. Upon completion of service at center $i$, a job of job class $j$ moves to center $k$ and becomes a job of job class $l$ with probability $p_{ij,kl}$. The nonzero elements of the routing matrix $\boldsymbol{P}$ are $p_{11,15} = p_{12,26} = p_{13,12} = p_{14,13} = p_{15,14} = p_{26,14} = 1$. Assume that service times are mutually independent and service time for a job of job class $j$ at center $i$ is distributed as a positive random variable $V(i,j)$ with finite mean but otherwise arbitrary distribution.

Set

$$X(t) = (C_5(t), C_4(t), C_3(t), C_2(t), C_1(t), S_1(t), Q_2(t), M(t)),$$

where

$$M(t) = \begin{cases} 1 & \text{if there is a preempted job waiting at center 1} \\ & \text{at time } t \\ 0 & \text{otherwise,} \end{cases}$$

$Q_2(t)$ is the number of jobs waiting or in service at center 2, $S_1(t)$ is the job class of the job in service at center 1 ($S_1(t)$ equals 0 if there is no job in service at center 1), and $C_j(t)$ is the number of jobs of job class $j$ in queue at center 1 at time $t$.

Let $T_n$ be the time of the $n$th service completion to a job of job class 1 at center 1. Show that $\{X(T_n)\colon n \geq 0\}$ is an irreducible DTMC with a finite state space, provided that service time to a job of job class 4 at center 1 is exponentially distributed as a random variable $V(1,4)$. Also show that the process $\{X(t)\colon t \geq 0\}$ is a regenerative process in continuous time and the expected time between regeneration points is finite.

4.2. (Finite-capacity queue with preemptive-resume priority) Consider a queueing system with one single-server center and two job types numbered 1 and 2. Jobs of type $i$ arrive from an external source, queue and receive service at the center, and depart to an external sink. A job of type $i$ that arrives when the system already contains $B_i - 1$ ($\geq 0$) jobs of type $i$ causes the arrival process for jobs of type $i$ to shut down; the arrival process for jobs of type $i$ remains shut down until the first subsequent departure of a job of type $i$. Jobs of the same type receive service in order of their arrival at the center, and jobs of type 1 have preemptive-resume priority over jobs of type 2. A job of type 2 that has been preempted receives additional service at the center before any other job of type 2 receives service. Assume that the interarrival times for jobs of type $i$ are i.i.d. as a positive random variable $A_i$ with finite mean and a density function that is continuous and positive on $(0, +\infty)$. Also assume that the service times for jobs of type $i$ are i.i.d. as a positive random variable $L_i$ with finite mean and a density function that is continuous and positive on $(0, +\infty)$.

Set

$$X(t) = (B_2(t), B_1(t), M(t), L(t)),$$

where

$$L(t) = \begin{cases} 1 & \text{if the job at the head of the type 2 queue} \\ & \text{at time } t \text{ has been preempted} \\ 0 & \text{otherwise,} \end{cases}$$

$$M(t) = \begin{cases} 2 & \text{if there is a job of type 2 in service at time } t \\ 1 & \text{if there is a job of type 1 in service at time } t \\ 0 & \text{if there is no job in service at time } t, \end{cases}$$

and $B_i(t)$ is the number of jobs of type $i$ waiting in queue at time $t$.

Show that $\{X(t): t \geq 0\}$ is a regenerative process in continuous time and the expected time between regeneration points is finite. Describe how you would use the regenerative method to obtain a strongly consistent point estimate and an asymptotic confidence interval for the limiting probability that there are at least $k$ jobs of type $i$ in the system, $k = 0, 1, \ldots, B_i$. Be explicit about the sequence of regeneration points, the quantities to be measured, and the construction of the point and interval estimates. State the ratio formula and central limit theorem that lead to the point and interval estimates.

4.3 (Preemptive cyclic queues) Consider a network of queues with two single-server centers that provides service to a fixed number $N$ of customers labelled $1, 2, \ldots, N$. After completion of service at center 1, a customer moves instantaneously from center 1 to center 2, and after completion of service at center 2 moves instantaneously back to center 1. Suppose that the queueing discipline at each center is LCFS. Also suppose that interruptions of service are of the premptive-resume type (service continues from the point of interruption when the preempted job goes back into service). Customers waiting in queue appear in the order of preemption at the center, the most recently preempted customer at the head of the queue. Assume that all service times are mutually independent. Also assume that service times at center 1 are identically and exponentially distributed as a random variable $V_1$ and service times at center 2 are identically distributed as a positive random variable $V_2$ with finite mean and a density function that is continuous and positive on $(0, +\infty)$.

Let $X(t)$ be the number of customers waiting or in service at center 2 at time $t$. Take event $e_j$ = "service completion at center 1 with $j$ customers at center 1" and event $e_{N+j}$ = "service completion at center 2 with $j$ customers at center 2" ($j = 1, 2, \ldots, N$). Formal specification of the process $\{X(t)\colon t \geq 0\}$ is as a GSMP with finite state space $S = \{0, 1, \ldots, N\}$ and event set $E = \{e_1, e_2, \ldots, e_{2N}\}$. Show that $\{X(t)\colon t \geq 0\}$ is a regenerative process in continuous time and the expected time between regeneration points is finite.

## Chapter 5

# Passage Times

Certain random time intervals correspond to delays in discrete-event stochastic systems. A "passage time" in a network of queues is the random time interval, computed as a random sum of waiting times, for a job to traverse a portion of the network. Formal specification of a sequence of passage times in a network of queues is in terms of an arbitrarily chosen, "marked" job by means of four subsets of the state space of the "augmented job stack process." The augmented job stack process records the position in the job stack of the marked job along with the number of jobs of each job class at each center. These subsets jointly define the random times at which passage times for the marked job start and terminate. The subsets in effect determine when to start and stop the clock to measure a passage time for the marked job. Additional conditions ensure that the start times and termination times for the specified passage time strictly alternate, so that there is at most one passage time for the marked job underway at any time. Regenerative methods for estimation of passage times are available when there exist special "single-states" of the job stack process or the augmented job stack process.

Estimation procedures for characteristics of limiting passage times in a network of queues can be based on measurement of individual passage times for one or more jobs. The "marked job method" provides point and interval estimates based on measurement of individual passage times for an arbitrarily chosen, distinguished job. The "labelled jobs method" prescribes measurement of individual passage times for all of the jobs in the network. When both methods are applicable, confidence intervals constructed using the labelled jobs method are shorter than those obtained from the marked job method. This result is consistent with intuition since the labelled jobs method extracts more information about passage times from a simulation run of fixed length.

## 5.1. The Augmented Job Stack Process

In order to measure individual passage times in a closed network of queues with priorities among job classes, it is convenient to augment the job stack used to describe the state of the system. A suitable and apparently minimal augmentation of the job stack uses the concept of an arbitrarily chosen, distinguished job. The idea is to keep track of the position in the job stack of the distinguished job and to measure passage times for this *marked job*.

As in Chapter 4, we consider closed networks of queues with a finite number $N$ of stochastically identical jobs (customers), a finite number $L$ of single-server or multiple-server centers numbered $1, 2, \ldots, L$, and a finite number $K$ of mutually exclusive job classes numbered $1, 2, \ldots, K$. At every point of continuous time each job is in exactly one job class, but jobs may change job class as they traverse the network. Upon completion of service at center $i$, a job of job class $j$ moves to center $k$ and becomes a job of job class $l$ with probability $p_{ij,kl}$. The routing matrix

$$\boldsymbol{P} = \{p_{ij,kl}\colon\ (i,j),(k,l) \in C\}$$

is a given irreducible stochastic matrix, and

$$C \subseteq \{1,2,\ldots,L\} \times \{1,2,\ldots,K\}$$

is the set of (center, class) pairs in the network. In accordance with the routing matrix $\boldsymbol{P}$, some centers may never see jobs of certain job classes. At each center, jobs queue and receive service according to a fixed priority scheme among job classes; the priority scheme may differ from center to center. Within a job class at a center, jobs receive service according to a fixed queue service discipline that depends only on the order of arrival of the jobs. According to a fixed procedure for each center, a job in service may or may not be preempted when a job of higher priority joins the queue at the center. A job may not be preempted by another job of the same job class. If more than one job subject to preemption is in service at a center when a job of higher priority joins the queue at the center, the preempted job is selected according to a fixed scheme. Any interruption of service is of the "preemptive-repeat" type in that service starts afresh when the preempted job goes back into service. A preempted job goes into service at a center before any other job of the same job class that is waiting in queue at the center. We assume that service times are mutually independent and the service time for a job of job class $j$ at center $i$, distributed as a positive random variable $V(i,j)$, has finite first and second moments and a density function that is continuous and positive on $(0, +\infty)$.

Let $S_i(t)$ be the job class of the job in service at center $i$ at time $t$, $i = 1, 2, \ldots, L$; by convention $S_i(t) = 0$ if at time $t$ there are no jobs at center $i$. If center $i$ has more than one server, we take $S_i(t)$ to be a vector that records the job class of the job in service from each server at center $i$. Specifically, we enumerate the servers at center $i$ as $1, 2, \ldots, m(i)$ and set

$$S_i(t) = (S_{i,1}(t), S_{i,2}(t), \ldots, S_{i,m(i)}(t)),$$

where $S_{i,m}(t)$ is the job class of the job in service from server $m$ at center $i$ at time $t$. If there is more than one available server when a job goes into service, the job receives service from the lowest numbered available server. We denote the job classes served at center $i$ and ordered by decreasing priority by $j_1(i), j_2(i), \ldots, j_{k(i)}(i)$, and we let $C_{j_1}^{(i)}(t), \ldots, C_{j_{k(i)}}^{(i)}(t)$ be the number of jobs waiting in queue at time $t$ of the various job classes served at center $i$.

As before, set

$$Z(t) = \big(C_{j_k(1)}^{(1)}(t), \ldots, C_{j_1}^{(1)}(t), S_1(t); \ldots; \\ C_{j_k(L)}^{(L)}(t), \ldots, C_{j_1}^{(L)}(t), S_L(t)\big). \tag{1.1}$$

The job stack at time $t$ then corresponds to the nonzero components in the vector $Z(t)$ and is an ordering of the jobs by job class at the individual centers. Within a job class at a particular center, jobs appear in the job stack in order of their arrival at the center, the latest to arrive being closest to the top of the job stack. A job that has been preempted appears at the head of the queue of jobs of its job class. The job stack process $\boldsymbol{Z} = \{Z(t)\colon\ t \geq 0\}$ has a finite state space $D^*$.

Take event $e_{i,m}$ = "service completion by server $m$ at center $i$." Formal specification of the job stack process is as a GSMP with finite state space $D^*$ and event set

$$E = \big\{e_{1,1}, \ldots, e_{1,m(1)}, \ldots, e_{L,1}, \ldots, e_{L,m(L)}\big\}. \tag{1.2}$$

For $z = (c_{j_k(1)}^{(1)}, \ldots, c_{j_1}^{(1)}, s_{1,1}, \ldots, s_{1,m(1)}; \ldots; c_{j_k(L)}^{(L)}, \ldots, c_{j_1}^{(L)}, s_{L,1}, \ldots, s_{L,m(L)}) \in D^*$, event $e_{i,m}$ is active in state $z$ ($e_{i,m} \in E(z)$) if and only if $s_{i,m} > 0$. According to this definition, event $e_{i,m} \in E(z)$ if and only if in state $z$ a job receives service from server $m$ at center $i$. All speeds $r(z, e)$ for active events are equal to 1.

Now let $N(t)$ be the position of an arbitrarily chosen, marked job in the job stack at time $t$, and set

$$X(t) = (Z(t), N(t)). \tag{1.3}$$

Recall that the job at the top of the job stack is said to be in position 1, the next job in position 2, etc. The process $\{X(t)\colon t \geq 0\}$ defined by (1.3) is called the *augmented job stack process.*

Formal specification of the augmented job stack process is as a GSMP with finite state space $G^* = D^* \times \{1, 2, \ldots, N\}$ and event set $E$ given by (1.2). For $x = (c^{(1)}_{j_{k(1)}}, \ldots, c^{(1)}_{j_1}, s_{1,1}, \ldots, s_{1,m(1)}; \ldots; c^{(L)}_{j_{k(L)}}, \ldots, c^{(L)}_{j_1}, s_{L,1}, \ldots, s_{L,m(L)}, n) \in G^*$, event $e_{i,m}$ = "service completion by server $m$ at center $i$" is active in state $x$ ($e_{i,m} \in E(x)$) if and only if $s_{i,m} > 0$. According to this definition $e_{i,m} \in E(x)$ if and only if in state $x$ a job receives service from server $m$ at center $i$. All speeds $r(x, e)$ for active events are equal to 1. Recall that by assumption, any interruption of service is of the preemptive-repeat type.

For $x \in G^*$, $(i, j), (k, l) \in C$, and $m \in \{1, 2, \ldots, m(i)\}$, denote by $q(x; i, j, k, l; m)$ the state of the augmented job stack process obtained when in state $x$ there is a service completion by server $m$ to a job of job class $j$ at center $i$ and the job moves to center $k$ and becomes a job of job class $l$. If $e = e_{i,m}$ and $s_{i,m} = j$, then the state transition probability $p(x'; x, e) = p_{ij,kl}$ when

$$x = (c^{(1)}_{j_{k(1)}}, \ldots, s_{1,m(1)}; \ldots; c^{(L)}_{j_{k(L)}}, \ldots, s_{L,m(L)}, n)$$
$$\text{with } s_{i,m} = j \text{ and } x' = q(x; i, j, k, l; m).$$

All other state transition probabilities $p(x'; x, e)$ are equal to 0. The clock-setting distribution functions are as follows. If the new event $e' = e_{i,m}$ and $s_{i,m} = j$, then $F(u; x', e', x, e^*) = P\{V(i, j) \leq u\}$.

We assume that the initial distribution $\mu$ of the underlying GSSMC satisfies

$$\mu(A) = q(x_0) \prod_{e_i \in E(x_0)} F_0(a_i; e_i, x_0)$$

for all sets

$$A = \{x_0\} \times \big\{(c_{0,1},\ldots,c_{0,M}) \in C(x_0)\colon\ 0 \le c_{0,i} \le a_i \\ \text{for } 1 \le i \le M\big\},$$

where $M = m(1)+m(2)+\cdots+m(L)$, $q$ is a probability distribution on $G^*$, and $F_0(\,\cdot\,;e,x)$ is a proper distribution function for $x \in G^*$ and $e \in E(x)$.

For $x, x' \in G^*$, we say that $x'$ is a *successor state* of $x$ and write $x \to x'$ if $p(x';x,e_{i,m}) > 0$ for some event $e_{i,m} \in E(x)$. State $x'$ is said to be *accessible* from $x$, and we write $x \rightsquigarrow x'$, if there exist states $x_0 = x, x_1, \ldots, x_{n-1}, x_n = x'$ of the augmented job stack process and (center, server) pairs $(i_0,m_0)$, $(i_1,m_1),\ldots,(i_{n-1},m_{n-1})$ such that

$$p(x_1;x,e_{i_0,m_0})p(x_2;x_1;e_{i_1,m_1})\cdots p(x_n;x_{n-1},e_{i_{n-1},m_{n-1}}) > 0.$$

When $x \rightsquigarrow x'$ and $x' \rightsquigarrow x$, states $x$ *and* $x'$ are said to *communicate.*

Passage times are specified in terms of the marked job by means of four subsets ($A_1$, $A_2$, $B_1$, and $B_2$) of the state space $G^*$ of the augmented job stack process. The subsets $A_1$, $A_2$ [resp., $B_1$, $B_2$] jointly define the random times at which passage times for the marked job start [resp., terminate]. The subsets $A_1$, $A_2$, $B_1$, and $B_2$ in effect determine when to start and stop the clock that measures a particular passage time for the marked job. We assume that for all $x' \in A_2$ [resp., $x' \in B_2$] there exists $x \in A_1$ [resp., $x \in B_1$] such that $x \to x'$ [resp., $x \to x'$].

For $x_1, x_2, \ldots, x_{k+2} \in G^*$ with $x_1 \to x_2$, $x_2 \to x_3,\ldots$, $x_{k+1} \to x_{k+2}$, and $k \ge 1$, we require that the sets $A_1$, $A_2$, $B_1$, and $B_2$ satisfy:

(i) if $x_1 \in A_1$, $x_2 \in A_2$, $x_{k+1} \in A_1$, and $x_{k+2} \in A_2$, then $x_{m+1} \in B_1$ and $x_{m+2} \in B_2$ for some $0 < m \le k$; and

(ii) if $x_1 \in B_1$, $x_2 \in B_2$, $x_{k+1} \in B_1$, and $x_{k+2} \in B_2$, then $x_{m+1} \in A_1$ and $x_{m+2} \in A_2$ for some $0 \le m \le k$.

These conditions ensure that the start times and termination times for the specified passage time strictly alternate.

In terms of the sets $A_1$, $A_2$, $B_1$, and $B_2$, we define two sequences $\{\mathcal{S}_j\colon j \ge 0\}$ and $\{\mathcal{T}_j\colon j \ge 1\}$ of random times, where $\mathcal{S}_{j-1}$ is the start time of the $j$th passage time for the marked job, and $\mathcal{T}_j$ is the termination time of this $j$th passage time. Assuming that the initial state of the augmented job stack process is such that a passage time for the marked job starts at time 0, set $\mathcal{S}_0 = 0$,

$$\mathcal{S}_j = \inf\{\zeta_n \ge \mathcal{T}_j\colon X(\zeta_n) \in A_2, X(\zeta_{n-1}) \in A_1\},$$

and

$$\mathcal{T}_j = \inf\{\zeta_n > \mathcal{S}_{j-1}\colon X(\zeta_n) \in B_2, X(\zeta_{n-1}) \in B_1\}$$

for $j \ge 1$. Then the $j$th passage time for the marked job is $P_j = \mathcal{T}_j - \mathcal{S}_{j-1}$.

**Example 1.4.** (Cyclic queues with feedback) Consider a network of queues with two single-server centers ($L = 2, m(1) = m(2) = 1$) and a single job class ($K = 1$) that provides service to $N$ ($\ge 2$) jobs. Take $C = \{(1,1),(2,1)\}$ to be the set of (center, class) pairs. Assume that both queueing disciplines are FCFS. With fixed probability $p$ ($0 < p < 1$), a job that completes service at center 1 joins the tail of the queue at center 1 and with probability $1-p$ moves to center 2. A job that completes service at center 2 moves to center 1. Set

$$X(t) = (Z(t), N(t)),$$

where $Z(t)$ is the number of jobs waiting or in service at center 1 at time $t$ and $N(t)$ is the position of the marked job in the job stack at time $t$. The state space of the augmented job stack process is

$$G^* = \{(i, n) \colon 0 \le i \le N, 1 \le n \le N\}.$$

Consider the passage time that starts when a job completes service at center 2 (and moves to center 1) and terminates when the job next moves to center 2. The subsets $A_1$ and $A_2$ that define the start times of individual passage times are

$$A_1 = \{(i, N) \colon 0 \le i < N\}$$

and

$$A_2 = \{(i, 1) \colon 0 < i \le N\}.$$

The subsets $B_1$ and $B_2$ that define the termination times of individual passage times are

$$B_1 = \{(i, i) \colon 0 < i \le N\}$$

and

$$B_2 = \{(i-1, i) \colon 0 < i \le N\}.$$

Figure 5.1 shows state transitions of the augmented job stack process and the sets $A_1$, $A_2$, $B_1$, and $B_2$ for $N = 2$ jobs. In the schematic representation of the job stack, jobs to the left of the vertical bar are at center 1, and jobs to the right of the bar are at center 2. The symbol $\times$ denotes the marked job and $\circ$ denotes an unmarked job.

Now consider the passage time that starts when a job completes service at center 2 (and moves to center 1) and terminates the next such time the job moves to center 1. The subsets that define the start times and termination times of individual passage times are

$$A_1 = B_1 = \{(i, N) \colon 0 \le i < N\}$$

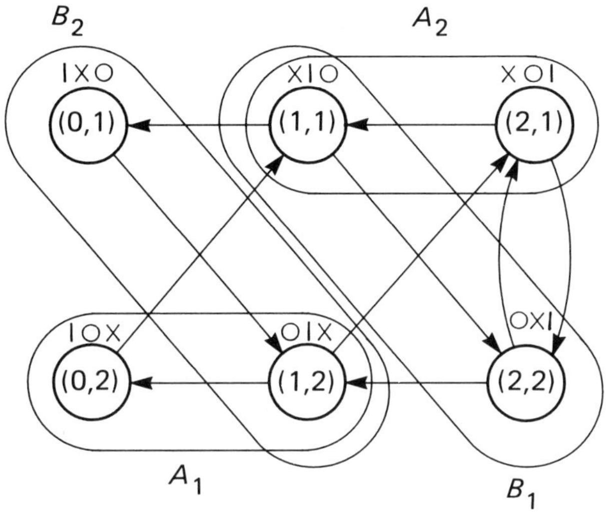


**Figure 5.1.** Subsets for passage time through center 1.

and

$$A_2 = B_2 = \{(i,1) \colon \ 0 < i \le N\}.$$

Figure 5.2 shows state transitions of the augmented job stack process and the sets $A_1$, $A_2$, $B_1$, and $B_2$ for $N = 2$ jobs.

Without further restrictions on the building blocks of a network of queues with priorities among job classes, the sequence of passage times for the marked job need not converge in distribution to a random variable independent of the initial conditions. Theorem 1.7 of Chapter 4 postulates the existence of a distinguished center $i_0$: either center $i_0$ serves only one job class or service at center $i_0$ to a job of the lowest priority job class $j_{k(i_0)}(i_0)$ served at center $i_0$ is preempted when any other job of higher priority joins the queue. Let $z^*_{i_0}$ be the state of the job stack process in which all $N$ jobs are of the lowest priority job class $j_{k(i_0)}(i_0)$ at center $i_0$. Then $z_{i_0}$ is a

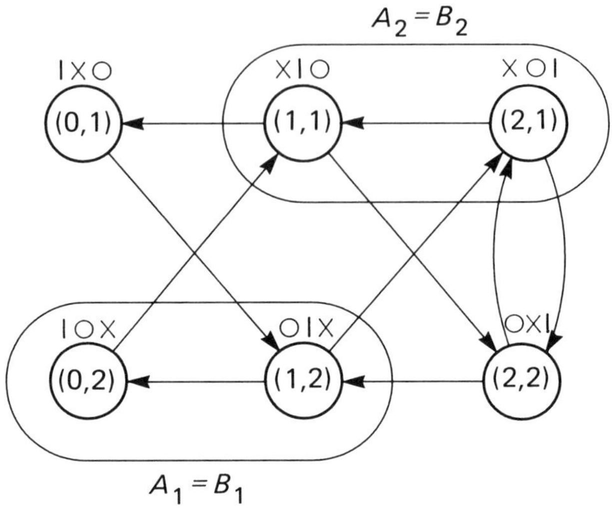


**Figure 5.2.** Subsets for passage time through centers 1 and 2.

target state for the job stack process, and the set

$$D = \{z \in D^*:\ z^*_{i_0} \rightsquigarrow z\}$$

of all states of the job stack process that are accessible from $z^*_{i_0}$ is irreducible.

For networks with more than one center ($L > 1$), the set

$$G = \{(z, n) \in G^*:\ z \in D\}$$

is also irreducible. The idea is to show that $x^*_{i_0} = (z^*_{i_0}, N)$ can serve as a target state of the augmented job stack process in the sense that $x \rightsquigarrow x^*_{i_0}$ for all $x \in G$; cf the proof of Theorem 1.7 of Chapter 4.

**Theorem 1.5.** *Assume that there are at least two centers ($L > 1$). Also assume that the routing matrix $\boldsymbol{P}$ is irreducible and for some center $i_0$, either $k(i_0) = 1$ or service to a job of job class $j_{k(i_0)}(i_0)$ at center $i_0$ is preempted when any other job of higher priority joins the queue. Then $x$ and $x'$ communicate for all $x, x' \in G$.*

Henceforth, we assume that for some $z^* \in D^*$ the sets $D = \{z \in D^*\colon z^* \rightsquigarrow z\}$ and $G = \{(z,n) \in G^*\colon z \in D\}$ are irreducible in the sense that $z$ and $z'$ communicate for all $z, z' \in D$ and $x$ and $x'$ communicate for all $x, x' \in G$. We also assume that the subsets $A_1$, $A_2$, $B_1$, and $B_2$ that define the start times and termination times of individual passage times for the marked job are subsets of $G$, and the initial distribution $q$ is concentrated on the set $G$ so that $X(0) \in G$.

Definition 1.6 identifies special "single-states" of the augmented job stack process. Recall that we write $h(z,n) = (i,j)$ when the job in position $n$ of the job stack associated with $z \in D$ is a job of job class $j$ at center $i$.

**Definition 1.6.** *An element $(z, N)$ of the set $A_1$ is said to be a single-state of the augmented job stack process for the passage time specified by the sets $A_1$, $A_2$, $B_1$, and $B_2$ if*

*(i) there exists $(z', n') \in A_2$ such that $(z, N) \to (z', n')$; and*
*(ii) $h(z,n) \in \{(i_0, j_1(i_0)), (i_0, j_2(i_0)), \ldots, (i_0, j_{k(i_0)}(i_0))\}$ for some single-server center $i_0$ and all $n = 1, 2, \ldots, N$.*

According to this definition, $(z, N)$ is a single-state of the augmented job stack process for the specified passage time if the marked job is in service at some single-server center ($i_0$), the other $N - 1$ jobs are in queue at the same center, and with positive probability a passage time for the marked job starts upon completion of the service in progress. Recall that the job classes served at center $i_0$ and ordered by decreasing priority are $j_1(i_0), j_2(i_0), \ldots, j_{k(i_0)}(i_0)$. Denote by $A'_1$ the set of all states that are single-states of the augmented job stack process in the sense of Definition 1.6, and set

$$S = \{(z, n, z', n') \colon (z, n) \in A_1, (z', n') \in A_2, \text{ and } (z, n) \to (z', n')\}.$$

Also set

$$A'_2 = \{(z', n') \in A_2 \colon (z, N, z', n') \in S \text{ for some } (z, N) \in A'_1\} \tag{1.7}$$

and assume that $A'_2 \neq \emptyset$. This assumption ensures there exists a single-state $(z, N)$ of the augmented job stack process and a suitable "return state" $(z', n')$ such that $(z, N) \to (z', n')$.

Observe that by Theorem 1.5 the augmented job stack process restricted to the set $G$ is irreducible in the sense that $x$ and $x'$ communicate for all $x, x' \in G$. Theorem 1.10 below asserts that the augmented job stack process is a regenerative process in continuous time and the expected time between regeneration points is finite. The argument rests on the fact that the augmented job stack process restricted to the set of states accessible from a fixed target state can be represented as an irreducible GSMP with state space $G$ and event set $E$ given by (1.2).

**Theorem 1.8.** *Assume that the conditions of Theorem 1.5 hold. Also assume that $A'_2 \neq \emptyset$. Then $P\{X(\mathcal{S}_n) = x' \text{ i.o.}\} = 1$ for any $x' \in A'_2$.*

*Proof.* Let $x' \in A'_2$. The definition of the set $A'_2$ implies that there exists $x$ such that $x \to x'$ and $x$ is a single-state of the augmented job stack process in the sense of Definition 1.6. Let $L(t)$ be the last state occupied by the augmented job stack process before jumping to $X(t)$, and set

$V(t) = (L(t), X(t))$. It is sufficient to show that the process $\{V(t)\colon t \geq 0\}$ hits state $(x, x')$ infinitely often with probability 1. To see this, observe that the process $\{V(t)\colon t \geq 0\}$ can be represented as a GSMP with finite state space $B = \{(u, w) \in G \times G\colon u \to w\}$ and event set $E$ given by (1.2). For $v = (u, w) \in B$ with $u, w \in G$, event $e_{i,m}$ = "service completion by server $m$ at center $i$" is active in state $v$ if and only if $e_{i,m} \in E(w)$. For $v = (u, w), v' = (w, y) \in B$, and $e \in E$, the state transition probability $p(v'; v, e) = p(y; w, e)$ and for $e' \in N(v'; v, e^*)$, the clock-setting distribution function $F(\,\cdot\,; v', e', v, e^*) = F(\,\cdot\,; y, e', w, e^*)$. Let $\zeta_n$ be the time of the $n$th state transition, $n \geq 0$.

Proposition 3.1 of Chapter 2 implies that $P\big\{V(\zeta_n) = (x, x') \text{ i.o.}\big\} = 1$ provided that

(i) the process $\{V(t)\colon t \geq 0\}$ is irreducible in the sense that $v$ and $v'$ communicate for all states $v, v'$ of the process; and

(ii) each clock-setting distribution has finite mean and a density function that is continuous and positive on $(0, +\infty)$.

Condition (i) holds as a consequence of Theorem 1.5 and Condition (ii) holds by assumption. □

**Example 1.9.** (Cyclic queues with feedback) As in Example 1.4, set $X(t) = (Z(t), N(t))$, where $Z(t)$ is the number of jobs waiting or in service at center 1 and $N(t)$ is the position of the marked job in the job stack at time $t$. Consider the passage time that starts when a job joins the queue at center 1 upon completion of service at center 2 and terminates when the job next joins the queue at center 2. The subsets $A_1$ and $A_2$ that define the start times of individual passage times are

$$A_1 = \{(i, N)\colon\ 0 \leq i < N\}$$

and

$$A_2 = \{(i, 1)\colon\ 0 < i \leq N\}.$$

Then $S = \{(i, N, i+1, 1)\colon\ 0 \leq i < N\}$, $(0, N)$ is a single-state of the augmented job stack process in the sense of Definition 1.6 and $A_2' = \{(1, 1)\}$.

Now select an element $x'$ from the set $A_2'$ of (1.7). By Theorem 1.8, the process $\{X(\mathcal{S}_n)\colon\ n \geq 0\}$ hits state $x'$ infinitely often with probability 1. Furthermore, at any such time $\mathcal{S}_{\beta_k}$ the clocks associated with the only (at most two) active events have just been set, the active events are the same, and the new clock readings have the same joint distribution. Thus, the future evolution of the augmented job stack process is independent of the history of the process before time $\mathcal{S}_{\beta_k}$ and has the same distribution as it does when $X(0) = x'$. It follows that the successive random times $\{\mathcal{S}_{\beta_k}\colon\ k \geq 0\}$ at which $\{X(\mathcal{S}_n)\colon\ n \geq 0\}$ hits state $x'$ form a sequence of regeneration points for the augmented job stack process. An argument similar to the one used in Theorem 2.3 of Chapter 4 shows that the expected time between regeneration points is finite.

**Theorem 1.10.** *Assume that the conditions of Theorem 1.5 hold. Also assume that $A_2' \neq \emptyset$. Then the augmented job stack process $\{X(t)\colon\ t \geq 0\}$ is a regenerative process in continuous time, and the expected time between regeneration points is finite.*

Theorem 1.10 holds even if each distribution of service time does not have a density function that is continuous and positive on $(0, +\infty)$, provided that $P\{X(\mathcal{S}_n) = x' \text{ i.o.}\} = 1$ and the expected value of the number $L_n$ of service completions in the time interval $[\mathcal{S}_{\beta_{n-1}}, \mathcal{S}_{\beta_n})$ is finite.

Next observe that the random indices $\{\beta_k\colon\ k \geq 0\}$ form a sequence of regeneration points for the discrete time process $\{(X(\mathcal{S}_n), P_{n+1})\colon\ n \geq 0\}$. This assertion follows from the fact that passage times for the marked job start from scratch when $X(\mathcal{S}_k) = x'$. Using properties of the GSSMC associated with state transitions of the augmented job stack process and the assumptions on the clock-setting distributions, it can be shown that $E[\beta_k - \beta_{k-1}] < \infty$.

**Theorem 1.11.** *Assume that the conditions of Theorem 1.5 hold and $A_2' \neq \emptyset$. Then the process $\{(X(\mathcal{S}_n), P_{n+1})\colon\ n \geq 0\}$ is a regenerative process in discrete time, and the expected time between regeneration points is finite.*

It follows from Theorem 1.11 and general results for regenerative processes that

$$(X(\mathcal{S}_n), P_{n+1}) \Rightarrow (X, P)$$

as $n \to \infty$, provided that $\beta_k - \beta_{k-1}$ is aperiodic; that is, there exist random variables $X$ and $P$ such that

$$\lim_{n\to\infty} P\{X(\mathcal{S}_n) = i, P_{n+1} \leq y\} = P\{X = i, P \leq y\} \stackrel{\text{def}}{=} F(i, y)$$

for all $i \in A_2$ and $y \in [0, +\infty)$ for which $F(i, \cdot\,)$ is continuous.

The sequence of passage times for any other job converges in distribution to the same random variable as the sequence of passage times for the marked job. To see this, label the jobs from 1 to $N$, and let $Z(t)$ be the vector in (1.1) that defines the job stack at time $t$. Set

$$X^i(t) = (Z(t), N^i(t)),$$

where $N^i(t)$ is the position of job $i$ in the job stack at time $t$, $i = 1, 2, \ldots, N$. We assume that for some $z^*$, the set $D$ of all states of the job stack process that are accessible from $z^*$ and the set $G = \{(z, n) \in G^*\colon\ z \in D, 1 \leq n \leq N\}$ are irreducible in the sense that all pairs of states communicate. Each of the processes $\{X^i(t)\colon t \geq 0\}$ is a GSMP (with state space $G$ and event set $E$ given by (1.2)) defined on a common underlying probability space $(\Omega, \mathcal{F}, P)$.

Observe that if the marked job is job $i$, then the process $\{X^i(t)\colon t \geq 0\}$ coincides with the augmented job stack process $\{X(t)\colon t \geq 0\}$ except possibly for the initial condition at $t = 0$. For each job, define two sequences of random times: the start times and termination times of the successive passage times for the job. Denote these sequences of random times for job $i$ by $\{\mathcal{S}_j^i\colon j \geq 0\}$ and $\{\mathcal{T}_j^i\colon j \geq 1\}$. Formal definition of these sequences of random times is in terms of the state transitions of the process $\{X^i(t)\colon t \geq 0\}$ and is completely analogous to what was done in terms of the state transitions of the process $\{X(t)\colon t \geq 0\}$. Then the $j$th passage time for job $i$ is $P_j^i = \mathcal{T}_j^i - \mathcal{S}_{j-1}^i$.

Let $X_j^i$ be the state of the GSMP $\{X^i(t)\colon t \geq 0\}$ when the $(j+1)$st passage time starts for job $i$: $X_j^i = X^i(S_j^i)$. Each of the discrete time processes $\{(X_j^i, S_j^i)\colon j \geq 0\}$ is defined on the probability space $(\Omega, \mathcal{F}, P)$.

**Theorem 1.12.** *For* $i = 1, 2, \ldots, N$, $P_j^i \Rightarrow P$ *as* $j \to \infty$.

*Proof.* Since the $N$ jobs are stochastically identical with respect to their service requirements and routing through the network, the probability mass functions and clock-setting distributions governing the processes $\{X^i(t)\colon t \geq 0\}$ are the same. This assertion implies that the processes $\{(X_j^i, \mathcal{S}_j^i)\colon j \geq 0\}$ have the same finite-dimensional distributions. In fact, for any particular job the only difference is that (with possibly one exception) the job does not start a passage time at $t = 0$. However, this difference does not alter limiting results. Since the process $\{X^i(t)\colon t \geq 0\}$ hits each state infinitely often with probability 1, job $i$ starts a passage time with probability 1 and the situation is then the same. Observe in particular that $\mathcal{S}_j^i \to +\infty$ with probability 1 for all $i$; thus, for every job there is always a next passage time. It follows that $P_j^i \Rightarrow P$ as $j \to \infty$.

## 5.2. The Marked Job Method

Point and interval estimates for characteristics of limiting passage times can be based on measurement of individual passage times for an arbitrarily chosen, distinguished job. Select an element $x'$ of the set $A_2'$ defined by (1.7). Begin the simulation of the augmented job stack process with $X(0) = x'$, and simulate the process in cycles defined by the successive random times $\{\mathcal{S}_{\beta_k}\colon k \geq 0\}$ at which $\{X(\mathcal{S}_n)\colon n \geq 0\}$ hits state $x'$. Observe that there may be more than one $x \in A_1$ such that $x \to x'$. Set $\alpha_k = \beta_k - \beta_{k-1}$ so that $\alpha_k$ is the length (that is, the number of state transitions) of the $k$th cycle of $\{X(\mathcal{S}_n)\colon n \geq 0\}$. Thus, $\alpha_k$ is the number of passage times for the marked job in the $k$th cycle. For a real-valued measurable function $f$, also set

$$Y_k(f) = \sum_{j=\beta_{k-1}+1}^{\beta_k} f(P_j).$$

Proposition 2.1 follows directly from the fact that the random indices $\{\beta_k\colon k \geq 0\}$ form a sequence of regeneration points for the process $\{(X(\mathcal{S}_n), P_{n+1})\colon n \geq 0\}$.

**Proposition 2.1.** *The sequence $\{(Y_k(f), \alpha_k)\colon k \geq 1\}$ consists of independent and identically distributed random vectors.*

The final step is to establish a ratio formula for $E[f(P)]$. We give a direct proof of the ratio formula. This proof requires the assumption that $P\{P \in D(f)\} = 0$, but does not use the key renewal theorem.

**Proposition 2.2.** *Provided that* $P\{P \in D(f)\} = 0$ *and* $E[\|f(P)\|] < \infty$,

$$E[f(P)] = \frac{E[Y_1(f)]}{E[\alpha_1]} \stackrel{\text{def}}{=} r(f).$$

*Proof.* Assume that $f \geq 0$ and $E[f(P)] < \infty$. Let $0 < a < \infty$ and set $f_a = \min(f, a)$. Since $P_{n+1} \Rightarrow P$ as $n \to \infty$, $f_a$ is bounded, and $P\{P \in D(f)\} = 0$ by assumption, $f_a(P_{n+1}) \Rightarrow f_a(P)$ as $n \to \infty$ and

$$\lim_{n\to\infty} E[f_a(P_{n+1})] = E[f_a(P)]. \tag{2.3}$$

Next apply Césaro summability to compute the limit of the sequence appearing in (2.3). Write

$$\frac{E[\sum_{n=0}^{m} f_a(P_{n+1})]}{m+1} = \frac{E[\sum_{k=1}^{l(m)+1} Y_k(f_a)]}{m+1} - \frac{E[Y'(m)]}{m+1}, \tag{2.4}$$

where $l(m) = \max\{k\colon\ \beta_k \leq m\}$, and set

$$Y'(m) = \sum_{n=m+1}^{\beta_{l(m)+1}} f_a(P_{n+1}).$$

Observe that $0 \leq Y'(m) \leq a(\beta_{l(m)+1} - m)$ since $0 \leq f_a \leq a$. Moreover, Wald's identity implies that

$$E[\beta_{l(m)+1}] = E[\alpha_1]E[l(m)+1]$$

and

$$E\Big[\sum_{k=1}^{l(m)+1} Y_k(f_a)\Big] = E[Y_1(f_a)]E[l(m)+1].$$

These equations together with the elementary renewal theorem imply that

$$\lim_{m\to\infty} \frac{E[\beta_{l(m)+1} - m]}{m+1} = 0$$

and

$$\lim_{m\to\infty} \frac{E[\sum_{k=1}^{l(m)+1} Y_k(f_a)]}{m+1} = \frac{E[Y_1(f_a)]}{E[\alpha_1]}.$$

Hence from (2.4),

$$\lim_{m\to\infty} \frac{E[\sum_{n=0}^{m} f_a(P_{n+1})]}{m+1} = \frac{E[Y_1(f_a)]}{E[\alpha_1]}. \tag{2.5}$$

From (2.3) and (2.5) conclude that

$$E[f_a(P)] = \frac{E[Y_1(f_a)]}{E[\alpha_1]}. \tag{2.6}$$

Now let $a \to \infty$ on both sides of (2.6) and use the assumption that $E[f(P)] < \infty$ to obtain

$$E[f(P)] = \frac{E[Y_1(f)]}{E[\alpha_1]}. \tag{2.7}$$

For a general $f$ function, write $f = f^+ - f^-$ and apply the above argument to both $f^+$ and $f^-$. Thus, (2.7) holds provided that $E[|f(P)|] < \infty$. □

From $n$ cycles, the regenerative method provides an estimator

$$\hat{r}(n) \stackrel{\text{def}}{=} \frac{\bar{Y}(n)}{\bar{\alpha}(n)} = \frac{\sum_{k=1}^{n} Y_k(f)}{\sum_{k=1}^{n} \alpha_k}$$

that is strongly consistent for $r(f)$. Asymptotic confidence intervals for $r(f)$ are based on the CLT

$$\frac{\sqrt{n}\,\{\hat{r}(n) - r(f)\}}{\sigma(f)/E[\alpha_1]} \Rightarrow N(0,1)$$

as $n \to \infty$, where $\sigma^2(f) \stackrel{\text{def}}{=} \text{Var}\,[Y_1(f) - r(f)\alpha_1]$.

**Algorithm 2.8.** *Marked job method*

1. *Select an element $x'$ of the set $A_2'$ defined by (1.7).*
2. *Set $X(0) = x'$, and observe a fixed number $n$ of cycles of the augmented job stack process defined by the successive random times $\{\mathcal{S}_{\beta_k}\colon k \geq 0\}$ at which (a passage time for the marked job starts and) the augmented job stack process hits state $x'$.*
3. *Measure $(Y_1(f), \alpha_1), (Y_2(f), \alpha_2), \ldots, (Y_n(f), \alpha_n)$.*
4. *Form the point estimate*

$$\hat{r}(n) = \frac{\bar{Y}(n)}{\bar{\alpha}(n)}$$

   *for $r(f)$.*

5. *Form the asymptotic $100(1-2\gamma)\%$ confidence interval*

$$\left[\hat{r}(n) - \frac{s(n)z_{1-\gamma}}{\bar{\alpha}(n)\sqrt{n}}, \hat{r}(n) + \frac{s(n)z_{1-\gamma}}{\bar{\alpha}(n)\sqrt{n}}\right]$$

   *for $r(f)$.*

To apply the marked job method it is necessary to select a return state $x' \in A_2' \subseteq G$. For complex networks of queues it is nontrivial to determine the set $G = \{(z, n) \in G^*\colon z \in D\}$ by inspection. It is sufficient to determine the elements of the set $D$. Let $z_i^*$ be the state of the job stack process in which all $N$ jobs are of job class $j_{k(i)}(i)$ at center $i$ $(i = 1, 2, \ldots, L)$.

**Proposition 2.9.** *Assume that the routing matrix $\boldsymbol{P}$ is irreducible and for some center $i_0$, either $k(i_0) = 1$ or service to a job of job class $j_{k(i_0)}(i_0)$ at center $i_0$ is preempted when any other job of higher priority joins the queue. Let $z \in D^*$ and $n \in \{1, 2, \ldots, N\}$. Then $x = (z, n) \in G$ if and only if $z_i^* \rightsquigarrow z$ for some $i = 1, 2, \ldots, L$.*

**Example 2.10.** (Cyclic queues with nonpreemptive priority) Consider a network of queues with two single-server centers ($L = 2, m(1) = m(2) = 1$) and two job classes ($K = 2$) that provides service to $N$ ($\geq 2$) jobs. Take $C = \{(1,1),(2,1),(2,2)\}$ to be the set of (center, class) pairs. Assume that jobs of job class 2 have nonpreemptive priority over jobs of job class 1 at center 2. A job (of job class 1) that completes service at center 1 moves to center 2 and continues to be a job of job class 1. A job of job class 1 that completes service at center 2 stays at center 2 and becomes a job of job class 2, and a job of job class 2 that completes service at center 2 moves to center 1. Thus, the nonzero elements of the irreducible routing matrix $\boldsymbol{P}$ are $p_{11,21} = 1$, $p_{21,22} = 1$, and $p_{22,11} = 1$.

Set

$$Z(t) = \left(Q_1(t), C_1^{(2)}(t), C_2^{(2)}(t), S_2(t)\right),$$

where $Q_1(t)$ is the number of jobs waiting or in service at center 1 at time $t$, $S_2(t)$ is the job class of the job receiving service at center 2, and $C_j^{(2)}(t)$ is the number of jobs of job class $j$ waiting in queue at center 2. Denote the state space of the process $\{Z(t)\colon t \geq 0\}$ by $D^*$ and set $D = \{z \in D^*\colon z_1^* \rightsquigarrow z\}$, where $z_1^* = (N,0,0,0)$. Observe that the set $D^* - D \neq \emptyset$. Take

$$A_1 = \left\{(q_1, c_1^{(2)}, c_2^{(2)}, s_2, N) \in G\colon\ s_2 = 2\right\}$$

and

$$A_2 = \left\{(q_1, c_1^{(2)}, c_2^{(2)}, s_2, 1) \in G\colon\ q_1 > 0\right\},$$

so that a passage time starts when a job of job class 2 completes service at center 2 and joins the queue at center 1 (as a job of job class 1). By Theorem 2.9, $z = (0, N-1, 0, 2) \in D$ since $z_2^* = (0, N-1, 0, 1)$ and $z_2^* \rightsquigarrow z$. Then $(0, N-1, 0, 2, N)$ is a single-state of the augmented job stack process, $(0, N-1, 0, 2, N, 1, N-2, 0, 2, 1) \in S$, and $(1, N-2, 0, 2, 1) \in A_2'$.

**Exercises**

2.1. (Cyclic queues with feedback and two job types) Consider a closed network of queues with two job types, two job classes, and two centers. that provides service to $N$ ($\geq 2$) jobs: $N_1$ jobs of type 1 and $N_2$ jobs of type 2. Upon completion of service at center 1, a job of type $i$ joins the queue at center 1 with fixed probability $p^{(i)}$ and with probability $1 - p^{(i)}$ joins the queue at center 2, Upon completion of service at center 2, jobs join the queue at center 1. At each center, jobs of type 1 have nonpreemptive priority over jobs of type 2. Jobs of the same type at a center receive service in order of their arrival at the center. All service times are mutually independent. Jobs of type $i$ at center $k$ receive service that is exponentially distributed with rate parameter $\lambda_k^{(i)}$. Center $k$ serves jobs of both types, but only serves jobs of job class $k$.

Set

$$Z(t) = (C_1^{(2)}(t), C_1^{(1)}(t), S_1(t), C_2^{(2)}(t), C_2^{(1)}(t), S_2(t)),$$

where $C_k^{(k)}(t)$ is the number of jobs of type $i$ waiting in queue at center $k$ at time $t$ and $S_k(t)$ is the type of job in service at center $k$ at time $t$.

Select a marked job of type 1 and a marked job of type 2, and let $N_i(t)$ be the position from the top of the marked job of type $i$ in the job stack defined by the vector $Z(t)$. Set

$$X(t) = (Z(t), N_1(t), N_2(t)).$$

Let $L(t)$ be the last state occupied by the process $\{X(t)\colon\ t \geq 0\}$ before jumping to $X(t)$, and set $V(t) = (L(t), X(t))$. Denote by $R(i)$ the limiting passage time for jobs of type $i$ that starts when a job of type $i$ completes service at center 2 (and joins the queue at center 1) and terminates when the job next completes service at center 2. Describe how you would use the regenerative method to obtain a strongly consistent point estimate and an asymptotic confidence interval for $E[R(1)]$ and

$P\{R(1) \leq x\}$ ($x$ fixed). Be explicit about the sequence of regeneration points, the quantities to be measured, and the construction of the point and interval estimates. State the ratio formula and central limit theorem that lead to the point and interval estimates. Also explain how you would obtain strongly consistent point estimates and asymptotic confidence intervals for $E[R(1)]$ and $E[R(2)]$ from a single sample path of the process $\{V(t)\colon\ t \geq 0\}$.

2.2. (Finite-capacity queue with preemptive-resume priority) Consider an open queueing system with one single-server center and two job types. Jobs of type $i$ arrive from an external source, queue and receive service at the center, and depart to an external sink. A job of type $i$ that arrives when the system already contains $B_i - 1$ ($\geq 0$) jobs of type $i$ causes the arrival process for jobs of type $i$ to shut down. The arrival process remains shut down until the first subsequent departure of a job of type $i$. Jobs of the same type receive service in order of their arrival at the center, and jobs of type 1 have "preemptive-resume" priority over jobs of type 2. A job of type 2 that has been preempted receives additional service at the center before any other job of type 2 receives service. Assume that the interarrival times for jobs of type $i$ are i.i.d. as a positive random variable $A_i$ with finite mean and a density function that is continuous and positive on $(0, +\infty)$. Also assume that the service times for jobs of type $i$ are i.i.d. as a positive random variable $V_i$ with finite mean and a density function that is continuous and positive on $(0, +\infty)$.

Set

$$X(t) = (B_2(t), B_1(t), M(t), L(t)),$$

where

$$L(t) = \begin{cases} 1 & \text{if the job at the head of the type 2 queue} \\ & \text{at time } t \text{ has been preempted} \\ 0 & \text{otherwise,} \end{cases}$$

$$M(t) = \begin{cases} 2 & \text{if there is a job of type 2 in service at time } t \\ 1 & \text{if there is a job of type 1 in service at time } t \\ 0 & \text{if there is no job in service at time } t, \end{cases}$$

and $B_i(t)$ is the number of jobs of type $i$ waiting in queue at time $t$. Show that the state transitions and state transition times determine the delay waiting for service experienced by jobs of type $i$. Also show that $\{X(t)\colon\ t \geq 0\}$ is a regenerative process in continuous time and the expected time between regeneration points is finite. Let $D(i)$ be the limiting delay experienced by jobs of type $i$. Can the regenerative method be used to obtain point and interval estimates for $P\{D(i) > x\}$? Explain your answer.

2.3. (Finite-capacity queue with nonpreemptive priority) Consider an open queueing system with one single-server center and two job types. Jobs of type $i$ arrive from an external source, queue and receive service at the center, and depart to an external sink. A job of type $i$ that arrives when the system already contains $B_i - 1$ $(\geq 0)$ jobs of type $i$ causes the arrival process for jobs of type $i$ to shut down; the arrival process remains shut down until the first subsequent departure of a job of type $i$. Jobs of the same type receive service in order of their arrival at the center, and jobs of type 1 have nonpreemptive priority over jobs of type 2. Assume that all interarrival times and service times are mutually independent and the interarrival times for job $i$ are identically distributed as a positive random variable $A_i$ with finite mean and a density function that is continuous and positive on $(0, +\infty)$. Also assume that service times for jobs of type $i$ are identically distributed as a positive random variable $V_i$ with finite mean and a density function that is continuous and positive on $(0, +\infty)$.

What stochastic process $\{X(t)\colon\ t \geq 0\}$ would you simulate to estimate the number of jobs of type $i$ in the system and the delay (waiting for service) experienced by jobs of type $i$? Specify the state space $S$ and the set $E$ of events associated with simulation of the process $\{X(t)\colon t \geq 0\}$. Show that the state transitions and state transition times determine the delay experienced by jobs of type $i$. Let $D(i)$ be the limiting delay experienced by jobs of type $i$. Can the regenerative method be used to obtain point and interval estimates for $E[D(i)]$? Explain your answer.

## 5.3. The Labelled Jobs Method

Passage times for all of the jobs in a network of queues can be measured by observing a "fully augmented job stack process" that maintains the position of each of the jobs in the job stack. Under a mild restriction on the priorities among job classes, it is possible to define a suitable discrete time process and to identify a sequence of regeneration points that leads to point and interval estimates. This "labelled jobs method" prescribes measurement of passage times for all of the jobs.

It is necessary to restrict the job stack process defined by (1.1) to a suitable subset of the state space $D^*$. We assume throughout that there exists a target state of the job stack process; that is, there exists $z^* \in D^*$ such that $z \rightsquigarrow z^*$ for all $z \in D^*$. This assumption ensures that the set $D = \{z \in D^*\colon z^* \rightsquigarrow z\}$ is irreducible in the sense that $z$ and $z'$ communicate for all $z, z' \in D$. Recall from Theorem 1.7 of Chapter 4 that in order for a target state to exist it is sufficient that for some center $i_0$, either $k(i_0) = 1$ or service at center $i_0$ to a job of the lowest priority job class $j_{k(i_0)}(i_0)$ at center $i_0$ is preempted when any other job of higher priority joins the queue. Let $z^*_{i_0} \in D^*$ be the state of the job stack process in which all $N$ jobs are of job class $j_{k(i_0)}(i_0)$ at center $i_0$. Then the set $D = \left\{z \in D^*\colon z^*_{i_0} \rightsquigarrow z\right\}$ of all states of the job stack process that are accessible from $z^*_{i_0}$ is irreducible. Also assume that the sets $A_1$, $A_2$, $B_1$, and $B_2$ that define the start times and termination times of individual passage times for an arbitrarily chosen, marked job are subsets of $G = \{(z, n) \in G^*\colon z \in D\}$.

Set

$$S = \{(z, n, z', n')\colon\ (z, n) \in A_1,\ (z', n') \in A_2, \\ \text{and } (z, n) \to (z', n')\} \tag{3.1}$$

and

$$T = \{(z, n, z', n')\colon\ (z, n) \in B_1,\ (z', n') \in B_2, \\ \text{and } (z, n) \to (z', n')\}. \tag{3.2}$$

Denote by $H$ the set of all (center, class) pairs $(i, j) \in C$ such that an arbitrarily chosen, marked job can be of job class $j$ at center $i$ when the passage time specified by the sets $A_1$, $A_2$, $B_1$, and $B_2$ is not underway. Formally, let $H^c \subset C$ be the set of (center, class) pairs $(i, j)$ such that there exists a sequence $(z_0, n_0), (z_1, n_1), \ldots, (z_m, n_m) \in G$ satisfying

(i) $(z_0, n_0) \to (z_1, n_1) \to \cdots \to (z_m, n_m)$;
(ii) $(z_0, n_0, z_1, n_1) \in S$, $(z_{m-1}, n_{m-1}, z_m, n_m) \in T$, and $(z_{k-1}, n_{k-1}, z_k, n_k) \notin T$ for $1 \le k \le m-1$; and
(iii) $h(z_k, n_k) = (i, j)$ for some $1 \le k \le m-1$.

Then set $H = C - H^c$. The labelled jobs method is applicable to passage times for which $S \cap T = \emptyset$ so that the set $H$ is nonempty. Definition 3.3 identifies special "single-states" of the job stack process.

**Definition 3.3.** *An element $z_0$ of the set $D$ is said to be a single-state of the job stack process for the passage time specified by the sets $A_1$, $A_2$, $B_1$, and $B_2$ if*
*(i) there exists $(z_1, m) \in B_1$ such that $(z_1, m) \to (z_0, n)$ for some $(z_0, n) \in B_2$; and*
*(ii) $h(z_0, n) \in H \cap \{(i_0, j_1(i_0)), (i_0, j_2(i_0)), \ldots, (i_0, j_{k(i_0)}(i_0))\}$ for some single-server center $i_0$ and all $n = 1, 2, \ldots, N$.*

According to this definition, $z_0$ is a single-state of the job stack process if there exists $z_1$ such that a passage time for some job terminates when there is a state transition from $z_1$ to $z_0$, and in state $z_0$ all jobs are at the same single-server center ($i_0$) with no passage times underway ($h(z_0, n) \in H$ for all $n$). We assume that there exists a single-state of the job stack process in the sense of Definition 3.3.

**Example 3.4.** (Cyclic queues with feedback) Let $Z(t)$ be the number of jobs waiting or in service at center 1 at time $t$. Then the set $H = \{(2,1)\}$, and $z_0 = 0$ is a single-state of the job stack processs for the passage times specified by the subsets

$$A_1 = \{(i, N)\colon\ 0 \le i < N\},$$

$$A_2 = \{(i, 1)\colon\ 0 < i \le N\},$$

$$B_1 = \{(i, i)\colon\ 0 < i \le N\},$$

and

$$B_2 = \{(i-1, i)\colon\ 0 < i \le N\}.$$

Observe that the set $S = \{(i, N, i+1, 1)\colon\ 0 \le i < N\}$ and the set $T = \{(i, i, i-1, i)\colon\ 0 < i \le N\}$, so that $S \cap T = \emptyset$.

Label the jobs from 1 to $N$, and denote by $N^i(t)$ the position of job $i$ in the job stack at time $t$ ($i = 1, 2, \ldots, N$). Then in terms of the vector $Z(t)$, set

$$X^0(t) = (Z(t), N^1(t), \ldots, N^N(t)).$$

The process $\{X^0(t)\colon\ t \ge 0\}$ is called the *fully augmented job stack process*. Formal specification of the fully augmented job stack process is as a GSMP with finite state space

$$G^0 = \{(z, n_1, \ldots, n_N) \in D^* \times \{1, 2, \ldots, N\}^N : n_i \ne n_j \text{ if } i \ne j\}$$

and event set $E$ given by (1.2).

The fully augmented job stack process need not be irreducible even if the set $G$ is irreducible.

**Example 3.5.** (Cyclic queues) Let $Z(t)$ be the number of jobs waiting or in service at center 1 at time $t$. The process $\{Z(t)\colon t \geq 0\}$ has state space $D^* = \{0, 1, \ldots, N\}$. The set of all states that are accessible from $z^* = N$ is $D = D^*$, and all pairs of states in the set $G = \{(z, n) \in G^*\colon z \in D, 1 \leq n \leq N\}$ communicate. Set $X^0(t) = (Z(t), N^1(t), N^2(t), \ldots, N^N(t))$, where $N^i(t)$ is the position of job $i$ in the job stack at time $t$. When there are $N = 3$ jobs in the network, the set $D = \{0, 1, 2, 3\}$, and the fully augmented job stack process has two irreducible, closed sets of states. All pairs of states in the set $\{(z, 1, 2, 3), (z, 2, 3, 1), (z, 3, 1, 2)\colon z \in D^*\}$ communicate as do all pairs in the set $\{(z, 1, 3, 2), (z, 3, 2, 1), (z, 2, 1, 3)\colon z \in D^*\}$.

Set $X^i(t) = (Z(t), N^i(t))$, where $N^i(t)$ is the position of job $i$ in the job stack at time $t$. Let $L^i(t)$ be the last state occupied by the process $\{X^i(t)\colon t \geq 0\}$ before jumping to $X^i(t)$. Set $V^i(t) = (L^i(t), X^i(t))$, and denote the state space of the process $\{V^i(t)\colon t \geq 0\}$ by $B^i$. Also set

$$S^i = \{(z, n, z', n') \in B^i\colon (z, n) \in A_1,\ (z', n') \in A_2\}$$

and

$$T^i = \{(z, n, z', n') \in B^i\colon (z, n) \in B_1,\ (z', n') \in B_2\}.$$

The random times at which the process $\{V^i(t)\colon t \geq 0\}$ hits the set $S^i$ correspond to the start times of individual passage times for job $i$, and the random times at which the process hits the set $T^i$ correspond to the termination times. Similarly, set $V^0(t) = (L^0(t), X^0(t))$ and denote the state space of $\{V^0(t)\colon t \geq 0\}$ by $B^0$. Also set

$$S^0 = \{(z, n_1, \ldots, n_N, z', n'_1, \ldots, n'_N) \in B^0\colon (z, n_k) \in A_1 \text{ and } (z', n'_k) \in A_2 \text{ for some } k\}$$

and

$$T^0 = \{(z, n_1, \ldots, n_N, z', n'_1, \ldots, n'_N) \in B^0\colon (z, n_k) \in B_1 \text{ and } (z', n'_k) \in B_2 \text{ for some } k\}.$$

The random times at which the process $\{V^0(t)\colon t \geq 0\}$ hits the set $S^0$ correspond to the start times of passage times irrespective of job identity, and the random times at which $\{V^0(t)\colon t \geq 0\}$ hits the set $T^0$ correspond to the termination times.

Let $z_0$ be a single-state of the job stack process as in Definition 3.3. Select $(z_0, n_1, \ldots, n_N) \in G^0$ and set $X^0(0) = (z_0, n_1, \ldots, n_N)$. Denote the successive passage times, irrespective of job identity and enumerated in termination order, by $\{P^0_n\colon n \geq 1\}$. Also let $\{T^0_n\colon n \geq 1\}$ be the corresponding sequence of termination times. Set $T^0_0 = 0$. The labelled jobs method rests on the fact that $\{(Z(T^0_n), P^0_{n+1})\colon n \geq 0\}$ is a regenerative process in discrete time and the expected time between regeneration points is finite. We first show that the job stack process hits a fixed single-state infinitely often with probability 1.

**Theorem 3.6.** *Assume that the routing matrix* $\boldsymbol{P}$ *is irreducible and for some center* $i_0$, *either* $k(i_0) = 1$ *or service to a job of job class* $j_{k(i_0)}(i_0)$ *at center* $i_0$ *is preempted when any other job of higher priority joins the queue. Then*

$$P\{Z(T^0_n) = z_0 \text{ i.o.}\} = 1$$

*for any single-state* $z_0$ *of the job stack process.*

*Proof.* Let $L(t)$ be the last state occupied by the job stack process before jumping to the state occupied at time $t$. Set

$Y(t) = (L(t), Z(t))$, and let $\zeta_n$ be the time of the $n$th state transition, $n \geq 0$. Then, by Theorem 2.2 of Chapter 4,

$$P\{Y(\zeta_k) = (z, z') \text{ i.o.}\} = 1 \tag{3.7}$$

for all $z, z' \in D$ such that $z \to z'$. Now let $z_0$ be a single-state of the job stack process as in Definition 3.3. Condition (i) of Definition 3.3 ensures that there exists $z_1$ with $z_1 \to z_0$ such that a passage time for some job terminates when the job stack process makes a state transition from $z_1$ to $z_0$. The desired result follows from (3.7) since every time $\zeta_k$ at which $Y(\zeta_k) = (z_1, z_0)$ is an element of $\{\mathcal{T}_n^0\colon n \geq 0\}$ and $Y(\mathcal{T}_n^0) = (z_1, z_0)$ implies that $Z(\mathcal{T}_n^0) = z_0$. □

**Theorem 3.8.** *Assume that the conditions of Theorem 3.6 hold. Then the process $\{(Z(\mathcal{T}_n^0), P_{n+1}^0)\colon n \geq 0\}$ is a regenerative process in discrete time, and the expected time between regeneration points is finite.*

*Proof.* Recall that $X^0(0) = (z_0, n_1, \ldots, n_N)$ for some single-state $z_0$ of the job stack process as in Definition 3.3 and some $(z_0, n_1, \ldots, n_N) \in G^0$. By Theorem 3.6, $P\{Z(\mathcal{T}_n^0) = z_0 \text{ i.o.}\} = 1$. Denote the successive random times at which (at passage time terminates and) the job stack process hits state $z_0$ by $\{\mathcal{T}_{\beta_k^0}^0\colon k \geq 1\}$. Set $\mathcal{T}_{\beta_0^0}^0 = 0$.

It is sufficient to show that $\{\beta_k^0\colon k \geq 0\}$ is an increasing sequence of a.s. finite random indices for which the post-$\beta_k^0$ process $\{(Z(\mathcal{T}_n^0), P_{n+1}^0),\ n \geq \beta_k^0;\ \beta_l - \beta_{l-1},\ l \geq k+1\}$ is independent of the pre-$\beta_k^0$ process $\{(Z(\mathcal{T}_n^0), P_{n+1}^0),\ 0 \leq n < \beta_k^0;\ \beta_l - \beta_{l-1},\ 1 \leq l \leq k\}$ and has the same distribution for each $k$. Set $\zeta(\beta_k^0) = \min\{\zeta_n\colon \zeta_n > \mathcal{T}_{\beta_k^0}^0\}$. At time $\mathcal{T}_{\beta_k^0}^0$, no passage times are underway. Now observe that each of the clocks running at time $\zeta(\beta_k^0)+$ was set at time $\zeta(\beta_k^0)$. Therefore,

$\{Z(t)\colon t \geq \zeta(\beta_k^0)\}$ determines the finite-dimensional distributions of $Z(T^0_{\beta_k^0+i})$, $P^0_{\beta_k^0+i+1}$ for $i \geq 1$ and the distribution of $\beta_{k+1}^0 - \beta_k^0$. The joint distribution of $Z(\zeta(\beta_k^0))$ and the clocks set at time $\zeta(\beta_k^0)$ depends on the past history of the job stack process only through $z_0$ and the event occurring at time $\zeta(\beta_k^0)$. This distribution is the same for all $\beta_k^0$, and therefore the post-$\beta_k^0$ process is independent of the pre-$\beta_k^0$ process and has the same distribution for all $k$. It can be shown using properties of the GSSMC associated with state transitions of the job stack process and the assumptions on the clock-setting distributions, that $E[\beta_{k+1}^0 - \beta_k^0] < \infty$. □

Observe that the successive random times at which $\{Z(T_k^0)\colon k \geq 0\}$ hits state $z_0$ do not form a sequence of regeneration points for the job stack process.

Assume that $\beta_k^0 - \beta_{k-1}^0$ is aperiodic. It then follows from Theorem 3.8 and general results for regenerative processes that

$$(Z(T_n^0), P_{n+1}^0) \Rightarrow (Z^0, P^0)$$

as $n \to \infty$. The limiting random variable $P^0$ has the same distribution as the limiting passage time $P$ (in Theorem 1.12) for any marked job. To see this, let $P_j^i$ be the $j$th passage time for job $i$. Also let $\alpha_1^i$ be the number of passage times for job $i$ in the $k$th block defined by the random times $\{T_{\beta_k^0}\colon k \geq 0\}$, and set $\beta_k^i = \alpha_1^i + \cdots + \alpha_k^i$ ($\beta_0^i = 0$). Next observe that for all $A$, the sequence $\{(\sum_{j=\beta_{k-1}^i+1}^{\beta_k^i} 1_A(P_j^i), \alpha_k^i)\colon k \geq 1\}$ consists of i.i.d. random vectors and, by the argument in the proof of Proposition 2.2,

$$E[1_A(P)] = \frac{E\left[\sum_{j=1}^{\alpha_1^i} 1_A(P_j^i)\right]}{E[\alpha_1^i]}.$$

Then

$$\begin{aligned} E[1_A(P^0)] &= \frac{E\left[\sum_{j=1}^{\alpha_1^0} 1_A(P_j^0)\right]}{E[\alpha_1^0]} \\ &= \frac{E\left[\sum_{i=1}^{N}\sum_{j=1}^{\alpha_1^i} 1_A(P_j^i)\right]}{\sum_{i=1}^{N} E[\alpha_1^i]} \\ &= \frac{\sum_{i=1}^{N} E\left[\sum_{j=1}^{\alpha_1^i} 1_A(P_j^i)\right]}{\sum_{i=1}^{N} E[\alpha_1^i]} \\ &= \frac{\sum_{i=1}^{N} E[1_A(P)]E[\alpha_1^i]}{\sum_{i=1}^{N} E[\alpha_1^i]} \\ &= E[1_A(P)]. \end{aligned}$$

Select a single-state $z_0$ of the job stack process as in Definition 3.3 and an initial state $(z_0, n_1, \ldots, n_N) \in G^0$ for the fully augmented job stack process $\{X^0(t)\colon t \geq 0\}$. Since the labelling of the jobs is arbitrary and the $N$ jobs are stochastically identical with respect to their service requirements and routing probabilities, the sequence of passage times irrespective of job identity associated with starting the job stack process in a fixed single-state converges in distribution to $P^0$.

We simulate the fully augmented job stack process in blocks defined by the successive random times $\{\mathcal{T}^0_{\beta_k^0}\colon k \geq 0\}$ at which (a passage time terminates and) the job stack process hits state $z_0$ ($\beta_0^0 = 0$ and $\mathcal{T}^0_{\beta_0^0} = 0$). Set

$$Y_k^0(f) = \sum_{j=\beta_{k-1}^0+1}^{\beta_k^0} f(P_j^0)$$

and $\alpha_k^0 = \beta_k^0 - \beta_{k-1}^0$ for $k \geq 1$.

Proposition 3.9 follows directly from Theorem 3.6, and the argument used in the proof of Proposition 2.2 establishes a ratio formula for $E[f(P^0)]$.

**Proposition 3.9.** *The sequence* $\{(Y_k^0(f), \alpha_k^0) \colon k \geq 1\}$ *consists of independent and identically distributed random vectors.*

**Proposition 3.10.** *Provided that* $P\{P^0 \in D(f)\} = 0$ *and* $E[\| f(P^0) \|] < \infty$,

$$E[f(P^0)] = \frac{E[Y_1^0(f)]}{E[\alpha_1^0]} \stackrel{\text{def}}{=} r^0(f).$$

From $n$ cycles, the regenerative method provides an estimator

$$\hat{r}^0(n) \stackrel{\text{def}}{=} \frac{\bar{Y}^0(n)}{\bar{\alpha}^0(n)} = \frac{\sum_{k=1}^n Y_k^0(f)}{\sum_{k=1}^n \alpha_k^0}$$

that is strongly consistent for $r^0(f)$. Asymptotic confidence intervals for $r^0(f)$ are based on the CLT

$$\frac{\sqrt{n}\,\{\hat{r}^0(n) - r(f)\}}{\sigma^0(f)/E[\alpha_1^0]} \Rightarrow N(0,1)$$

as $n \to \infty$, where $(\sigma^0(f))^2 \stackrel{\text{def}}{=} \text{Var}\,[Y_1^0(f) - r^0(f)\alpha_1^0]$.

**Algorithm 3.11.** *Labelled jobs method*

1. *Select a single-state* $z_0$ *of the job stack process as in Definition 3.3 and an initial state* $(z_0, n_1, \ldots, n_N)$ *for the fully augmented job stack process.*

2. *Set* $X^0(0) = (z_0, n_1, \ldots, n_N)$, *and observe a fixed number* $n$ *of blocks of the fully augmented job stack process defined by the successive random times* $\{\mathcal{T}^0_{\beta^0_k}\colon k \geq 0\}$ *at which (a passage time terminates and) the job stack process hits state* $z_0$.
3. *Measure* $(Y^0_1(f), \alpha^0_1), (Y^0_2(f), \alpha^0_2), \ldots, (Y^0_n(f), \alpha^0_n)$.
4. *Form the point estimate*

$$\hat{r}^0(n) = \frac{\bar{Y}^0(n)}{\bar{\alpha}^0(n)}$$

*for* $r^0(f)$.

5. *Form the asymptotic* $100(1-2\gamma)\%$ *confidence interval*

$$\left[\hat{r}^0(n) - \frac{s(n)z_{1-\gamma}}{\bar{\alpha}^0(n)\sqrt{n}}, \hat{r}^0(n) + \frac{s(n)z_{1-\gamma}}{\bar{\alpha}^0(n)\sqrt{n}}\right]$$

*for* $r^0(f)$.

**Example 3.12.** (Cyclic queues with nonpreemptive priority) Set $Z(t) = (Q_1(t), C_1^{(2)}(t), C_2^{(2)}(t), S_2(t))$ as in Example 2.10, and suppose that there are $N = 3$ jobs in the network. Since center 1 serves only one job class, take $i_0 = 1$ and $z^*_{i_0} = (3,0,0,0)$. All states in $D^*$ except (0,0,2,2) and (0,1,1,1) are accessible from $z^*_1$. Therefore,

$$D = D^* - \{(0,0,2,2), (0,1,1,1)\}$$

and any two elements of the set $D$ communicate.

Consider the passage time specified by the subsets

$$A_1 = \{(1,0,1,2,1), (1,1,0,1,1), (2,0.0,0,2)\},$$

$$A_2 = \{(0,1,1,2,1), (0,2,0,1,1), (1,1,0,1,2)\},$$

$$B_1 = \{(0,1,1,2,2),(1,0,1,2,2)\},$$

and

$$B_2 = \{(1,1,0,1,1),(2,0,0,0,1),(1,0,1,2,1)\}.$$

The passage time starts when a job joins the queue at center 2 as a job of job class 1 and terminates when the job completes service at center 2 as a job of job class 2. Observe that $S \cap T = \emptyset$. Also observe that because $(0,0,2,2) \notin D$, $(0,0,2,2,2) \notin B_1$. For this passage time, $H = \{(1,1)\}$ and $z_0 = (3,0,0,0)$ is a single-state of the job stack process. Clearly, $z_0$ satisfies Condition (i) of Definition 3.3. Condition (ii) also holds because $(1,0,1,2,2) \to (2,0,0,0,1)$.

Now consider the passage time specified by the sets

$$A_1 = \{(1,0,1,2,2),(0,0,2,2,2),(0,1,1,2,2)\},$$

$$A_2 = \{(2,0,0,1,1),(1,0,1,2,1),(1,1,0,1,1)\},$$

$$B_1 = \{(1,1,0,1,2),(0,2,0,1,2)\},$$

and

$$B_2 = \{(1,0,1,2,2),(0,1,1,2,2)\}.$$

The passage time starts when a job completes service at center 2 as a job of job class 2 and terminates when the job completes service at center 2 as a job of job class 1. For this passage time, the set $H = \{(2,2)\}$. There is at least one passage time underway unless the configuration of the job stack is (0,0,2,2). Since $(0,0,2,2) \notin D$ there is no single-state, and the labelled jobs method does not apply.

**Exercises**

3.1. (Central server) Consider a network of queues with $L$ $(> 1)$ single-server centers $(m(1) = m(2) = 1)$ and a single job class $(K = 1)$. Take $C = \{(1,1),(2,1),\ldots,(L,1)\}$ to be the set of (center, class) pairs. With fixed probability $p_i$ ($\sum_{i=2}^{L} p_i = 1$) a job that completes service at center 1 moves to center $i$ ($2 \leq i \leq L$). A job that completes service at center $i$ moves to center 1. Assume that all queueing disciplines are FCFS. Also assume that service times are mutually independent and the service time at center $i$ is exponentially distributed with rate parameter $\lambda(i)$.

The *cycle time* is the passage time that starts when a job moves to center 1 and terminates when the job next moves to center 1. Describe how you would use the regenerative method to obtain a strongly consistent point estimate and an asymptotic confidence interval for the limiting expected cycle time. Be explicit about the sequence of regeneration points, the quantities to be measured, and the construction of the point and interval estimates. State the ratio formula and central limit theorem that lead to the point and interval estimates.

## 5.4. Comparison of Statistical Efficiency

The marked job method prescribes observation of passage times for an arbitrarily chosen, distinguished job and provides point and interval estimates for characteristics of limiting passage times. The half length of the confidence interval obtained from a simulation of fixed length for the expected value of a general function $f$ of the limiting passage time is proportional to a certain quantity $e(f)$. The labelled jobs method is applicable to passage times that correspond to the random times for a job to traverse a subnetwork, and observed passage times for all of the jobs enter into the point and interval estimates. With the same constant of proportionality, the half length of the confidence interval for the labelled jobs method is proportional to a quantity $e^0(f)$. Since $e(f)$ and $e^0(f)$ are independent of the blocks of the underlying regenerative process, these quantities are appropriate measures of the statistical efficiency of the estimation procedures.

For closed networks of queues with exponential service times, theoretical values for expected passage times and the associated variance constants appearing in CLT's used to form confidence intervals can be computed. Such computations provide a quantitative assessment of the relative statistical efficiency of the marked job and labelled jobs methods for Markovian networks. For networks of queues with general service times, there is little hope of computing the needed theoretical values, even for expected passage times. An argument using central limit theorems and the continuous mapping theorem shows that $e^0(f) \leq e(f)$ for any function $f$ and all numbers of jobs in the network. It follows that confidence intervals constructed using the labelled jobs method are shorter than those obtained from the marked job method. This result is consistent with intuition since the labelled jobs method extracts more information about passage times from a simulation run of fixed length.

Assume that for the specified passage time the set $A_2'$ of (1.7) is nonempty so that there exists a single-state of the augmented job stack process as in Definition 1.6. Also assume that the sets $S$ and $T$ of (3.1) and (3.2) are disjoint and there exists a single-state of the job stack process as in Definition 3.3. These assumptions ensure that both the marked job and labelled jobs methods are applicable. Let $L(t)$ be the last state occupied by the augmented job stack process before jumping to $X(t)$. Set $V(t) = (L(t), X(t))$, and denote the state space of the process $\{V(t)\colon\ t \geq 0\}$ by $B$. Similarly, let $L^0(t)$ be the last state occupied by the fully augmented job stack process before jumping to $X^0(t)$. Set $V^0(t) = (L^0(t), X^0(t))$, and denote the state space of $\{V^0(t)\colon\ t \geq 0\}$ by $B^0$. We work with the process $\{V^0(t)\colon\ t \geq 0\}$ and take job 1 to be the marked job. Select a single-state $z_0$ as in Definition 3.3. Also select $v_0 = (z_0', m_1', z_0, m_1) \in T$ and $v_0^0 = (z_0', m_1', m_2', \ldots, m_N', z_0, m_1, m_2, \ldots, m_N) \in B^0$. Take $V^0(0) = v_0^0$.

Lemma 4.1 is the basis for construction of CLT's in continuous time for sequences of passage times. Let $\{\gamma_k^*\colon k \geq 0\}$ be a renewal process defined on the same probability space as $\{P_n^0\colon n \geq 1\}$, and let $\{P_{n_j}^0\colon j \geq 1\}$ be a subsequence of $\{P_n^0\colon n \geq 1\}$. Set $\delta_{k+1}^* = \gamma_{k+1}^* - \gamma_k^*$, and denote by $m^*(t)$ the number of $P_{n_j}^0$ completed in $(0, t]$. Also set

$$\alpha_k^* = m^*(\gamma_k^*) - m^*(\gamma_{k-1}^*)$$

and

$$Y_k^*(f) = \sum_{j=m^*(\gamma_{k-1}^*)+1}^{m^*(\gamma_k^*)} f(P_{n_j}^0).$$

**Lemma 4.1.** *Assume that the sequence $\{(Y_k^*(f), \alpha_k^*)\colon k \geq 1\}$ consists of independent and identically distributed random vectors and*

$$E[f(P)] = \frac{E[Y_1^*(f)]}{E[\alpha_1^*]} \stackrel{\text{def}}{=} r(f).$$

*In addition, assume that $E[\delta_1^*] < \infty$, $E\left[(\alpha_1^*)^2\right] < \infty$, and $E\left[(Y_1^*(\mid f \mid))^2\right] < \infty$. Then*

$$\frac{\sqrt{t}\left(\frac{1}{m^*(t)} \sum_{j=1}^{m^*(t)} f(P_{n_j}^0) - r(f)\right)}{\sqrt{E[\delta_1^*]}\sigma^*(f)/E[\alpha_1^*]} \Rightarrow N(0,1)$$

*as $t \to \infty$, where $(\sigma^*(f))^2 \stackrel{\text{def}}{=}$ Var $[Y_1^*(f) - r(f)\alpha_1^*] < \infty$.*

*Proof.* Set

$$A(t) = \sum_{j=1}^{m^*(t)} f(P_{n_j}^0) - r(f)m^*(t),$$

and let $\{n^*(t)\colon\ t \geq 0\}$ be the counting process associated with the renewal process $\{\gamma_k^*\colon\ k \geq 0\}$. Then the Doeblin decomposition is

$$A(t) = \sum_{k=1}^{n^*(t)} [Y_k^*(f) - r(f)\alpha_k^*]$$
$$+ \sum_{j=m^*(\gamma^*_{n^*(t)})+1}^{m^*(t)} f(P^0_{n_j}) - r(f)\left[m^*(t) - m^*(\gamma^*_{n^*(t)})\right]. \tag{4.2}$$

Also set $Z_k^*(f) = Y_k^*(f) - r(f)\alpha_k^*$. By assumption, $E[Z_k^*(f)] = 0$ and the sequence $\{Z_k^*(f)\colon\ k \geq 1\}$ consists of i.i.d. random variables. The weak law for renewal counting processes yields $n^*(t)/t \Rightarrow 1/E[\delta_1^*]$ as $t \to \infty$. These facts, together with the CLT for random partial sums and the continuous mapping theorem, yields the CLT

$$\frac{\sum_{k=1}^{n^*(t)} Z_k^*(f)}{\sqrt{t}\sigma^*(f)/\sqrt{E[\delta_1^*]}} \Rightarrow N(0,1) \tag{4.3}$$

as $t \to \infty$. Equation (4.3) implies that (suitably normalized) the first term on the right side of (4.2) converges weakly to $N(0,1)$. The last two terms on the right side of (4.2) are remainder terms that, when divided by $\sqrt{t}$, converge weakly to 0. This argument is standard, but requires $E[(\alpha_1^*)^2]$ and $E[(Y_1^*(\mid f \mid))^2]$ to be finite. The converging together lemma and (4.3) ensure that

$$\frac{A(t)}{\sqrt{t}\,\sigma^*(f)/\sqrt{E[\delta_1^*]}} \Rightarrow N(0,1) \tag{4.4}$$

as $t \to \infty$. Finally, dividing the numerator and denominator of the left side of (4.4) by $m^*(t)$ and using the continuous mapping theorem along with the fact that $m^*(t)/t \Rightarrow E[\alpha_1^*]/E[\delta_1^*]$ as $t \to \infty$, we obtain the desired result. □

The labelled jobs method prescribes selection of a single-state $z_0$ of the job stack process with respect to the specified passage time (along with $v_0 = (z_0', m_1', z_0, m_1) \in T$ and $v_0^0 = (z_0', m_1', m_2', \ldots, m_N', z_0, m_1, m_2, \ldots, m_N) \in B^0$). We simulate the process $\{V^0(t)\colon\ t \geq 0\}$ in blocks defined by the successive random times at which the process leaves the set $\{(z_0', n_1', n_2', \ldots, n_N', z_0, n_1, n_2, \ldots, n_N) \in B^0\colon\ (z_0', n_i', z_0, n_i) = v_0 \text{ for some } i\}$. Let $\delta_1^0$ be the length of a block and $m^0(t)$ be the number of passage times, irrespective of job identity, completed in the interval $(0, t]$. Lemma 4.1 implies that

$$\frac{\sqrt{t}\left(\frac{1}{m^0(t)}\sum_{n=1}^{m^0(t)} f(P_n^0) - r(f)\right)}{\sqrt{E[\delta_1^0]}\sigma^0(f)/E[\alpha_1^0]} \Rightarrow N(0,1)$$

as $t \to \infty$, where $\left(\sigma^0(f)\right)^2 = \text{ Var } [Y_1^0(f) - r(f)\alpha_1^0]$ and $r(f) = E[Y_1^0(f)]/E[\alpha_1^0]$. This CLT implies that the half length of the confidence interval obtained from a simulation of fixed length is proportional to the quantity $e^0(f)$ defined by

$$e^0(f) = \sqrt{E[\delta_1^0]}\sigma^0(f)/E[\alpha_1^0].$$

(We assume that $E[(\alpha_1^0)^2] < \infty$ and $E[(Y_1^0(|\ f\ |))^2] < \infty$. These assumptions make it possible to apply Lemma 4.1.) Since the numerator in this CLT and the limit ($N(0,1)$) is independent of the state ($v_0$) selected from $T$, so is the denominator. This assertion is a consequence of the convergence of types theorem. Thus, $e^0(f)$ is an appropriate measure of the statistical efficiency of the labelled jobs method.

The marked job method prescribes selection of a state $x' = (z', n') \in A_2'$. We simulate the process $\{V(t)\colon\ t \geq 0\}$ in blocks defined by the successive random times at which the process hits the set $\{(x, x^*) \in S\colon\ x^* = x'\}$. Let $\delta_1$ be the length of a block and $m(t)$ be the number of passage times for the marked job completed in the interval $(0, t]$. By Lemma 4.1

$$\frac{\sqrt{t}\left(\frac{1}{m(t)}\sum_{n=1}^{m(t)} f(P_n) - r(f)\right)}{\sqrt{E[\delta_1]}\sigma(f)/E[\alpha_1]} \Rightarrow N(0,1) \qquad (4.5)$$

as $t \to \infty$, where $\sigma^2(f) = \text{Var}\,[Y_1(f) - r(f)\alpha_1]$ and $r(f) = E[Y_1(f)]/E[\alpha_1]$. Using an argument similar to the one for the labelled jobs method, an appropriate measure of the statistical efficiency of the marked job method is the quantity

$$e(f) = \sqrt{E[\delta_1]}\sigma(f)/E[\alpha_1].$$

Theorem 4.11 below asserts that $e^0(f) \leq e(f)$ for all functions $f$, so that the confidence intervals produced by the labelled jobs method are shorter than those produced by the marked job method.

In terms of the single-state $z_0$ and the selected state $v_0 = (z_0', m_1', z_0, m_1) \in T$, define subsets $U^1, U^2, \ldots, U^N$ of $B^0$ as follows:

$$U^i = \big\{(z', n_1', \ldots, n_N', z, n_1, \ldots, n_N) \in B^0\colon\; z' = z_0', z = z_0,$$
$$n_1' = m_1', \ldots, n_i' = m_i', n_1 = m_1, \ldots, n_i = m_i\big\}$$

for $1 \leq i \leq N$. According to this definition, the times at which $\{V^0(t)\colon\; t \geq 0\}$ hits the set $U^i$ correspond to termination times of a passage time for job 1 (in position $m_1$ of the job stack associated with $z_0$) with jobs $2, \ldots, i$ in fixed positions $m_2, \ldots, m_i$ of the job stack associated with $z_0$. Recall that $v_0^0 = (z_0', m_1', m_2', \ldots, m_N', z_0, m_1, m_2, \ldots, m_N) \in B^0$ and $V^0(0) = v_0^0$. Observe that the random times at which $\{V^0(t)\colon\; t \geq 0\}$ hits the set $U^i$ are a subsequence of the random times at which the process hits the set $U^{i-1}$. Comparison of the marked job and labelled jobs methods rests on CLT's in continuous time for $U^i$-blocks defined by the successive random times at which the process $\{V^0(t)\colon\; t \geq 0\}$ leaves the set $U^i$.

Denote by $\delta_k^i$ the length of the $k$th $U^i$-block of the process $\{V^0(t)\colon t \geq 0\}$, $k \geq 1$. Let $\alpha_k^{ji}$ be the number of passage times for job $j$ in the $k$th $U^i$-block, and let $Y_k^{ji} = Y_k^{ji}(f)$ be the sum of the values of the function $f$ for these passage times for $1 \leq j \leq i$. Then set

$$e^{ji} = \sqrt{E[\delta_1^i]}\sigma^{ji}/E\left[\alpha_1^{ji}\right],$$

where $(\sigma^{ji})^2 = \text{ Var }[Y_k^{ji} - r(f)\alpha_k^{ji}]$ for all $k$. Denote the analogous quantities for jobs $1, 2, \ldots, j$ by $\alpha_k^{(j)i}$ , $Y_k^{(j)i}$, and

$$e^{(j)i} = \sqrt{E\left[\delta_1^{(j)i}\right]}\sigma^{(j)i}/E\left[\alpha_1^{(j)i}\right],$$

respectively, and set $(\sigma^{(j)i})^2 = \text{ Var }[Y_k^{(j)i} - r(f)\alpha_k^{(j)i}]$.

We require three lemmas. The first lemma asserts that the statistical efficiency of a simulation based on observation of passage times for job 1 (the marked job) in $U^1$-blocks is the same as that of a simulation based on observation of passage times for job $i$ in $U^i$-blocks $(1 \leq i \leq N)$.

**Lemma 4.6.** $e^{ii} = e^{11}$.

*Proof.* Recall that $N^i(t)$ is the position of job $i$ in the job stack at time $t$. Set $X^i(t) = (Z(t), N^i(t))$, and consider the process $\{X^i(t)\colon t \geq 0\}$. Let $L^i(t)$ be the last state occupied by the process $\{X^i(t)\colon t \geq 0\}$ before jumping to $X^i(t)$ and set $V^i(t) = (L^i(t), X^i(t))$. Also let $\{P_n^i\colon n \geq 0\}$ be the sequence of passage times for job $i$ enumerated in termination order. Observe that the successive random times at which the process $\{V^i(t)\colon t \geq 0\}$ hits state $v_0 = (z_0', m_1', z_0, m_1)$ form a sequence of regeneration points. Let $m^i(t)$ be the number of passage times for job $i$ completed in $(0, t]$. Based on these $v_0$-cycles Lemma 4.1 implies that, for some $e_i$,

$$\frac{\sqrt{t}\left(\frac{1}{m^i(t)}\sum_{n=1}^{m^i(t)} f(P_n^i) - r(f)\right)}{e^i} \Rightarrow N(0,1) \qquad (4.7)$$

as $t \to \infty$. Similarly, based on $U^i$-blocks of $\{V^i(t)\colon\ t \geq 0\}$, for some $e^{ii}$ we have

$$\frac{\sqrt{t}\left(\frac{1}{m^i(t)}\sum_{n=1}^{m^i(t)} f(P_n^i) - r(f)\right)}{e^{ii}} \Rightarrow N(0,1) \qquad (4.8)$$

as $t \to \infty$. Since the numerators and the limits in (4.7) and (4.8) are the same, the denominators must be the same; that is, $e^{ii} = e^i$.

To complete the proof, we show that $e^i = e^{11}$. Set $\zeta_1^i = \inf\{t > 0\colon\ V^i(t) = v_0\}$ and $W^i(t) = V^i(\zeta_1^i + t)$. Formal specification of each of the processes $\{V(t)\colon\ t \geq 0\}$ and $\{W^i(t)\colon\ t \geq 0\}$ is as a GSMP. These two processes have the same probability mass functions and clock-setting distribution functions since the jobs are stochastically identical and these building blocks are functions only of the routing matrix and of the parameters of the distributions of service time. It follows that the processes $\{V(t)\colon\ t \geq 0\}$ and $\{W^i(t)\colon\ t \geq 0\}$ have the same finite-dimensional distributions. Next enumerate the passage times for job $i$ from time $\zeta_1^i$ as $\{R_n^i\colon\ n \geq 0\}$, and let $m^i(t)$ be the number of passage times for job $i$ completed in $(\zeta_1^i, \zeta_1^i + t]$. Denote the sequence of passage times for job 1 by $\{P_n^1\colon\ n \geq 0\}$, and let $m^1(t)$ be the number of passage times for job 1 completed in $(0, t]$. Now set

$$C^1(t) = \frac{1}{m^1(t)} \sum_{n=1}^{m^1(t)} f(P_n^1) - r(f)$$

and consider the functional $\{C^1(t): t \geq 0\}$. Also set

$$C^i(t) = \frac{1}{m^i(t)} \sum_{n=1}^{m^i(t)} f(P_n^i) - r(f),$$

and observe that $\{C^i(t): t \geq 0\}$ is the same functional of the process $\{W^i(t): t \geq 0\}$. It follows that for all $t > 0$, $C^1(t)$ and $C^i(t)$ have the same distribution. Lemma 4.1 implies that

$$\frac{\sqrt{t}C^1(t)}{e^{11}} \Rightarrow N(0,1)$$

and

$$\frac{\sqrt{t}C^i(t)}{e^i} \Rightarrow N(0,1)$$

as $t \to \infty$, so that $e^i = e^{11}$. □

Lemma 4.9 asserts that simulations based on observation of passage times for jobs $1, 2, \ldots, j$ in $U^i$-blocks and in $U^{i+1}$-blocks are equally efficient.

**Lemma 4.9.** *Let $l = i + 1$. Then $e^{(j)i} = e^{(j)l}$ for $j = 1, 2, \ldots, i$.*

*Proof.* Set $v_0^{(k)} = (z_0', m_1', m_2', \ldots, m_k', z_0, m_1, m_2, \ldots, m_k)$ and $X^{(k)}(t) = (Z(t), N^1(t), \ldots, N^k(t))$ for $1 \leq k \leq N$. Let $L^{(k)}(t)$ be the last state occupied by the process $\{X^{(k)}(t): t \geq 0\}$ before jumping to $X^{(k)}(t)$. Also set $V^{(k)}(t) = (L^{(k)}(t), X^{(k)}(t))$. The successive random times at which $\{V^{(i)}(t): t \geq 0\}$ hits state $v_0^{(i)}$ form a sequence of regeneration points for the process, and the successive random times at which $\{V^{(l)}(t): t \geq 0\}$ hits state $v_0^{(l)}$ form a sequence of regeneration points for the process. Let $\{P_n^{(j)}: n \geq 0\}$ be the sequence of passage times for jobs $1, 2, \ldots, j$ enumerated in termination order, and let $m^{(j)}(t)$ be the number of passage times for jobs $1, 2, \ldots, j$ completed in $(0, t]$. Then

$$\frac{\sqrt{t}\left(\frac{1}{m^{(j)}(t)} \sum_{n=1}^{m^{(j)}(t)} f(P_n^{(j)}) - r(f)\right)}{e^{(j)i}} \Rightarrow N(0,1)$$

and

$$\frac{\sqrt{t}\left(\frac{1}{m^{(j)}(t)} \sum_{n=1}^{m^{(j)}(t)} f(P_n^{(j)}) - r(f)\right)}{e^{(j)l}} \Rightarrow N(0,1)$$

as $t \to \infty$. Since the numerators and the limits are the same, the denominators must be the same. □

Lemma 4.10 asserts that a simulation based on observation of passage times for jobs 1 and 2 in $U^2$-blocks is at least as efficient as a simulation based on passage times for job 1 in $U^1$-blocks. The idea is to develop a bivariate CLT for passage times of jobs 1 and 2 in $U^2$-blocks, apply the continuous mapping theorem to obtain a one-dimensional CLT, and make an identification of the resulting variance constant.

**Lemma 4.10.** $e^{(2)2} \leq e^{11}$.

*Proof.* Set $W_k^1 = Y_k^{12} - r(f)\alpha_k^{12}$ and $W_k^2 = Y_k^{22} - r(f)\alpha_k^{22}$, and let $\boldsymbol{W}_k$ be the column vector with components $W_k^1$ and $W_k^2$. Set $\Sigma = \{\sigma_{ij}\}$ and $\sigma_{ij} = E\left[W_1^i W_1^j\right]$ for $i, j = 1, 2$. The vectors $\{\boldsymbol{W}_k \colon k \geq 0\}$ are i.i.d. with mean 0 and it follows that

$$\frac{1}{\sqrt{n}} \sum_{k=1}^{n} \boldsymbol{W}_k \Rightarrow N(0, \Sigma),$$

as $n \to \infty$. Now set

$$\sigma^2 = \sum_i \sigma_{ii} + 2 \sum_{i<j} \sigma_{ij}.$$

An application of the continuous mapping theorem using the mapping $h(w_1, w_2) = w_1 + w_2$ yields

$$\frac{1}{\sigma\sqrt{n}} \sum_{k=1}^{n} \sum_{i=1}^{2} W_k^i \Rightarrow N(0,1)$$

as $n \to \infty$. Furthermore since $W_k^1 + W_k^2$ corresponds to jobs 1 and 2, it is clear from the form of this CLT that $\sigma^2 = (\sigma^{(2)2})^2$, $\sigma_{11} = (\sigma^{12})^2$, and $\sigma_{22} = (\sigma^{22})^2$.

Next observe that $e^{22} = e^{11}$ by Lemma 4.9 and $e^{12} = e^{11}$ by Lemma 4.9. Using the definitions of $e^{12}, e^{11}$, and $e^{22}$, these results imply

$$\sigma_{11} = \frac{(\sigma^{11})^2 E[\delta_1^1] E^2[\alpha_1^{12}]}{E[\delta_1^2] E^2[\alpha_1^{11}]}$$

and

$$\sigma_{22} = \frac{(\sigma^{11})^2 E[\delta_1^1] E^2[\alpha_1^{22}]}{E[\delta_1^2] E^2[\alpha_1^{11}]}.$$

From the Cauchy-Schwarz inequality,

$$2 \mid \sigma_{12} \mid \leq 2(\sigma_{11}\sigma_{22})^{1/2} = \frac{(\sigma^{11})^2 E[\delta_1^1] \left(2E[\alpha_1^{22}] E[\alpha_1^{12}]\right)}{E[\delta_1^2] E^2[\alpha_1^{11}]}.$$

Hence

$$\begin{aligned} (\sigma^{(2)2})^2 &= \sigma_{11} + \sigma_{22} + 2\sigma_{12} \\ &\leq \frac{(\sigma^{11})^2 E[\delta_1^1] \left(E[\alpha_1^{22}] + E[\alpha_1^{12}]\right)^2}{E[\delta_1^2] E^2[\alpha_1^{11}]}. \end{aligned}$$

Since

$$E\left[\alpha_1^{(2)2}\right] = E[\alpha_1^{22}] + E[\alpha_1^{12}],$$

it follows that $e^{(2)2} e^{11} \leq 1$. □

**Theorem 4.11.** *For all functions $f$, $e^0(f) \le e(f)$.*

*Proof.* It is sufficient to show that $e^{(N)N} \le e^{11}$. To see this, let $m^0(t)$ be the number of passage times, irrespective of job identity, completed in $(0, t]$. Observe, again by Lemma 4.1, that

$$\frac{\sqrt{t}\left(\frac{1}{m^0(t)}\sum_{n=1}^{m^0(t)} f(P_n^0) - r(f)\right)}{e^{(N)N}} \Rightarrow N(0,1)$$

as $t \to \infty$ and

$$\frac{\sqrt{t}\left(\frac{1}{m^0(t)}\sum_{n=1}^{m^0(t)} f(P_n^0) - r(f)\right)}{e^0(f)} \Rightarrow N(0,1)$$

as $t \to \infty$; therefore $e^{(N)N} = e^0(f)$. But $e(f) = e^{11}$; this assertion follows from the CLT's of (4.5) and (4.8) (with $i = 1$).

To complete the proof, we show that $e^{(N)N} \le e^{11}$. We first show that $e^{(3)3} \le e^{11}$. Set $W_k^1 = Y_k^{(2)3} - r(f)\alpha_k^{(2)3}$ and $W_k^2 = Y_k^{33} - r(f)\alpha_k^{33}$, and let $\boldsymbol{W}_k$ be the column vector with components $W_k^1$ and $W_k^2$. Set $\Sigma = \{\sigma_{ij}\}$ and $\sigma_{ij}] = E\left[W_1^i W_1^j\right]$. The vectors $\{\boldsymbol{W}_k\colon\ k \ge 0\}$ are i.i.d. with mean 0 and therefore

$$\frac{1}{\sqrt{n}}\sum_{k=1}^{n} W_k \Rightarrow N(0,\Sigma),$$

as $n \to \infty$. By the argument used in the proof of Lemma 4.10, $\sigma_{11} = (\sigma^{(2)3})^2$ and $\sigma_{22} = (\sigma^{33})^2$. By Lemma 4.9, $e^{33} = e^{11}$. Therefore,

$$\sigma_{22} = \frac{(\sigma^{11})^2 E[\delta_1^1] E^2[\alpha_1^{33}]}{E[\delta_1^3] E^2[\alpha_1^{11}]}.$$

Next observe that $e^{(2)2} \le e^{11}$ by Lemma 4.10 and $e^{(2)3} = e^{(2)2}$ by Lemma 4.9. Thus,

$$\sigma_{11} \le \frac{(\sigma^{11})^2 E[\delta_1^1] E^2\left[\alpha_1^{(2)3}\right]}{E[\delta_1^3] E^2\left[\alpha_1^{11}\right]}.$$

By the Cauchy-Schwarz inequality,

$$(\sigma^{(3)2})^2 \leq \sigma_{11} + \sigma_{22} + 2(\sigma_{11}\sigma_{22})^{1/2}.$$

Using the fact that $E\left[\alpha_1^{(3)3}\right] = E\left[\alpha_1^{33}\right] + E\left[\alpha_1^{(2)3}\right]$, it follows that $e^{(3)3}/e^{11} \leq 1$. The same argument using $e^{(3)3} \leq e^{11}$ shows that $e^{(4)4} \leq e^{11}$. Continuing in this way, it follows that $e^{(N)N} \leq e^{11}$. □

## 5.5. Abstract Passage Times

Passage times in a GSMP are random time intervals that correspond to delays in discrete-event stochastic systems. We first consider sequences of passage times that terminate in the order in which they start; for example, sequences such that at most one passage time can be underway at any time. Such sequences of passage times can be specified in terms of the underlying GSSMC $\{(S_n, C_n)\colon n \geq 0\}$ of the GSMP by means of four nonempty subsets ($A_1$, $A_2$, $B_1$, and $B_2$) of the state space $S$. The sets $A_1$ and $A_2$ jointly define the random times at which passage times start, and the sets $B_1$ and $B_2$ jointly define the random times at which passage times terminate. Formally, let $A_1$, $A_2$, $B_1$, and $B_2$ be nonempty subsets of $S$ such that

(i) for all $s' \in A_2$ there exists $s \in A_1$ such that $s \to s'$; and
(ii) for all $s' \in B_2$ there exists $s \in B_1$ such that $s \to s'$.

(We write $s \to s'$ if $p(s'; s, e)r(s, e) > 0$ for some event $e \in E(s)$.)

In terms of the sets $A_1$, $A_2$, $B_1$, and $B_2$, we define two sequences $\{\zeta_{\alpha(n)}\colon n \geq 0\}$ and $\{\zeta_{\beta(n)}\colon n \geq 0\}$ of random times; $\zeta_{\alpha(n)}$ is the start time of the $n$th passage time, and $\zeta_{\beta(n)}$ is the termination time of this $n$th passage time. Assuming that initially there are no passage times underway, set

$$\alpha(n) = \inf\{k > \alpha(n-1)\colon S_k \in A_2, S_{k-1} \in A_1\} \qquad (5.1)$$

and

$$\beta(n) = \inf\{k > \beta(n-1)\colon\ S_k \in B_2, S_{k-1} \in B_1\} \tag{5.2}$$

for $n \geq 0$. (Take $\alpha(-1) = \beta(-1) = 0$.) Then the $n$th passage time is

$$P_n = \zeta_{\beta(n)} - \zeta_{\alpha(n)}.$$

Thus, the $n$th passage time starts when the process $\{S_k\colon\ k \geq 0\}$ makes a state transition from the set $A_1$ to the set $A_2$ for the $n$th time. The $n$th passage time terminates when the process makes a state transition from the set $B_1$ to the set $B_2$ for the $n$th time. Observe that there can be more than one passage time underway simultaneously. (If, initially there are $n(S_0) > 0$ passage times underway, take $\beta(n)$ to be the index of the $(n(S_0)+n)$th state transition from the set $B_1$ to the set $B_2$).

We assume that there exists a set $\{n(s)\colon\ s \in S\}$ of non-negative integers such that for $s_0, s_1, \ldots, s_k \in S$ with $s_0 \to s_1$, $s_1 \to s_2, \ldots, s_{k-1} \to s_k$, and $k \geq 1$,

$$n(s_0) + \sum_{n=1}^{k} 1_A(s_{n-1}, s_n) - \sum_{n=1}^{k} 1_B(s_{n-1}, s_n) = n(s_k),$$

where

$$A \stackrel{\text{def}}{=} \{(s_1, s_2) \in A_1 \times A_2\colon\ s_1 \to s_2\}$$

and

$$B \stackrel{\text{def}}{=} \{(s_1, s_2) \in B_1 \times B_2\colon\ s_1 \to s_2\}.$$

Thus, for $s \in S$, $s_0, s_1, \ldots, s_k = s \in S$ with $s_0 \to s_1$, $s_1 \to s_2, \ldots, s_{k-1} \to s_k$, and $k \geq 1$,

$$n(s_0) + \sum_{n=1}^{k} 1_A(s_{n-1}, s_n) - \sum_{n=1}^{k} 1_B(s_{n-1}, s_n) = n(s),$$

and for $s \in S$, $s = s_0, s_1, \ldots, s_k \in S$ with $s_0 \to s_1$, $s_1 \to s_2, \ldots,$ $s_{k-1} \to s_k$, and $k \geq 1$,

$$n(s) + \sum_{n=1}^{k} 1_A(s_{n-1}, s_n) - \sum_{n=1}^{k} 1_B(s_{n-1}, s_n) = n(s_k).$$

These conditions ensure that a passage time starts every time that the process $\{S_k\colon\ k \geq 0\}$ makes a state transition from the set $A_1$ to the set $A_2$ and a passage time terminates every time that there is a state transition from the set $B_1$ to the set $B_2$. Moreover, the state of the GSMP determines the number of passage times underway simultaneously. We assume throughout that

$$P_\mu\{\,\alpha(n) < \infty\,\} = P_\mu\{\,\beta(n) < \infty\,\} = 1$$

for $n \geq 1$.

**Example 5.3.** (Transportation system) Consider a transportation system with $N$ terminals numbered $1, 2, \ldots, N$ in a circular configuration and a fleet of $K$ ($\geq 2$) vehicles that travel from terminal to terminal in a strictly defined order: $1, 2, \ldots, N, 1, 2, \ldots$. Passengers that wish to travel to the next terminal arrive at terminal $i$ ($i = 1, 2, \ldots, N$) according to a renewal process and are served in the order of their arrival by vehicles that arrive at the terminal. The interarrival times at terminal $i$ are i.i.d. as a positive random variable $D_i$ with absolutely continuous distribution. Boarding occurs instantaneously as free units of vehicle capacity become available. Upon arrival at a terminal, a vehicle remains idle until fully loaded, at which time it transports its passengers to the next terminal leaving behind all passengers unable to board. Travel times are independent of the passenger arrival processes. The time for a vehicle to travel from terminal $i$ to the next terminal is distributed as a positive random variable $T_i$ with absolutely continuous distribution. There is no dependence among the travel times for different vehicles so vehicles are free to pass one another. When a vehicle arrives at a terminal, all passengers disembark instantaneously and exit from the system. Passengers waiting at the terminal board the vehicle, subject to vehicle capacity and a FCFS queueing discipline. If there is more than one vehicle that is idle at a terminal, passengers that arrive at the terminal for transportation to the next terminal board vehicles in the order that the vehicles arrive. There is no limit on the number of passengers that may wait at a terminal, and there are no passenger defections. Assume that vehicle capacity equals 1.

Set

$$X(t) = (Z_1(t), \dots, Z_N(t), M_1(t), \dots, M_K(t), N_1(t), \dots, N_K(t)),$$

where $Z_i(t)$ is the number of passengers waiting at terminal $i$ for transportation to the next terminal at time $t$ ($i = 1, 2, \dots, N$),

$$M_k(t) = \begin{cases} 1 & \text{if vehicle } k \text{ is not idle at time } t \\ -j & \text{if vehicle } k \text{ is idle at some terminal at time } t \\ & \text{with } j-1 \text{ vehicles ahead at the terminal,} \end{cases}$$

$N_k(t) = 1$ if vehicle $k$ is idle at terminal $N$ or vehicle $k$ is travelling from terminal $N$ to terminal 1 at time $t$, and (for $j = 2, 3, \dots, N$) $N_k(t) = j$ if vehicle $k$ is idle at terminal $j-1$ or vehicle $k$ is travelling from terminal $j-1$ to terminal $j$ at time $t$ ($k = 1, 2, \dots, K$).

Take event $e_{N+k}$ = "arrival of vehicle $k$ at terminal" ($k = 1, 2, \dots, K$) and $e_i$ = "arrival of passenger at terminal $i$ for transportation to the next terminal" ($i = 1, 2, \dots, N$). Formal specification of the process $\{X(t)\colon\ t \geq 0\}$ is as a GSMP with countably infinite state space

$$S = \{(z_1, \ldots, z_N, m_1, \ldots, m_K, n_1, \ldots, n_K) \in \{0, 1, \ldots\}^N \times \{-K, \ldots, -2, -1, 1\}^K \times \{1, 2, \ldots, N\}^K : m_k = m_l \text{ only if } m_k = m_l = 1, n_k \neq n_l, \text{ or } k = l; m_k < 0 \text{ only if } z_k = 0 \text{ for } k = 1, 2, \ldots, K\}$$

and event set $E = \{e_1, e_2, \ldots, e_{N+K}\}$. For $s = (z_1, \ldots, z_N, m_1, \ldots, m_K, n_1, \ldots, n_K) \in S$, event $e_{N+k}$ = "arrival of vehicle $k$ at terminal" $\in E(s)$ if and only if $m_k = 1$ and $e_i$ = "arrival of passenger at terminal $i$ for transportation to the next terminal" $\in E(s)$ for all $s \in S$. All speeds $r(s, e)$ for active events are equal to 1.

If $e = e_i$ = "arrival of passenger at terminal $i$ for transportation to the next terminal," then the state transition probability $p(s'; s, e) = 1$ when

$$s = (z_1, \ldots, z_N, m_1, \ldots, n_K) \text{ with } m_k \neq 1 \text{ or } n_k \neq i \text{ for all } k \text{ and } s' = (z_1, \ldots, z_{i-1}, z_i + 1, z_{i+1}, \ldots, z_N, m_1, \ldots, n_K)$$

and when

$$s = (z_1, \ldots, z_N, m_1, \ldots, n_K) \text{ with } m_k = -1, n_k = i \text{ and } s' = (z_1, \ldots, z_N, m_1, \ldots, m_{k-1}, 1, m_{k+1}, \ldots, m_K, n_1, \ldots, n_{k-1}, i + 1, n_{k+1}, \ldots, n_K).$$

If $e = e_{N+k}$ = "arrival of vehicle $k$ at terminal," then $p(s'; s, e) = 1$ when

$$s = (z_1, \ldots, z_N, m_1, \ldots, m_K, n_1, \ldots, n_K) \text{ with } n_k = i, z_i > 0, n_l \neq i \text{ or } m_l = 1 \text{ for } l \neq k \text{ and } s' = (z_1, \ldots, z_{i-1}, z_i - 1, z_{i+1}, \ldots, z_N, m_1, \ldots, m_{i-1}, 1, m_{i+1}, \ldots, m_K, n_1, \ldots, n_{i-1}, i + 1, n_{i+1}, \ldots, n_K),$$

when

$$s = (z_1, \ldots, z_N, m_1, \ldots, m_K, n_1, \ldots, n_K) \text{ with } n_k = i, z_i = 0,$$
$$n_l = i \text{ for some } l$$
$$\text{and } s' = (z_1, \ldots, z_{i-1}, z_i, z_{i+1}, \ldots, z_N, m_1, \ldots, m_{i-1},$$
$$m - 1, m_{i+1}, \ldots, m_K, n_1, \ldots, n_{i-1}, i, n_{i+1}, \ldots, n_K)$$
$$\text{with } m = \min_{\{l:\ n_l = i\}} m_l,$$

and when

$$s = (z_1, \ldots, z_N, m_1, \ldots, m_K, n_1, \ldots, n_K) \text{ with } n_k = i, z_i = 0,$$
$$n_l \neq i \text{ or } m_l = 1 \text{ for } l \neq k$$
$$\text{and } s' = (z_1, \ldots, z_{i-1}, z_i, z_{i+1}, \ldots, z_N, m_1, \ldots, m_{i-1}, -1,$$
$$m_{i+1}, \ldots, m_K, n_1, \ldots, n_{i-1}, i, n_{i+1}, \ldots, n_K).$$

All other state transition probabilities $p(s'; s, e)$ are equal to 0. The clock-setting distribution functions are as follows. If the new event $e' = e_{N+k}$ and $s' = (z_1, \ldots, z_N, m_1, \ldots, m_K, n_1, \ldots, n_K)$, then $F(x; s', e', s, e^*) = P\{T_{n_k - 1} \leq x\}$ if $n_k > 1$ and equals $P\{T_N \leq x\}$ if $n_k = 1$. If $e' = e_i$, then $F(x; s', e', s, e^*) = P\{D_i \leq x\}$.

The *delay times* experienced by passengers at terminal $i$ are the random time intervals from arrival of a passenger at terminal $i$ until the passenger boards a vehicle. The delay time experienced by a passenger at terminal $i$ starts when (event $e_i$ = "arrival of passenger at terminal $i$ for transportation to the next terminal" occurs and) the process $\{X(t)\colon t \geq 0\}$ makes a state transition from some state $s = (z_1, \ldots, z_{i-1}, z_i, z_{i+1}, \ldots, n_K)$ to state $s' = (z_1, \ldots, z_{i-1}, z_i + 1, z_{i+1}, \ldots, n_K)$ or from some state $s = (z_1, \ldots, z_{i-1}, 0, z_{i+1}, \ldots, z_N, m_1, \ldots, m_{k-1}, -1, m_{k+1}, \ldots, n_1, \ldots, n_{k-1}, i, n_{k+1}, \ldots, n_K)$ to state $s' = (z_1, z_2, \ldots, m_{k-1}, 1, m_{k+1}, n_1, \ldots, n_{k-1}, i + 1, n_{k+1}, \ldots, n_K)$. The delay time ends when for some $k$ (event $e_{N+k}$ = "arrival of vehicle $k$ at terminal" occurs and) the process makes a state transition from some state $s = (z_1, \ldots, z_N, m_1, \ldots, m_{k-1}, 1, m_{k+1}, \ldots, n_{k-1}, i, n_{k+1}, \ldots, n_K)$ with $z_i > 0$ to state $s' = (z_1, \ldots, z_{i-1}, z_i - 1, z_{i+1}, \ldots, m_{k-1}, 1, m_{k+1}, \ldots, n_{k-1}, i + 1, n_{k+1}, \ldots, n_K)$.

The sets that define the start times of individual delay times experienced by passengers at terminal $i$ are

$$A_1 = \big\{(z_1, z_2, \ldots, z_N, m_1, \ldots, m_K, n_1, \ldots, n_K) \in S\colon \text{ either } m_k = 1 \text{ or } n_k \neq i + 1 \text{ for all } k \text{ or } z_i = 0, m_k = -1, n_k = i \text{ for some } k\big\}$$

and

$$A_2 = \big\{(z_1, z_2, \ldots, z_N, m_1, \ldots, m_K, n_1, \ldots, n_K) \in S\colon \text{ either } z_i > 0 \text{ or } m_k = 1, n_k = i + 1 \text{ for some } k\big\}.$$

The sets that define the termination times of individual delay times experienced by passengers at terminal $i$ are

$$B_1 = \big\{(z_1, z_2, \ldots, z_N, m_1, \ldots, m_K, n_1, \ldots, n_K) \in S\colon z_i > 0, m_k = -1, n_k = i \text{ for some } k\big\}$$

and

$$B_2 = \big\{(z_1, z_2, \ldots, z_N, m_1, \ldots, m_K, n_1, \ldots, n_K) \in S\colon m_k = 1, n_k = i + 1 \text{ for some } k\big\}.$$

Assume that initially there are no delay times underway. Then the $n$th delay time experienced by passengers at terminal $i$ is $\zeta_{\beta(n)} - \zeta_{\alpha(n)}$, where

$$\alpha(n) = \inf\{k > \alpha(n-1)\colon\ S_k \in A_2, S_{k-1} \in A_1\}$$

and

$$\beta(n) = \inf\big\{k \geq \beta(n-1)\colon\ S_k \in B_2, S_{k-1} \in B_1\big\}.$$

We now consider sequences of passage times that need not terminate in the order in which they start. Such sequences of passage times can be specified in terms of the underlying GSSMC $\{(S_n, C_n)\colon n \geq 0\}$ of the GSMP by means of two nonempty subsets $(A_1, A_2)$ of the state space $S$, a partition $\boldsymbol{A}$ of the set

$$A \stackrel{\text{def}}{=} \{(s, s') \in A_1 \times A_2\colon s \to s'\},$$

and two collections $(\{B_1(A^*)\colon A^* \in \boldsymbol{A}\}, \{B_2(A^*)\colon A^* \in \boldsymbol{A}\})$ of nonempty subsets of $S$ such that for $A^* \in \boldsymbol{A}$ the sets

$$B(A^*) \stackrel{\text{def}}{=} \{(s, s') \in B_1(A^*) \times B_2(A^*)\colon (s, s') \in A^*\}$$

are pairwise disjoint. The sets $A_1$ and $A_2$ jointly define the random times at which passage times start. The sets $\{B_1(A^*)\colon A^* \in \boldsymbol{A}\}$ and $\{B_2(A^*)\colon A^* \in \boldsymbol{A}\}$ jointly define the random times at which passage times terminate and provide the link between the start times and the termination times for individual passage time. A passage time that starts when the process $\{S_k\colon k \geq 0\}$ makes a state transition from some $s$ to some $s'$ with $(s, s') \in A^*$ is said to be of *type* $A^*$. We assume that

(i) for all $s' \in A_2$ there exists $s \in A_1$ such that $s \to s'$; and
(ii) for all $A^* \in \boldsymbol{A}$ and $s' \in B_2(A^*)$, there exists $s \in B_1(A^*)$ such that $s \to s'$.

In terms of the sets $A_1$, $A_2$, $\{B_1(A^*)\colon A^* \in \boldsymbol{A}\}$, and $\{B_2(A^*)\colon A^* \in \boldsymbol{A}\}$, we define two sequences $\{\zeta_{\alpha(n)}\colon n \geq 0\}$ and $\{\zeta_{\beta(n)}\colon n \geq 0\}$ of random times; $\zeta_{\alpha(n)}$ is the start time of the $n$th passage time, and $\zeta_{\beta(n)}$ is the termination time of this $n$th passage time. Set

$$\alpha(n) = \inf\{k > \alpha(n-1)\colon S_k \in A_2, S_{k-1} \in A_1\} \tag{5.4}$$

and

$$\beta(n) = \inf\{k > \alpha(n)\colon S_k \in B_2(A^*), S_{k-1} \in B_1(A^*)\} \tag{5.5}$$

for $n \geq 0$, where $S_{\alpha(n)-1} = s$, $S_{\alpha(n)} = s'$, and $(s, s') \in A^*$. (Take $\alpha(-1) = 0$.) Then the $n$th passage time is

$$P_n = \zeta_{\beta(n)} - \zeta_{\alpha(n)}.$$

Thus, the $n$th passage time starts when the process $\{S_k\colon k \geq 0\}$ makes a state transition from the set $A_1$ to the set $A_2$ for the $n$th time. The $n$th passage time terminates when the process next makes a state transition from the set $B_1(A^*)$ to the set $B_2(A^*)$, where $S_{\alpha(n)-1} = s$, $S_{\alpha(n)} = s'$, and $(s, s') \in A^*$. Observe that there can be more than one passage time underway simultaneously. Also observe that at most one passage time may start at a state transition and at most one passage time may terminate at a state transition.

We assume throughout that

$$P_\mu\{\,\alpha(n) < \infty\,\} = P_\mu\{\,\beta(n) < \infty\,\} = 1$$

for $n \geq 1$. For $A^* \in \boldsymbol{A}$, $(s, s'), (\bar{s}, \bar{s}') \in A^*$, $s_1, s_2, \ldots, s_{k+2} \in S$ with $s_1 \to s_2$, $s_2 \to s_3, \ldots, s_{k+1} \to s_{k+2}$, and $k \geq 1$, we require that the sets $A_1$, $A_2$, $B_1(A^*)$, and $B_2(A^*)$ satisfy:

(i) if $s_1 = s$, $s_2 = s'$, $s_{k+1} = \bar{s}$, and $s_{k+2} = \bar{s}'$, then $(s_{m+1}, s_{m+2}) \in B(A^*)$ for some $0 < m \leq k$; and

(ii) if $(s_1, s_2) \in B(A^*)$ and $(s_{k+1}, s_{k+2}) \in B(A^*)$, then $(s_{m+1}, s_{m+2}) \in A^*$ for some $0 \leq m < k$.

These alternation conditions ensure that a passage time starts every time that the process $\{S_k\colon k \geq 0\}$ makes a state transition from the set $A_1$ to the set $A_2$ and, for $A^* \in \boldsymbol{A}$, a passage time terminates every time that there is a state transition from the set $B_1(A^*)$ to the set $B_2(A^*)$. Moreover, the state of the GSMP determines the number of passage times underway simultaneously and the type of each passage time underway. We denote by $n(s, A^*)$ the number of passage times of type $A^*$ underway when the state is $s$.

**Example 5.6.** (Patrolling repairman) Assume that the time to repair and restart machine $j$ is distributed as a positive random variable $R_j$ and the walking time from machine $j$ to the next machine is a positive constant $W_j$. Also assume that the lifetimes for machine $j$ are i.i.d. as a positive random variable $L_j$ with absolutely continuous distribution. Set

$$X(t) = (Z_1(t), \ldots, Z_N(t), M(t), N(t)),$$

where

$$Z_j(t) = \begin{cases} 1 & \text{if machine } j \text{ is waiting for repair at time } t \\ 0 & \text{otherwise,} \end{cases}$$

$$M(t) = \begin{cases} j & \text{if machine } j \text{ is under repair at time } t \\ 0 & \text{if no machine is under repair at time } t, \end{cases}$$

$N(t) = 1$ if machine $N$ is under repair or the repairman is walking to machine 1 at time $t$, and $N(t) = j$ if machine $j-1$ is under repair or the repairman is walking to machine $j$ at time $t$ $(j = 2, 3, \ldots, N)$. Take event $e_{N+2}$ = "arrival of repairman at machine," $e_{N+1}$ = "end of repair," and $e_j$ = "stoppage of machine $j$" $(j = 1, 2, \ldots, N)$. Recall that formal specification of the process $\{X(t)\colon t \geq 0\}$ is as a GSMP with a finite state space $S$ and event set $E = \{e_1, e_2, \ldots, e_{N+2}\}$.

The *waiting times for repair* experienced by machine 1 are the random time intervals from stoppage of machine 1 until the arrival of the repairman at machine 1. The sets that define the start times of individual waiting times for repair are

$$A_1 = \{(z_1, \ldots, z_N, m, n) \in S\colon\ z_1 = 0, m \neq 1\}$$

and

$$A_2 = \{(z_1, \ldots, z_N, m, n) \in S\colon\ z_1 = 1, m \neq 1\}.$$

Take $\boldsymbol{A} = \{A\}$, where $A = \{(s, s') \in A_1 \times A_2\colon\ s \to s'\}$. The sets $B_1(A)$ and $B_2(A)$ that define the termination times of individual waiting times for repair and the link between the start times and the termination times are

$$B_1(A) = \{(z_1, \ldots, z_N, m, n) \in S\colon\ z_1 = 1, m = 0\}$$

and

$$B_2(A) = \{(z_1, \ldots, z_N, m, n) \in S\colon\ z_1 = 0, m = 1\}\,.$$

Then the $n$th waiting time for repair experienced by machine 1 is $\zeta_{\beta(n)} - \zeta_{\alpha(n)}$, where

$$\alpha(n) = \inf\{k > \alpha(n-1)\colon\ S_k \in A_2, S_{k-1} \in A_1\}$$

and

$$\beta(n) = \inf\bigl\{k \geq \alpha(n)\colon\ S_k \in B_2(A), S_{k-1} \in B_1(A)\bigr\}$$

for $n \geq 0$; $S_{\alpha(n)-1} = s$, $S_{\alpha(n)} = s'$, and $(s, s') \in A$.

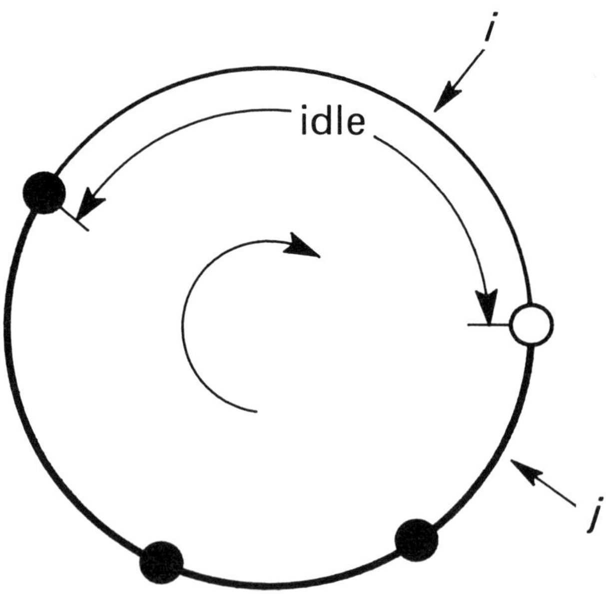


**Figure 5.3.** Token ring.

**Example 5.7.** (Token ring) Consider a unidirectional ring network with a fixed number of ports, labelled $1, 2, \ldots, N$ in the direction of signal propagation; see Figure 5.3. Message packets arrive at each port for transmission according to a random process and queue externally. A single token (that is, a distinguished bit pattern denoted by $\circ$) circulates around the ring from one port to the next. The time for the token to propagate from port $j$ to the next port is a positive constant $R_j$. When a port observes the token and there is a packet queued for transmission, the port converts the token ("holds the token") to a connector (another distinguished bit pattern, denoted by $\bullet$) and transmits a packet; the token continues to propagate if there is no packet queued for transmission. By destroying the connector prefix the port removes the transmitted packet when it returns around the ring. A port transmits the token pattern ("releases the token") after it has finished transmitting a packet and has observed the return around the ring of the lead bit of its packet; see Figure 5.4. A port that finishes transmission before it observes the return of the lead bit transmits the token pattern even if there is another packet waiting for transmission at the port. This rule ensures that all ports have an opportunity to transmit.

Assume that the time for port $j$ to transmit a packet is distributed as a positive random variable $L_j$ with finite mean. Also assume that packets arrive at individual ports randomly in the sense that the time from end of transmission by port $j$ until the arrival of the next packet for transmission by port $j$ is distributed as a positive random variable $D_j$ with finite mean and absolutely continuous distribution. Observe that there is at most one packet queued for transmission at any time at any particular port. Figure 5.5 shows a time-line diagram for the token ring with $N = 4$ ports.

Set

$$X(t) = (Z_1(t), \ldots, Z_N(t), M(t), K(t), N(t)),$$

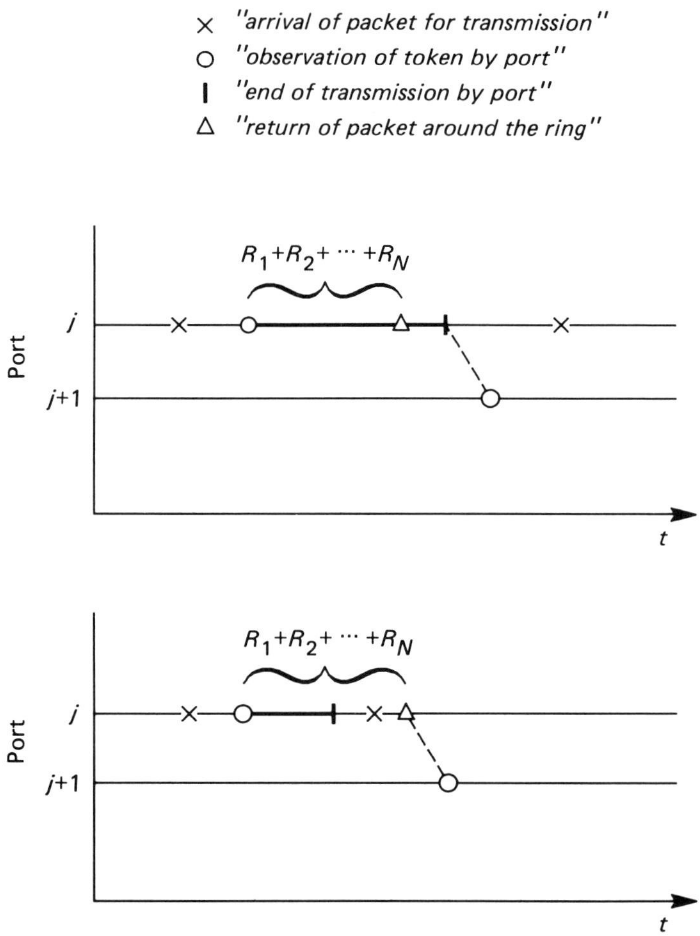


**Figure 5.4.** Release of token in token ring.

where

$$Z_j(t) = \begin{cases} 1 & \text{if there is a packet queued for transmission at} \\ & \text{port } j \text{ at time } t \\ 0 & \text{otherwise,} \end{cases}$$

$$M(t) = \begin{cases} j & \text{if port } j \text{ is transmitting a packet at time } t \\ 0 & \text{if no port is transmitting a packet at time } t, \end{cases}$$

$$K(t) = \begin{cases} 1 & \text{if some port is transmitting a packet that} \\ & \text{has not returned around the ring at time } t \text{ or} \\ & \text{no port is transmitting a packet but} \\ & \text{some port is holding the token at time } t \\ 0 & \text{otherwise,} \end{cases}$$

$N(t) = 1$ if port $N$ is transmitting a packet or the token is propagating to port 1 at time $t$, and $N(t) = j$ if port $j-1$ is transmitting a packet or the token is propagating to port $j$ at time $t$ ($j = 2, 3, \ldots, N$). Observe that $(M(t), K(t))$ equals $(j, 1)$ if port $j$ is transmitting a packet at time $t$ and the packet has not returned around the ring, equals $(j, 0)$ if port $j$ is transmitting a packet at time $t$ and the packet has returned around the ring, equals (0,1) if no port is transmitting a packet at time $t$ and the token is not propagating, and equals (0,0) if no port is transmitting a packet at time $t$ and the token is propagating.

Take event $e_{N+3}$ = "return of packet around the ring," $e_{N+2}$ = "observation of token by port," $e_{N+1}$ = "end of transmission by port," and $e_j$ = "arrival of packet for transmission by port $j$" ($j = 1, 2, \ldots, N$). Formal specification of the process $\{X(t)\colon\ t \geq 0\}$ is as a GSMP with finite state space

$$\begin{aligned} S = &\{(z_1, \ldots, z_N, m, k, n) \in \{0,1\}^N \times \{0, 1, \ldots, N\} \times \{0,1\} \\ &\times \{1, 2, \ldots, N\}\colon\ \ n = m + 1 \text{ when } 0 < m < N;\ n = 1 \\ &\text{when } m = N;\ m = j \text{ only if } z_j = 0 \text{ for } j = 1, 2, \ldots, N\} \end{aligned}$$

and event set $E = \{e_1, e_2, \ldots, e_{N+3}\}$. For $s = (z_1, \ldots, z_N, m, \ldots, k, n) \in S$, event $e_{N+3}$ = "return of packet around the ring" $\in E(s)$ if and only if $m > 0$ and $k = 1$, event $e_{N+2}$ = "observation of token by port" $\in E(s)$ if and only if $m = 0$ and $k = 0$, and $e_{N+1}$ = "end of transmission by port" $\in E(s)$ if and only if $m > 0$. Event $e_j$ = "arrival of packet for transmission

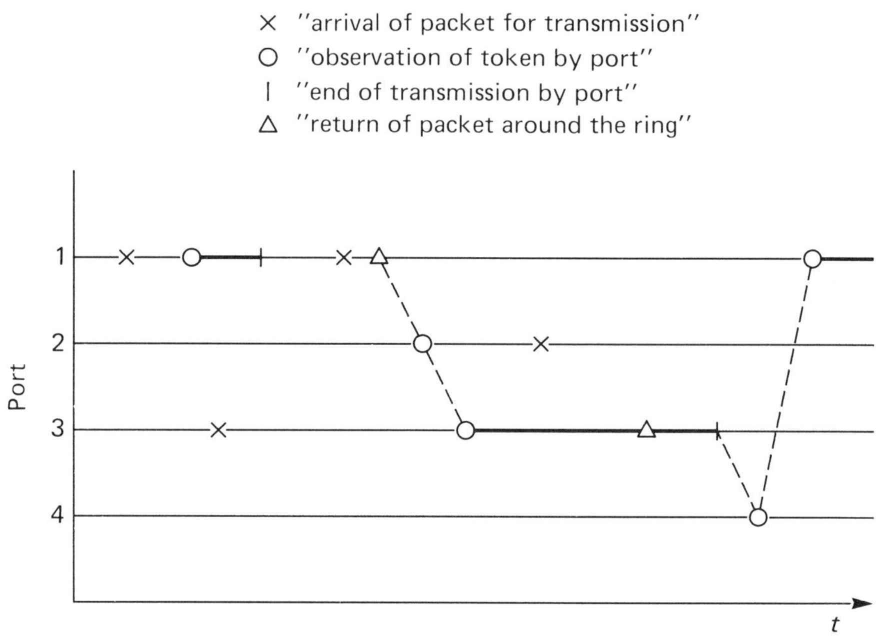


**Figure 5.5.** Time-line diagram for token ring.

by port $j$" $\in E(s)$ if and only if $z_j = 0$ and $m \neq j$. All speeds for active events are equal to 1.

If $e = e_{N+3} =$ "return of packet around the ring," then the state transition probability $p(s'; s, e) = 1$ when

$$s = (z_1, \ldots, z_N, m, 1, n) \text{ and } s' = (z_1, \ldots, z_N, m, 0, n).$$

If $e = e_{N+2} =$ "observation of token by port," then $p(s'; s, e) = 1$ when

$$s = (z_1, \ldots, z_{n-1}, 1, z_{n+1}, \ldots, z_N, 0, 0, n) \text{ with } n < N$$
$$\text{and } s' = (z_1, \ldots, z_{n-1}, 0, z_{n+1}, \ldots, z_N, n, 1, n+1),$$

when

$$s = (z_1, \ldots, z_{N-1}, 1, 0, 0, N) \quad \text{and } s' = (z_1, \ldots, z_{N-1}, 0, N, 1, 1),$$

when

$$s = (z_1, \ldots, z_{n-1}, 0, z_{n+1}, \ldots, z_N, 0, 0, n) \text{ with } n < N \quad \text{and } s' = (z_1, \ldots, z_{n-1}, 0, z_{n+1}, \ldots, z_N, 0, 0, n+1),$$

and when

$$s = (z_1, \ldots, z_{N-1}, 0, 0, 0, N) \quad \text{and } s' = (z_1, \ldots, z_{N-1}, 0, 0, 0, 1).$$

If $e = e_{N+1}$ = "end of transmission by port," then $p(s'; s, e) = 1$ when

$$s = (z_1, \ldots, z_N, m, k, m+1) \text{ with } 0 < m < N \quad \text{and } s' = (z_1, \ldots, z_N, 0, k, m+1)$$

and when

$$s = (z_1, \ldots, z_N, N, k, 1) \text{ and } s' = (z_1, \ldots, z_N, 0, k, 1).$$

If $e = e_j$ = "arrival of packet for transmission by port $j$," then $p(s'; s, e) = 1$ when

$$s = (z_1, \ldots, z_{j-1}, 0, z_{j+1}, \ldots, z_N, m, k, n) \quad \text{and } s' = (z_1, \ldots, z_{j-1}, 1, z_{j+1}, \ldots, z_N, m, k, n).$$

All other state transition probabilities $p(s'; s, e)$ are equal to 0. The clock-setting distribution functions are as follows. If the new event $e' = e_{N+3}$ = "return of packet around the ring," then $F(x; s', e', s, e^*) = 1_{[R_1 + \cdots + R_N, \infty)}(x)$. If $e' = e_{N+2}$ = "observation of token by port" and $s' = (z_1, \ldots, z_N, m, k, n)$, then $F(x; s', e', s, e^*) = 1_{[R_{n-1}, \infty)}(x)$ if $n > 1$ and equals

$1_{[R_N,\infty)}(x)$ if $n = 1$. If $e' = e_{N+1}$ = "end of transmission by port" and $s' = (z_1, \ldots, z_N, m, k, n)$, then $F(x; s', e', s, e^*) = P\{L_m \leq x\}$. If $e' = e_j$ = "arrival of packet for transmission by port $j$," then $F(x; s', e', s, e^*) = P\{D_j \leq x\}$.

The *access times* for port $j$ are the random time intervals from the arrival of a packet for transmission by port $j$ until the start of transmission by port $j$. The sets that define the start times of individual access times for port $j$ are

$$A_1 = \{(z_1, \ldots, z_N, m, k, n) \in S\colon\ z_j = 0, m \neq j, k = 0\}$$

and

$$A_2 = \{(z_1, \ldots, z_N, m, k, n) \in S\colon\ z_j = 1, m \neq j, k = 0\}.$$

Take $\boldsymbol{A} = \{A\}$, where $A = \{(s, s') \in A_1 \times A_2\colon\ s \to s'\}$. The sets $B_1(A)$ and $B_2(A)$ that define the termination times of individual access times for port $j$ and the link between the start times and the termination times are

$$B_1(A) = \big\{(z_1, \ldots, z_N, m, k, n) \in S\colon\ z_j = 1, m = 0, k = 0, \\ n = j\big\}$$

and

$$B_2(A) = \{(z_1, \ldots, z_N, m, k, n) \in S\colon\ z_j = 0, m = j, k = 1\}.$$

Then the $n$th access time for port $j$ is $\zeta_{\beta(n)} - \zeta_{\alpha(n)}$, where

$$\alpha(n) = \inf\{k > \alpha(n-1)\colon\ S_k \in A_2, S_{k-1} \in A_1\}$$

and

$$\beta(n) = \inf\big\{k \geq \alpha(n)\colon\ S_k \in B_2(A), S_{k-1} \in B_1(A)\big\}$$

for $n \geq 0$; $S_{\alpha(n)-1} = s$, $S_{\alpha(n)} = s'$, and $(s, s') \in A$.

**Example 5.8.** (Slotted ring). Consider a ring network with a fixed number $K$ of equal size slots, and a fixed number of equally spaced ports, labelled $1, 2, \ldots, N$ in the direction of signal propagation; see Figure 5.6. At each port constant (slot size) length message packets arrive according to a random process and queue externally. The propagation delay from one port to the next is a positive constant $R$. Assume that the number $N$ of ports is a multiple of $K$ and, so that there is no loss of utilization due to "unused bits," that the time to transmit a message packet is equal to $NR/K$. The lead "full/empty" ($\boldsymbol{F}/\boldsymbol{E}$) bit maintains the status of each slot. Subject to the restriction that no port can hold more than one slot simultaneously, a port that has a packet queued for transmission and observes the status bit of an empty slot sets the bit to 1 (full) and starts transmission. Transmission ends when the slot contains the entire packet. When the status bit of the filled slot propagates back to the sending port, it resets the bit to 0 (empty) and releases the slot. The port releases the slot even if it has another packet queued for transmission. This rule ensures that all ports have an opportunity to transmit. A port holds a slot from the time that it begins filling the slot until it releases the slot.

Assume that the time from end of transmission by port $j$ until the arrival of the next packet for transmission by port $j$ is distributed as a positive random variable $D_j$ with finite mean and absolutely continuous distribution. Observe that there is at most one packet queued for transmission at any time at each of the ports.

Set

$$X(t) = (Z_1(t), \ldots, Z_N(t), M_1(t), \ldots, M_K(t), N_1(t), \ldots, N_K(t)),$$

where

$$Z_j(t) = \begin{cases} 1 & \text{if there is a packet queued for transmission at} \\ & \text{port } j \text{ at time } t \\ 0 & \text{otherwise,} \end{cases}$$

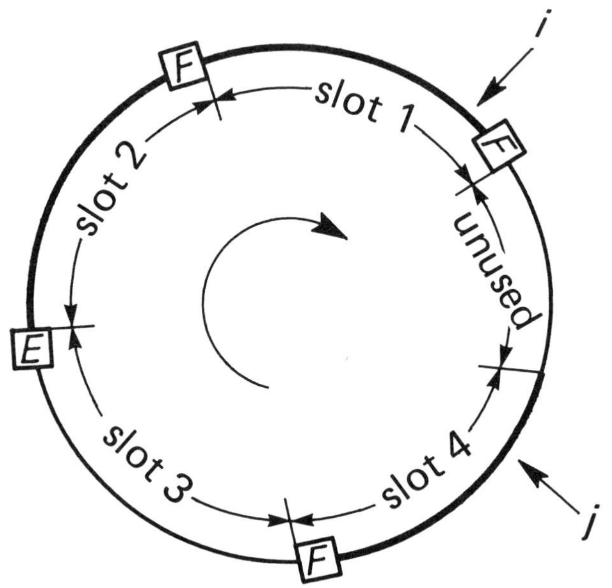


**Figure 5.6.** Slotted ring.

$$M_i(t) = \begin{cases} j & \text{if port } j \text{ holds slot } i \text{ at time } t \\ 0 & \text{otherwise,} \end{cases}$$

and $N_i(t) = j$ if the status bit of slot $i$ is propagating to port $j$ at time $t$ ($j = 1, 2, \ldots, N$). For any $i$ ($1 \leq i \leq K$) the vector $(Z_1(t), \ldots, Z_N(t), M_1(t), \ldots, M_K(t), N_i(t))$ contains the same information about the system as the vector $X(t)$. Incorporation of all the components $N_1(t), \ldots, N_K(t)$ into the state vector facilitates generation of sample paths of the process.

Take event $e_{N+1}$ = "observation of status bits by ports" and $e_j$ = "arrival of packet for transmission by port $j$" ($j = 1, 2, \ldots, N$). Formal specification of the process $\{X(t)\colon\ t \geq 0\}$ is as a GSMP with a finite state space $S$ and event set $E =$

$\{e_1, e_2, \ldots, e_{N+1}\}$. For $s = (z_1, \ldots, z_N, m_1, \ldots, m_K, n_1, \ldots, n_K) \in S$, event $e_{N+1}$ ="observation of status bits by ports"$\in E(s)$ for all $s \in S$. Event $e_j$ ="arrival of message for transmission by port $j$"$\in E(s)$ if and only if $z_j = 0$ and for each $i$ either (i) $m_i \neq j$ or (ii) $m_i = j$ and $n_i - 1 = j - 1 + l \pmod N$ for some integer $l$ such that $N/K < l \leq N$. Observe that the ends of transmission coincide with the occurrence of particular "observation of status bits by ports" events. Suppose, for example, that there are $N = 4$ ports and $K = 2$ slots and $s = (0, 0, 0, 0, 1, 0, 3, 1)$; that is, port 1 is transmitting a packet in slot 1, slot 2 is empty, the status bit of port 1 is propagating to port 3, and the status bit of slot 2 is propagating to port 4. Then the occurrence of the event "observation of status bits by ports" in state $s$ corresponds to an end of transmission by port 4.

The *access times* for port 1 are the random time intervals from arrival of a packet for transmission by port 1 until the start of transmission. The sets that define the start times of individual access times for port 1 are

$$A_1 = \{(z_1, z_2, \ldots, z_N, m_1, \ldots, m_K, n_1, \ldots, n_K) \in S\colon\ z_1 = 0\}$$

and

$$A_2 = \{(z_1, z_2, \ldots, z_N, m_1, \ldots, m_K, n_1, \ldots, n_K) \in S\colon\ z_1 = 1\}.$$

Take $\boldsymbol{A} = \{A\}$, where $A = \{(s, s') \in A_1 \times A_2\colon\ s \to s'\}$. The sets $B_1(A)$ and $B_2(A)$ that define the termination times of individual access times for port 1 and the link between the start times and the termination times are

$$B_1(A) = \big\{(z_1, z_2, \ldots, z_N, m_1, \ldots, m_K, n_1, \ldots, n_K) \in S\colon \\ z_1 = 1,\ m_i = 0,\ \text{and } n_i = 1 \text{ for some } i\big\}$$

and

$$B_2(A) = \big\{(z_1, z_2, \ldots, z_N, m_1, \ldots, m_K, n_1, \ldots, n_K) \in S\colon \\ z_1 = 0,\ m_i = 1,\ \text{and } n_i = 2 \text{ for some } i\big\}.$$

Then the $n$th access time for port 1 is $\zeta_{\beta(n)} - \zeta_{\alpha(n)}$, where

$$\alpha(n) = \inf\{k > \alpha(n-1)\colon\ S_k \in A_2, S_{k-1} \in A_1\}$$

and

$$\beta(n) = \inf\{k \geq \alpha(n)\colon\ S_k \in B_2(A), S_{k-1} \in B_1(A)\}$$

for $n \geq 0$; $S_{\alpha(n)-1} = s$, $S_{\alpha(n)} = s'$, and $(s, s') \in A$.

**Example 5.9.** (Cyclic queues with feedback) Label the jobs from 1 to $N$. Set

$$X(t) = (Z(t), N_1(t), N_2(t), \dots, N_N(t)),$$

where $Z(t)$ is the number of jobs waiting or in service at center 1 at time $t$ and $N_j(t)$ is the position of job $j$ in the job stack at time $t$. Take event $e_i$ = "service completion at center $i$." Formal specification of the process $\{X(t)\colon\ t \geq 0\}$ is as a GSMP with a finite state space $S$ and event set $E = \{e_1, e_2\}$.

The *sojourn times* at center 1 are the random time intervals from completion of service to a job at center 2 until the job next moves to center 2. The sets that define the start times of individual sojourn times at center 1 are

$$A_1 = \{(z, n_1, n_2, \dots, n_N) \in S\colon\ 0 \leq z < N\}$$

and

$$A_2 = \{(z, n_1, n_2, \dots, n_N) \in S\colon\ 0 < z \leq N\}.$$

Take $\boldsymbol{A} = \{A_1^*, A_2^*, \dots, A_N^*\}$, where

$$A_j^* = \{(z, n_1, \dots, n_N, z', n_1', \dots, n_N') \in A\colon\ n_j = N\}$$

for $1 \leq j \leq N$. For $s = (z, n_1, \dots, n_{j-1}, N, n_{j+1}, \dots, n_N) \in A_1$ and $s' = (z+1, n_1', \dots, n_{j-1}', 1, n_{j+1}', \dots, n_N') \in A_2$ with

$s \to s'$, the sets $B_1(A_j^*)$ and $B_2(A_j^*)$ that define the termination times of individual sojourn times at center 1 and the link between the start times and the termination times are

$$B_1(A_j^*) = \{(z, n_1, \ldots, n_{j-1}, z, n_{j+1}, \ldots, n_N) \in S\colon\ 0 < z \le N\}$$

and

$$B_2(A_j^*) = \{(z-1, n_1, \ldots, n_{j-1}, z, n_{j+1}, \ldots, n_N) \in S\colon \\ 0 < z \le N\}.$$

Then the $n$th sojourn time at center 1 is $\zeta_{\beta(n)} - \zeta_{\alpha(n)}$, where

$$\alpha(n) = \inf\{k > \alpha(n-1)\colon\ S_k \in A_2, S_{k-1} \in A_1\}$$

and

$$\beta(n) = \inf\{k \ge \alpha(n)\colon\ S_k \in B_2(A_j^*), S_{k-1} \in B_1(A_j^*)\}$$

for $n \ge 0$; $S_{\alpha(n)-1} = s$, $S_{\alpha(n)} = s'$, and $(s, s') \in A_j^*$.

Theorem 5.16 provides conditions under which a sequence of passage times $\{P_n\colon\ n \ge 0\}$ is a regenerative process in discrete time with finite cycle-length moments. The idea is to define a suitable "single-state" of the GSMP and to show that the sequence of state transitions that occur in such a single-state decomposes the sequence of passage times into i.i.d. cycles. We use the following definition of a regenerative process in discrete time. Let $\{X_n\colon\ n \ge 0\}$ be a stochastic process taking values in a set $S$.

**Definition 5.10.** *The process $\{X_n\colon\ n \ge 0\}$ is said to be a regenerative process in discrete time if there exists an increasing sequence $\{\gamma(k)\colon\ k \ge 0\}$ of a.s. finite nonnegative random* *indices such that*

$$\begin{aligned} P\{\, & X_{\gamma(k)} \in A_0, \ldots, X_{\gamma(k)+n} \in A_n, \gamma(k+1) - \gamma(k) = m_1, \ldots, \\ & \gamma(k+l) - \gamma(k+l-1) = m_l \mid X_i,\ 0 \leq i < \gamma(k); \\ & \gamma(j) - \gamma(j-1),\ 1 \leq j \leq k \,\} \\ = P\{\, & X_{\gamma(0)} \in A_0, \ldots, X_{\gamma(n)} \in A_n, \gamma(1) - \gamma(0) = m_1, \ldots, \\ & \gamma(l) - \gamma(l-1) \leq m_l \,\} \text{ a.s.} \end{aligned}$$

*for* $n, k \geq 0$, $l \geq 1$, $A_0, \ldots, A_n \subseteq S$, *and* $m_1, \ldots, m_l \geq 0$.

Definition 5.10 identifies special "single-states" of a GSMP with state space $S$. Denote by $\bar{S}$ the set of states such that, starting in the state, all passage times underway terminate at the next state transition. Formally, when the sequence of passage times is defined by (5.1) and (5.2), $\bar{S}$ is the set of all states $s \in S$ such that

(i) $n(s) = 0$; or
(ii) $n(s) = 1$ and $(s, s') \in B$ for each $s' \in S$ such that $s \to s'$.

When the sequence of passage times is defined by (5.4) and (5.5), $\bar{S}$ is the set of all states $s \in S$ such that, for each $A^* \in A$,

(i) $n(s, A^*) = 0$; or
(ii) $n(s, A^*) = 1$ and $(s, s') \in B(A^*)$ for each $s' \in S$ such that $s \to s'$.

**Definition 5.10.** *An element* $\bar{s}$ *of the set* $S$ *is said to be a single-state of the generalized semi-Markov process for the passage time specified by the sets* $A_1$, $A_2$, $\{B_1(A^*)\colon\ A^* \in \mathbf{A}\}$, *and* $\{B_2(A^*)\colon\ A^* \in \mathbf{A}\}$ *if* $\bar{s} \in \bar{S}$ *and* $|E(\bar{s})| = 1$.

For a fixed single-state $\bar{s}$, set $\theta(-1) = 0$ and

$$\theta(k) = \theta(k; \bar{s}) = \inf\{n > \theta(k-1)\colon\ S_{n-1} = \bar{s}\} \tag{5.11}$$

for $n \geq 0$. According to this definition, $\theta(k)$ is the index of the $k$th state transition that occurs in state $\bar{s}$. Observe that if $\theta(k) \leq \alpha(n) < \theta(k+1)$ for some $n, k \geq 0$, then $\theta(k) \leq \beta(n) \leq \theta(k+1)$. Define a sequence $\{\gamma(k)\colon\ k \geq 0\}$ of random indices for passage times by setting $\gamma(-1) = -1$ and

$$\gamma(k) = \inf\bigl\{n > \gamma(k-1)\colon\ \alpha(n-1) < \theta(m) \leq \alpha(n) \\ \text{for some } m \geq 0\bigr\} \tag{5.12}$$

for $k \geq 0$. The random indices $\{\theta(k)\colon\ k \geq 0\}$ of (5.11) effectively decompose the sequence of passage times into cycles. Passage times $P_n$ and $P_m$ are in the same cycle if and only if $\theta(j-1) \leq \alpha(n), \alpha(m) < \theta(j)$ for some $j \geq 1$; $\gamma(k)$ is the index of the first passage time in the $k$th cycle.

**Theorem 5.13.** *Assume that there exists a single-state $\bar{s} \in \bar{S}$ and $P_\mu\{S_n = \bar{s}$ i.o.$\} = 1$. Then the random indices $\bigl\{\gamma(k)\colon\ k \geq 0\bigr\}$ form a sequence of regeneration points for $\{P_n\colon\ n \geq 0\}$. Moreover, if $E\bigl[\bigl(\theta(1) - \theta(0)\bigr)^r\bigr] < \infty$ for some $r > 0$ then $E\bigl[\bigl(\gamma(1) - \gamma(0)\bigr)^r\bigr] < \infty$.*

*Proof.* For ease of exposition, suppose that for each (sample path and) $k \geq 1$, there is at least one passage time with start index between $\theta(k-1)$ and $\theta(k)$; that is,

$$\sum_{n=0}^{\infty} 1_{[\theta(k-1),\theta(k))}\bigl(\alpha(n)\bigr) > 0.$$

Fix $\bar{c} \in C(\bar{s})$, and set $\delta_k = \gamma(k) - \gamma(k-1)$ for $k \geq 1$. Using the strong Markov property and the fact that $\bar{s}$ is a single-state, it can be shown that

$$\begin{aligned} P_\mu\bigl\{ P_{\gamma(k)} \leq x_0, \ldots, P_{\gamma(k)+n} \leq x_n, \delta_{k+1} = m_1, \ldots,\\ \delta_{k+l} = m_l \bigm| (S_j, C_j),\ 0 \leq j \leq \theta(k) - 1 \bigr\}\\ = P_{(\bar{s},\bar{c})}\bigl\{ P_{\gamma(0)} \leq x_0, \ldots, P_{\gamma(0)+n} \leq x_n, \delta_1 = m_1, \ldots,\\ \delta_l = m_l \bigr\} \text{ a.s.} \end{aligned}$$

for $k, n, l \geq 0$, $x_0, \ldots, x_n \in \Re$, and $m_1, \ldots, m_l \geq 0$. (When $\mu(\{(s,c)\}) = 1$ for some fixed $(s,c) \in \Sigma$, we write $P_{(s,c)}$ instead of $P_\mu$.) Since $\{(S_n, C_n)\colon\ 0 \leq n \leq \theta(k) - 1\}$ determines both $\{\, P_0, P_1, \ldots, P_{\gamma(k)-1} \,\}$ and $\{\delta_1, \delta_2, \ldots, \delta_k\}$, it follows that

$$\begin{aligned} P_\mu\{\, & P_{\gamma(k)} \leq x_0, \ldots, P_{\gamma(k)+n} \leq x_n, \delta_{k+1} = m_1, \ldots, \\ &\qquad \delta_{k+l} = m_l \;\big|\; P_i,\ 0 \leq i < \gamma(k);\ \delta_j,\ 1 \leq j \leq k \,\} \\ &= P_{(\bar{s},\bar{c})}\{\, P_{\gamma(0)} \leq x_0, \ldots, P_{\gamma(0)+n} \leq x_n, \delta_1 = m_1, \ldots, \\ &\qquad \delta_l = m_l \,\} \text{ a.s.} \end{aligned}$$

for $k, n, l \geq 0$, $x_0, \ldots, x_n \in \Re$, and $m_1, \ldots, m_l \geq 0$. This result implies that the random indices $\{\gamma(k)\colon\ k \geq 0\}$ form a sequence of regeneration points for $\{P_n\colon\ n \geq 0\}$. The second assertion of the theorem follows easily in this case since $\gamma(1) - \gamma(0) \leq \theta(1) - \theta(0)$. A similar argument establishes the theorem when, for at least one sample path and index $k$, there is no passage time with start index between $\theta(k-1)$ and $\theta(k)$. □

The assumption that $|E(\bar{s})| = 1$ in Theorem 5.13 ensures that the clock-reading vector has the same distribution whenever there is a state transition in state $\bar{s}$. It is possible to relax the requirement that $|E(\bar{s})| = 1$ if some of the clock-setting distributions for events in $E(\bar{s})$ are exponential. Also observe that the random times $\{\zeta_{\theta(k)}\colon\ k \geq 0\}$ form a sequence of regeneration points for the process $\{X(t)\colon\ t \geq 0\}$.

**Example 5.14.** (Machine repair) Consider a group of $N$ ($\geq 1$) machines numbered $1, 2, \ldots, N$ under the care of a single repairman. The repairman provides service to (repairs) the machines one at a time in the order in which they fail. Successive repair times for machine $j$ are i.i.d. as a positive random variable $R_j$ with finite mean and a density function that is continuous and positive on $(0, +\infty)$. The machines stop randomly in the sense that the times (lifetimes) between end of repair and the next stoppage of machine $j$ are i.i.d. as a positive random variable $L_j$ with finite mean and a density function that is continuous and positive on $(0, +\infty)$. Machine lifetimes are independent of repair times and quantities associated with distinct machines are independent. Set

$$X(t) = (X_1(t), X_2(t), \dots, X_N(t)),$$

where

$$X_j(t) = \begin{cases} 0 & \text{if machine } j \text{ is running at time } t \\ 1 & \text{if machine } j \text{ is under repair at time } t \\ k & \text{if machine } j \text{ is waiting for repair at time } t \text{ with} \\ & k-1 \text{ machines ahead at the repair station.} \end{cases}$$

Take event $e_{N+1}$ = "end of repair" and $e_j$ = "failure of machine $j$" ($j = 1, 2, \dots, N$). Formal specification of the process $\{X(t)\colon t \geq 0\}$ is as a GSMP with finite state space

$$\begin{aligned} S = \Big\{&(x_1, x_2, \dots, x_N) \in \{0, 1, \dots, N\}^N : x_i = x_j \text{ only if } i = j \\ &\text{or } x_i = x_j = 0; x_i = n \text{ only if } 1_{\{0\}}(x_1) + \cdots + 1_{\{0\}}(x_N) \\ &\leq N - n \text{ for } n = 1, 2, \dots, N\Big\} \end{aligned}$$

and event set $E = \{e_1, e_2, \dots, e_N, e_{N+1}\}$. For $s = (x_1, x_2, \dots, x_N) \in S$, event $e_{N+1}$ = "end of repair" $\in E(s)$ if and only if $x_1 + \dots + x_N > 0$ and event $e_j$ = "failure of machine $j$" $\in E(s)$ if and only if $x_j = 0$. All speeds $r(s, e)$ for active events are equal to 1.

If $e = e_{N+1}$ = "end of repair," then $p(s'; s, e) = 1$ when

$$\begin{aligned} s = (&x_1, x_2, \dots, x_{j-1}, 1, x_{j+1}, \dots x_N) \\ &\text{and } s' = (x'_1, \dots, x'_{j-1}, 0, x'_{j+1}, \dots, x'_N) \end{aligned}$$

where $x'_i = \max(0, x_i - 1)$. If $e = e_j$ = "failure of machine $j$," then $p(s'; s, e) = 1$ when

$$\begin{aligned} s = (&x_1, x_2, \dots, x_{j-1}, 0, x_{j+1}, \dots, x_N) \\ &\text{and } s' = (x_1, x_2, \dots, x_{j-1}, \max(x_1, x_2, \dots, x_N) + 1, \\ &\qquad x_{j+1}, \dots, x_N). \end{aligned}$$

All other state transition probabilities $p(s'; s, e)$ are equal to 0. The clock-setting distribution functions are as follows. If the new event $e' = e_{N+1}$, then $F(x; s', e', s, e^*) = P\{R_j \le x\}$,. where $s' = (x'_1, x'_2, \ldots, x'_N)$ and $x'_j = 1$. If $e' = e_j$, then $F(x; s', e', s, e^*) = P\{L_j \le x\}$.

The *sojourn times* for machine $j$ are the random time intervals from stoppage of machine $j$ until end of repair. A sojourn time for machine $j$ starts when (event $e_j$ = "failure of machine $j$" occurs and) the process $\{X(t)\colon\ t \ge 0\}$ makes a state transition from some state $s = (x_1, x_2, \ldots, x_{j-1}, 0, x_{j+1}, \ldots, x_N)$ to state $s' = (x_1, x_2, \ldots, x_{j-1}, \max(x_1, x_2, \ldots, x_N) + 1, x_{j+1}, \ldots, x_N)$ and ends when (event $e_{N+1}$ = "end of repair" occurs and) the process makes a state transition from some state $s = (x_1, x_2, \ldots, x_{j-1}, 1, x_{j+1}, \ldots, x_N)$ to state $s' = (x_1, x_2, \ldots, x_{j-1}, 0, x_{j+1}, \ldots, x_N)$. The sets $A_1$ and $A_2$ that define the start times of individual sojourn times for machine $j$ are

$$A_1 = \{(x_1, \ldots, x_N) \in S\colon\ x_j > 0\}$$

and

$$A_2 = \{(x_1, \ldots, x_N) \in S\colon\ x_j = 0\}\,.$$

Take $\boldsymbol{A} = \{A\}$. The sets $B_1(A)$ and $B_2(A)$ that define the termination times of individual sojourn times for machine $j$ and the link between the start times and the termination times are

$$B_1(A) = \{(x_1, \ldots, x_N) \in S\colon\ x_j = 1\}$$

and

$$B_2(A) = \{(x_1, \ldots, x_N) \in S\colon\ x_j = 0\}\,.$$

Then the $n$th sojourn time for machine $j$ is $\zeta_{\beta(n)} - \zeta_{\alpha(n)}$, where

$$\alpha(n) = \inf\{k > \alpha(n-1)\colon\ S_k \in A_2, S_{k-1} \in A_1\}$$

and

$$\beta(n) = \inf\{k \geq \alpha(n)\colon\ S_k \in B_2(S_{\alpha(n)-1}, S_{\alpha(n)}),\\ S_{k-1} \in B_1(S_{\alpha(n)-1}, S_{\alpha(n)})\}$$

for $n \geq 0$; $S_{\alpha(n)-1} = s$, $S_{\alpha(n)} = s'$, and $(s, s') \in A$.

Since all clock speeds for active events are equal to 1, the GSMP is irreducible if for all $s, s' \in S$ state $s'$ is accessible from state $s$ in the sense that there exists a finite sequence of states $s = s_0, s_1, \ldots, s_{n-1}, s_n = s'$ and events $e_{i_0}, e_{i_1}, \ldots, e_{i_{n-1}} \in E$ such that $p(s_1; s_0, e_{i_0})p(s_2; s_1, e_{i_1}) \cdots p(s_n; s_{n-1}, e_{i_{n-1}}) > 0$. To show this, set $s^* = (0, 0, \ldots, 0)$. Observe that state $s^*$ is accessible from every state $s$, and every state $s$ is accessible from state $s^*$.

Take $\bar{s} = (1, 2, \ldots, N)$ and $\bar{s}' = (0, 1, 2, \ldots, N-1)$. Observe that $\bar{s}$ is a single-state and the set $E(\bar{s})$ of events that are active in state $\bar{s}$ is $E(\bar{s}) = \{e_{N+1}\}$. In state $\bar{s}$ machine 1 is under repair with machines $2, 3, \ldots, N$ waiting for repair (in this order). When event $e_{N+1}$ occurs, there is an end of repair for machine 1, and the GSMP makes a state transition from $\bar{s}$ to $\bar{s}'$. Since $S$ is finite and each clock-setting distribution has finite mean and a density function that is continuous and positive on $(0, +\infty)$, $P_\mu\{\, S_n = \bar{s} \text{ i.o.}\,\} = 1$. The sequence of passage times $\{P_n\colon\ n \geq 0\}$ is a regenerative process in discrete time, and the expected time between regeneration points is finite. The random indices $\{\gamma(k)\colon\ k \geq 0\}$ defined by (5.12) correspond to the successive passage times that terminate when there is a state transition from $\bar{s}$ to $\bar{s}'$. These random indices form a sequence of regeneration points for $\{P_n\colon\ n \geq 0\}$.

**Example 5.15.** (Token ring) Consider a unidirectional ring network as in Example 5.3 but now suppose that when a port observes the token and there is a packet queued for transmission, the port converts the token to a connector, transmits a packet, and then transmits the token pattern; the token continues to propagate if there is no packet queued for transmission.

Assume that the time for port $j$ to transmit a packet is distributed as a positive random variable $L_j$ with finite mean. Also assume that packets arrive at individual ports randomly in the sense that the time from end of transmission by port $j$ until the arrival of the next packet for transmission by port $j$ is distributed as a positive random variable $D_j$ with finite mean and absolutely continuous distribution. The time for the token to propagate from port $j$ to the next port is a positive constant $R_j$.

Set

$$X(t) = (Z_1(t), \ldots, Z_N(t), M(t), N(t)),$$

where

$$Z_j(t) = \begin{cases} 1 & \text{if there is a packet queued for transmission at} \\ & \text{port } j \text{ at time } t \\ 0 & \text{otherwise,} \end{cases}$$

$$M(t) = \begin{cases} j & \text{if port } j \text{ is transmitting a packet at time } t \\ 0 & \text{if no port is transmitting a packet at time } t, \end{cases}$$

$N(t) = 1$ if port $N$ is transmitting a packet or the token is propagating to port 1 at time $t$, and $N(t) = j$ if port $j-1$ is transmitting a packet or the token is propagating to port $j$ at time $t$ ($j = 2, 3, \ldots, N$). Take event $e_{N+2}$ = "observation of token by port," $e_{N+1}$ = "end of transmission by port," and $e_j$ = "arrival of packet for transmission by port $j$" ($j = 1, 2, \ldots, N$). Formal specification of the process $\{X(t)\colon\ t \geq 0\}$ is as a GSMP with finite state space

$$\begin{aligned} S = \big\{&(z_1, \ldots, z_N, m, n) \in \{0,1\}^N \times \{0, 1, \ldots, N\} \times \\ &\{1, 2, \ldots, N\}\colon\ \ n = m + 1 \text{ when } 0 < m < N;\ n = 1 \\ &\text{when } m = N;\ m = j \text{ only if } z_j = 0 \text{ for } j = 1, 2, \ldots, N\big\} \end{aligned}$$

and event set $E = \{e_1, e_2, \ldots, e_{N+2}\}$.

Take $\bar{s}' = (0, 1, \ldots, 1, 1, 2)$, and set $e^* = e_{N+2} =$ "observation of token by port." The process $\{X(t)\colon t \geq 0\}$ hits state $\bar{s}'$ when $e^*$ is the trigger event only if the event occurs in state $\bar{s} = (1, \ldots, 1, 0, 1)$. Observe that when event $e^*$ occurs and there is a state transition from $\bar{s}$ to $\bar{s}'$, port 1 starts transmission of a packet and there is a packet queued for transmission at ports $2, 3, \ldots, N$. Set

$$S^* = \{(z_1, \ldots, z_N, m, n) \in S\colon\ m = 0 \text{ and } n = 1\}.$$

Also let $T_n$ be the time of the $n$th state transition triggered by event $e^*$ in some state $s^* \in S^*$ so that $T_n$ is the time of the $n$th observation of the token by port 1, $n \geq 0$. Assume that the distribution of each $D_j$ is NBU and

$$\delta_j \stackrel{\text{def}}{=} P\{D_j \leq R_N\} > 0$$

for $j = 1, 2, \ldots, N$. Then

$$P\{X(T_n) = \bar{s}') \mid X(T_{n-1}), \ldots, X(T_0)\} \geq \delta \text{ a.s.},$$

where

$$\delta = \prod_{j=1}^{N} \delta_j = \prod_{j=1}^{N} P\{D_j \leq R_N\}.$$

It follows that $P_\mu\{\, S_n = \bar{s} \text{ i.o.}\,\} = 1$.

The *access times* for port $j$ are the random time intervals from the arrival of a packet for transmission by port $j$ until the start of transmission by port $j$. The sets that define the start times of individual access times for port $j$ are

$$A_1 = \{(z_1, \ldots, z_N, m, n) \in S\colon\ z_j = 0, m \neq j\}$$

and

$$A_2 = \{(z_1, \ldots, z_N, m, n) \in S\colon\ z_j = 1, m \neq j\}.$$

Take $\boldsymbol{A} = \{A\}$, where $A = \{(s, s') \in A_1 \times A_2\colon\ s \to s'\}$. The sets $B_1(A)$ and $B_2(A)$ that define the termination times of individual access times for port $j$ and the link between the start times and the termination times are

$$B_1(A) = \{(z_1, \ldots, z_N, m, n) \in S\colon\ z_j = 1, m = 0, n = j\},$$

and

$$B_2(A) = \{(z_1, \ldots, z_N, m, n) \in S\colon\ z_j = 0, m = j\}.$$

Then the $n$th access time for port $j$ is $\zeta_{\beta(n)} - \zeta_{\alpha(n)}$, where

$$\alpha(n) = \inf\{k > \alpha(n-1)\colon\ S_k \in A_2, S_{k-1} \in A_1\}$$

and

$$\beta(n) = \inf\bigl\{k \geq \alpha(n)\colon\ S_k \in B_2(A), S_{k-1} \in B_1(A)\bigr\}$$

for $n \geq 0$; $S_{\alpha(n)-1} = s$, $S_{\alpha(n)} = s'$, and $(s, s') \in A$. The random indices $\{\gamma(k)\colon\ k \geq 0\}$ defined by (5.12) correspond to the successive passage times that terminate when event $e^*$ occurs and there is a state transition from $\bar{s}$ to $\bar{s}'$. These random indices form a sequence of regeneration points for $\{P_n\colon\ n \geq 0\}$. Moreover, since

$$E[T_n - T_{n-1}] \leq R_1 + \cdots + R_N + \sum_{j=1}^{N} E[L_j] < \infty$$

for all $n \geq 1$, the expected time between regeneration points is finite.

Conditions for time-average convergence and convergence in distribution along with a central limit theorem for a sequence of passage times follow from Theorem 5.13 and standard results for regenerative processes. Under the conditions of Theorem 5.13, the random indices $\{\gamma(k)\colon\ k \ge 0\}$ form a sequence of regeneration points for $\{P_n\colon\ n \ge 0\}$. As before, we set $\delta_k = \gamma(k) - \gamma(k-1)$ for $k \ge 1$, so that $\delta_k$ is the number of passage times in the $k$th cycle. For a real-valued measurable function $f$, set

$$Y_k(f) = \sum_{n=\gamma(k-1)}^{\gamma(k)-1} f(P_n)$$

for $k \ge 0$. The regenerative property implies that the sequence $\{(Y_k(f), \delta_k)\colon\ k \ge 1\}$ consists of i.i.d. random vectors.

**Theorem 5.16.** *Assume that $E[\delta_1] < \infty$. Then*

$$\lim_{n\to\infty} \frac{1}{n} \sum_{j=0}^{n-1} f(P_j) = \frac{E[Y_1(f)]}{E[\delta_1]} \stackrel{\text{def}}{=} r(f) \text{ a.s.}$$

*for any real-valued measurable function $f$ such that $Y_0(|f|) < \infty$ a.s. and $E[Y_1(|f|)] < \infty$.*

**Theorem 5.17.** *Assume that $\delta_1$ is aperiodic with $E[\delta_1] < \infty$. Then $P_n \Rightarrow P$ as $n \to \infty$ and*

$$E[f(P)] = \frac{E[Y_1(f)]}{E[\delta_1]}$$

*for any real-valued measurable function $f$ such that $E[Y_1(|f|)] < \infty$ or $E[|f(P)|] < \infty$.*

Theorem 5.17 and the continuous mapping theorem together imply that

$$\lim_{n\to\infty} E[f(P_n)] = E[f(P)]$$

for any real-valued measurable function $f$ such that (i) either $E[Y_1(|f|)] < \infty$ or $E[|f(P)|] < \infty$ and (ii) $P\{P \in D(f)\} = 0$, provided that $\delta_1$ is aperiodic with $E[\delta_1] < \infty$. Set

$$\sigma^2(f) = \text{Var}\,[Y_1(f) - r(f)\delta_1].$$

**Theorem 5.18.** *Assume that $\delta_1$ is aperiodic with $E[\delta_1^2] < \infty$. Then*

$$\frac{\sum_{j=0}^{n-1} f(P_j) - nr(f)}{\sqrt{n}\sigma(f)/E[\delta_1]} \Rightarrow N(0,1)$$

*as $n \to \infty$ for any real-valued measurable function $f$ such that $Y_0(|f|) < \infty$ a.s. and $E[Y_1^2(|f|)] < \infty$.*

From $n$ cycles, the regenerative method provides an estimator

$$\hat{r}(n) \stackrel{\text{def}}{=} \frac{\bar{Y}(n)}{\bar{\delta}(n)} = \frac{\sum_{k=1}^{n} Y_k(f)}{\sum_{k=1}^{n} \delta_k}$$

that is strongly consistent for $r(f)$ along with an asymptotic confidence interval.

**Algorithm 5.19.** *Regenerative method for passage times*

1. *Select a single-state $\bar{s}$ as in Definition 5.10 such that $P_\mu\{S_n = \bar{s} \text{ i.o.}\} = 1$.*
2. *Set $X(0-) = \bar{s}$, and observe a fixed number $n$ of cycles of the sequence of passage times defined by the successive random indices $\{\gamma(k)\colon\ k \geq 0\}$ in (5.12).*
3. *Measure $(Y_1(f), \delta_1), (Y_2(f), \delta_2, \ldots, (Y_n(f), \delta_n))$.*
4. *Form the point estimate*

   $$\hat{r}(n) = \frac{\bar{Y}(n)}{\bar{\delta}(n)}$$

   *for $r(f)$.*
5. *Form the asymptotic $100(1-2\gamma)\%$ confidence interval*

   $$\left[\hat{r}(n) - \frac{s(n)z_{1-\gamma}}{\bar{\delta}(n)\sqrt{n}}, \hat{r}(n) + \frac{s(n)z_{1-\gamma}}{\bar{\delta}(n)\sqrt{n}}\right]$$

   *for $r(f)$.*

Table 5.1 gives simulation results for the token ring of Example 5.15 with $N = 4$ ports. The estimates for the expected value and percentiles of the access times experienced by port 1 are based on cycles defined by the successive random times at which event $e^* =$ "observation of token" triggers a state transition from (1,0,0,0,0,1) to (0,0,0,0,1,2). At these random times port 1 starts transmission of a packet (termination of a passage time) and there is no packet queued for transmission at any other port. Theoretical values for the expected access time are in parentheses.

Table 5.2 gives simulation results for the slotted ring of Example 5.8 with $N = 4$ ports and $K = 2$ slots. The estimates for the expected value and percentiles of the access times experienced by port 1 are based on cycles defined by the successive random times at which event $e^* =$ "observation of status bits by ports" triggers a state transition from (1,0,0,0,0,0,1,3) to (0,0,0,0,1,0,2,4). At these random times port 1 starts transmission of a packet in slot 1, slot 2 is empty, and no packet is queued for transmission at any of the ports.

## Exercises

5.1. (Road traffic) Consider a single lane of one-way traffic proceeding along a main road. At a fixed merge point on the main road, there is a side road. Upon arriving at the merge point, side-road cars attempt to merge with the stream of traffic on the main road. Side-road cars arrive and join a waiting line at the merge point according to a Poisson process of rate $\lambda > 0$. Main-road cars arrive at the merge point according to a renewal process $\{A_n\colon n \geq 0\}$; that is, the $n$th main-road car arrives at the merge point at time $A_n$, and the interarrival times are i.i.d. as a positive random variable with absolutely continuous distribution.

There is a gap-acceptance function $L$ that determines the manner in which (instantaneous) merging takes place. Assume that

$$L(x) = \begin{cases} 1 & \text{if } x > \Delta_1 \\ 0 & \text{if } x \le \Delta_1, \end{cases}$$

where $\Delta_1 > 0$ and $L(x)$ is the probability that the side-road car at the head of the line merges given that the time until the next main-road car passes the merge point is $x$. When a side-road car merges, $\Delta_2$ time units elapse before another side-road car may merge.

Set

$$X(t) = (Q(t), G(t), B(t)),$$

where $Q(t)$ is the number of cars waiting on the side road at time $t$,

$$B(t) = \begin{cases} 0 & \text{if } G(t) = 2 \text{ and a car has merged in } [t - \Delta_2, t) \\ 1 & \text{otherwise,} \end{cases}$$

and

$$G(t) = \begin{cases} 2 & \text{if } A_n \le t < A_{n+1},\ A_{n+1} - A_n > \Delta_1, \\ & \quad A_{n+1} - t > \Delta_1 \\ 1 & \text{if } A_n \le t < A_{n+1},\ A_{n+1} - A_n > \Delta_1, \\ & \quad A_{n+1} - t \le \Delta_1 \\ 0 & \text{if } A_n \le t < A_{n+1},\ A_{n+1} - A_n \le \Delta_1. \end{cases}$$

The *delay times* experienced by side-road cars are the random time intervals from arrival of a side-road car until the car merges. Specify the sets that define the start times of individual delay times experienced by side-road cars. Also specify the sets that define the termination times of the individual delay times and the link between the start times and the termination times.

5.2. (Producer/consumer with finite buffers). Consider a system consisting of two producers, two consumers, and two buffers, each numbered 1 and 2. The producers share a single channel for transmission one at a time of items from buffers to consumers. Producer $i$ creates items for consumer $i$ one at a time; items created but not yet transmitted are placed in buffer $i$ for transmission. Buffer $i$ has finite capacity $B_i > 0$; that is, an item created by producer $i$ when the system already contains $B_i - 1$ ($\geq 0$) items for consumer $i$ causes the process of creation of items for consumer $i$ to shut down. This process remains shut down until the first subsequent end of transmission to consumer $i$. Producer/consumer pair 1 has nonpreemptive priority over producer/consumer pair 2 for use of the channel. Items created by producer $i$ are transmitted in the order in which they are created. Assume that the successive times required by producer $i$ to create an item are i.i.d. as a positive random variable $D_i$ with finite mean and a density function that is continuous and positive on $(0, +\infty)$. Also assume that the successive times to transmit an item from buffer $i$ to consumer $i$ are i.i.d. as a positive random variable $L_i$ with finite mean and a density function that is continuous and positive on $(0, +\infty)$. Creation times are independent of transmission times, and random quantities associated with producer/consumer pair 1 are independent of random quantities associated with producer/consumer pair 2.

Set

$$X(t) = (B_1(t), B_2(t), M(t)),$$

where

$$M(t) = \begin{cases} 2 & \text{if producer 2 is transmitting at time } t \\ 1 & \text{if producer 1 is transmitting at time } t \\ 0 & \text{if neither producer is transmitting at time } t \end{cases}$$

and $B_i(t)$ is the number of items waiting for transmission to consumer $i$ at time $t$.

The *sojourn times* for items created by producer $i$ are the random time intervals from creation of an item until end of transmission to consumer $i$. Describe how you would use the regenerative method to obtain point and interval estimates for the limiting expected sojourn time for items created by producer 1.

5.3. (Producer/consumer with preemptive-repeat priority) Consider the producer/consumer system with finite buffers of Exercise 5.2, but now suppose that producer/consumer pair 1 has preemptive-repeat priority over producer/consumer pair 2 for use of the channel. Assume that the successive times required by producer $i$ to create an item are i.i.d. as a positive random variable $D_i$ with finite mean and a density function that is continuous and positive on $(0, +\infty)$. Also assume that the successive times to transmit an item from buffer $i$ to consumer $i$ are i.i.d. as a positive random variable $L_i$ with finite mean and a density function that is continuous and positive on $(0, +\infty)$. Creation times are independent of transmission times and random quantities associated with producer/consumer pair 1 are independent of random quantities associated with producer/consumer pair 2.

Set

$$X(t) = \big(B_2(t), B_1(t), M(t)\big),$$

where

$$M(t) = \begin{cases} 2 & \text{if producer 2 is transmitting at time } t \\ 1 & \text{if producer 1 is transmitting at time } t \\ 0 & \text{otherwise} \end{cases}$$

and $B_i(t)$ is the number of items waiting for transmission to consumer $i$ at time $t$.

The *waiting times* experienced by items created by producer $i$ are the random time intervals from creation of an item until start of transmission to consumer $i$. Can the regenerative method be used to obtain point and interval estimates for the probability that the limiting waiting time experienced by items created by producer $i$ is less than or equal to $x$ ($x$ fixed)? Explain your answer.

5.4. (Shuttle train) Consider a train that shuttles back and forth between two stations and has seats for $K$ ($\geq 1$) passengers. Passengers that wish to board the train and ride to the other station arrive at station $i$ according to a Poisson process of rate $\lambda_i$, queue, and subsequently board the train in the order in which they arrive at the station. Assume that passengers board the train instantaneously and all passengers that board at station $i$ disembark instantaneously when the train arrives at the other station. The time required for the train to travel from station $i$ to the other station is distributed as a positive random variable $L_i$ with finite mean. Travel times are independent of interarrival times, and the arrival process of passengers for boarding at station 1 is independent of the arrival process of passengers for boarding at station 2. Assume that $(\lambda_1 + \lambda_2)(E[L_1] + E[L_2]) < K$.

Set

$$X(t) = (K_1(t), K_2(t), B(t), D(t)),$$

where $K_i(t)$ is the number of passengers waiting to board the train at station $i$ at time $t$, $B(t)$ is the number of passengers on board at time $t$, and

$$D(t) = \begin{cases} 1 & \text{if the train is moving to station 1 at time } t \\ 2 & \text{if the train is moving to station 2 at time } t. \end{cases}$$

The *waiting times* experienced by passengers that board at station 1 are the random time intervals from arrival of a passenger at station 1 until the passenger boards the train. Specify the sets that define the start times of the individual waiting times experienced by passengers boarding at station 1. Also specify the sets that define the termination times of the individual waiting times and the link between the start times and the termination times. Explain how you would use the regenerative method to obtain a strongly consistent point estimate and an asymptotic confidence interval for the probability that the limiting waiting time experienced by passengers boarding at station 1 exceeds $x$ ($x$ fixed). Be explicit about the sequence of regeneration points, the quantities to be measured, and the construction of the point and interval estimates. State the ratio formulas and central limit theorems on which the estimates are based.

5.5. (Airport shuttle) Consider an airport shuttle that provides transportation service to $N$ stations numbered $1, 2, \ldots, N$. The shuttle has seats for $K$ ($\geq 1$) passengers and moves from station to station in a strictly defined order: $1, 2, \ldots, N, N-1, \ldots, 1, 2, \ldots$. Passengers that wish to board the shuttle at station $i$ in the "northbound" (from station 1 to station $N$) direction arrive at station $i$ according to a Poisson process of rate $\lambda_i$ and disembark at station $j$ ($> i$) with probability $p_{i,j}$ ($i = 1, 2, \ldots, N-1$). Passengers that wish to board the shuttle at station $i+1$ in the "southbound" (from station $N$ to station 1) direction arrive at station $i+1$ according to a Poisson process of rate $\lambda_{2N-(i+1)}$ and disembark at station $j$ ($< i+1$) with probability $p_{2N-(i+1),j}$ ($i = 1, 2, \ldots, N-1$). Passengers that arrive for boarding at station $i$ in each direction queue and subsequently board the shuttle in the order in which they arrive at the station. At each station passengers disembark before any waiting passengers board the shuttle. Assume that passengers board the shuttle instantaneously and passengers disembark instantaneously. The time required for the shuttle to travel (in the northbound direction) from station $i$ to station $i+1$ is distributed as a positive random variable $L_i$ with finite first and second moments, and the time required for the shuttle to travel (in the southbound direction) from station $i+1$to station $i$ is distributed as a positive random variable $L_{2N-(i+1)}$ with finite first and second moment. Travel times, interarrival times, and destinations are mutually independent. Assume that $\sum_{k=1}^{2(N-1)} \lambda_k \sum_{k=1}^{2(N-1)} E[L_k] < K$.

Set

$$X(t) = (D_{1,2}(t), D_{1,3}(t), \ldots, D_{1,N}(t), D_{2,3}(t), \ldots, D_{N-1,N}(t), \\ B_1(t), \ldots, B_{2(N-1)}(t), N(t)),$$

where

$$N(t) = \begin{cases} i & \text{if the shuttle is moving from station } i \text{ to station } i+1 \text{ in the northbound direction at time } t \\ 2N-(i+i) & \text{if the shuttle is moving from station } i+1 \text{ to station } i \text{ in the southbound direction at time } t, \end{cases}$$

$D_{i,j}(t)$ is the number of passengers on board the shuttle at time $t$ that boarded at station $i$ in the northbound direction and wish to disembark at station $j$ or boarded at station $j$ in the southbound direction and wish to disembark at station $i$ $(i = 1, 2, \ldots, N-1,\ j = 2, 3, \ldots, N,\ i < j)$, $B_i(t)$ is the number of passengers waiting at time $t$ to board the shuttle at station $i$ in the northbound direction, and $B_{2N-(i+1)}(t)$ is the number of passengers waiting at time $t$ to board the shuttle at station $i+1$ in the southbound direction $(i = 1, 2, \ldots, N-1)$.

The *sojourn times* experienced by passengers that board the shuttle at station $i$ in the northbound direction and disembark at station $j$ $(> i)$ are the random time intervals from arrival of a passenger for boarding until the passenger disembarks. Explain how you would use the regenerative method to obtain a strongly consistent point estimate and an asymptotic confidence interval for the probability that the limiting sojourn time experienced by passengers that board at station $i$ and disembark at station $j$ exceeds $x$ ($x$ fixed). Be explicit about the sequence of regeneration points, the quantities to be measured, and the construction of the point and interval estimates. State the ratio formulas and central limit theorems on which the estimates are based.

5.6. (Preemptive cyclic queues) Consider a network of queues with two single-server centers that provides service to a fixed number $N$ of customers labelled $1, 2, \ldots, N$. After completion of service at center 1, a customer moves instantaneously from center 1 to center 2, and after completion of service at center 2 moves instantaneously back to center 1. Suppose that the queueing discipline at each center is LCFS. Also suppose that interruptions of service are of the preemptive-resume type. Customers waiting in queue appear in the order of preemption at the center, the most recently preempted customer at the head of the queue. Assume that all service times are mutually independent. Also assume that service times at center 1 are identically and exponentially distributed as a random variable $V_1$ and service times at center 2 are identically distributed as a positive random variable $V_2$ with finite mean and a density function that is continuous and positive on $(0, +\infty)$.

Let $X(t)$ be the number of customers waiting or in service at center 2 at time $t$. A *sojourn time* starts when a customer completes service at center 1 and moves to center 2; the sojourn time ends when the customer next moves to center 1. Explain how you would use the regenerative method to obtain a strongly consistent point estimate and an asymptotic confidence interval for the probability that the limiting sojourn time exceeds $x$ ($x$ fixed). Be explicit about the sequence of regeneration points, the quantities to be measured, and the construction of the point and interval estimates. State the ratio formulas and central limit theorems on which the estimates are based.

**Table 5.1**

Simulation results for access times in token ring

$$N = 4,\ R_1 = R_2 = R_3 = R_4 = 0.20$$

$$E[D_1] = E[D_2] = E[D_3] = E[D_4] = 66.667$$

$$E[L_1] = E[L_2] = E[L_3] = E[L_4] = 5.00$$

Exponentially distributed interarrival times for packets
Constant packet transmission times for all ports
Cycles defined by state transitions from
(1,0,0,0,0,1) to (0,0,0,0,1,2)

Number of cycles

| | 200 | 400 | 600 | 800 | 1000 |
|---|---|---|---|---|---|
| Simulated time | 15301.02 | 32621.59 | 47202.65 | 62304.96 | 78420.69 |
| State transitions/ cycle | 286.18 | 305.24 | 294.14 | 291.40 | 294.9 |
| Access times/ cycle | 1.0150 | 1.0350 | 1.0367 | 1.0362 | 1.0380 |
| $E[P]$ (5.1452) | 1.2454 | 1.2702 | 1.2401 | 1.1775 | 1.1586 |
| | $\pm 0.2124$ | $\pm 0.1511$ | $\pm 0.1197$ | $\pm 0.1030$ | $\pm 0.0921$ |
| $P\{P \leq 0.2838\}$ | 0.2512 | 0.2609 | 0.2605 | 0.2702 | 0.2717 |
| (0.2773) | $\pm 0.0498$ | $\pm 0.0350$ | $\pm 0.0288$ | $\pm 0.0254$ | $\pm 0.0228$ |
| $P\{P \leq 0.5676\}$ | 0.5271 | 0.5459 | 0.5370 | 0.5597 | 0.5703 |
| (0.5558) | $\pm 0.0575$ | $\pm 0.0397$ | $\pm 0.0324$ | $\pm 0.0282$ | $\pm 0.0253$ |
| $P\{P \leq 5.1452\}$ | 0.7635 | 0.7657 | 0.7717 | 0.7937 | 0.7967 |
| (0.7974) | $\pm 0.0495$ | $\pm 0.0303$ | $\pm 0.0274$ | $\pm 0.0229$ | $\pm 0.0206$ |
| $P\{P \leq 6.2705\}$ | 0.8325 | 0.8237 | 0.8328 | 0.8456 | 0.8468 |
| (0.8408) | $\pm 0.0426$ | $\pm 0.0303$ | $\pm 0.0242$ | $\pm 0.0204$ | $\pm 0.0184$ |
| $P\{P \leq 5.6409\}$ | 0.9064 | 0.9082 | 0.9084 | 0.9156 | 0.9162 |
| (0.9297) | $\pm 0.0332$ | $\pm 0.0230$ | $\pm 0.01187$ | $\pm 0.0157$ | $\pm 0.0141$ |

## Table 5.2

Simulation results for access times in slotted ring

$$N = 4,\ K = 2,\ R_1 = R_2 = R_3 = R_4 = 0.84$$

$$E[D_1] = E[D_2] = E[D_3] = E[D_4] = 8.40$$

Exponentially distributed interarrival times for packets
Cycles defined by state transitions from
(1,0,0,0,0,0,1,3) to (0,0,0,0,1,0,2,4)

Number of cycles

| | 200 | 400 | 600 | 800 | 1000 |
|---|---|---|---|---|---|
| Simulated time | 18510.24 | 37336.84 | 55390.94 | 75037.41 | 93907.85 |
| State transitions/ cycle | 139.23 | 140.35 | 138.92 | 141.12 | 141.25 |
| Access times/ cycle | 7.360 | 7.375 | 7.325 | 7.452 | 7.404 |
| $E[P]$ (3.7627) | 6.6352 | 6.5946 | 6.5974 | 6.5967 | 6.6048 |
| | ±0.0896 | ±0.0618 | ±0.0509 | ±0.0434 | ±0.0382 |
| $P\{P \leq 0.6656\}$ | 0.1501 | 0.1546 | 0.1597 | 0.1624 | 0.1603 |
| | ±0.0145 | ±0.0101 | ±0.0087 | ±0.0077 | ±0.0068 |
| $P\{P \leq 5.3314\}$ | 0.3505 | 0.3471 | 0.3436 | 0.3428 | 0.3417 |
| | ±0.0200 | ±0.0142 | ±0.0120 | ±0.0102 | ±0.0090 |
| $P\{P \leq 6.6627\}$ | 0.5713 | 0.5864 | 0.5829 | 0.5815 | 0.5785 |
| | ±0.0215 | ±0.0152 | ±0.0124 | ±0.0107 | ±0.0095 |
| $P\{P \leq 3.3254\}$ | 0.8845 | 0.8898 | 0.8896 | 0.8883 | 0.9162 |
| | ±0.0154 | ±0.0104 | ±0.0084 | ±0.0069 | ±0.0061 |

## Chapter 6

# Simulations With Simultaneous Events

For most discrete-event simulations on a finite or countably infinite state space, a "natural" definition of the state of the system leads to a stochastic process that can be specified as a GSMP. However, a natural state definition can lead to a stochastic process that either cannot be specified as a GSMP at all or cannot be specified as a GSMP using a natural event set. Often this is because the natural state of the system, although adequate for the quantities to be estimated, does not determine the set of events that are scheduled to occur. That is, there is no $(s \mapsto E(s))$ event-set mapping. A simple example is a multi-server queue for which the natural definition of the state of the system is the number of customers waiting or in service. In this example, of course, the state of the system can be defined to be the number of customers waiting or in service along with the indices of the servers that are busy. This state definition leads to a stochastic process that can be specified as a GSMP.

The formulation of a GSMP in Chapter 1 assumes that scheduled events occur one at a time so that there is a unique trigger event associated with each state transition. However, there are discrete-event simulations in which "natural" events can occur simultaneously. The occurrence of simultaneous events is advantageous, and in fact may be unavoidable, for concise specification of some discrete-event simulations.

It is possible to extend the definition of a GSMP so that two or more events can occur simultaneously and trigger a state transition. Formal definition of a "GSMP with simultaneous trigger events" is in terms of a GSSMC that describes the process at successive state transition times. There are analogues for GSMP's with simultaneous trigger events of the conditions under which a GSMP with unique trigger events is a regenerative process with finite cycle-length moments.

## 6.1. Simultaneous Trigger Events

The probability that two or more events ever occur simultaneously is zero if all clock-setting distributions are absolutely continuous. However, events can occur simultaneously and trigger a state transition if some of the clock-setting distributions have discrete components.

**Example 1.1.** (Token ring with packet destinations) Consider a unidirectional ring network with a fixed number of ports, labelled $1, 2, \ldots, N$ in the direction of signal propagation. Message packets arrive at each port for transmission to a unique destination according to a random process and queue externally. A single token (that is, a distinguished bit pattern) circulates around the ring from one port to the next. The time for the token to propagate from port $j$ to the next port is a positive constant $R_j$. When a port observes the token and there is a packet queued for transmission, the port converts the token to a connector (another distinguished bit pattern) and transmits a packet followed by the token pattern; the token continues to propagate if there is no packet queued for transmission. By destroying the connector prefix, the port removes the transmitted packet when it returns around the ring.

Assume that the successive times for port $j$ to transmit a packet are i.i.d. as a positive random variable $L_j$. Also assume that the successive times from end of transmission by

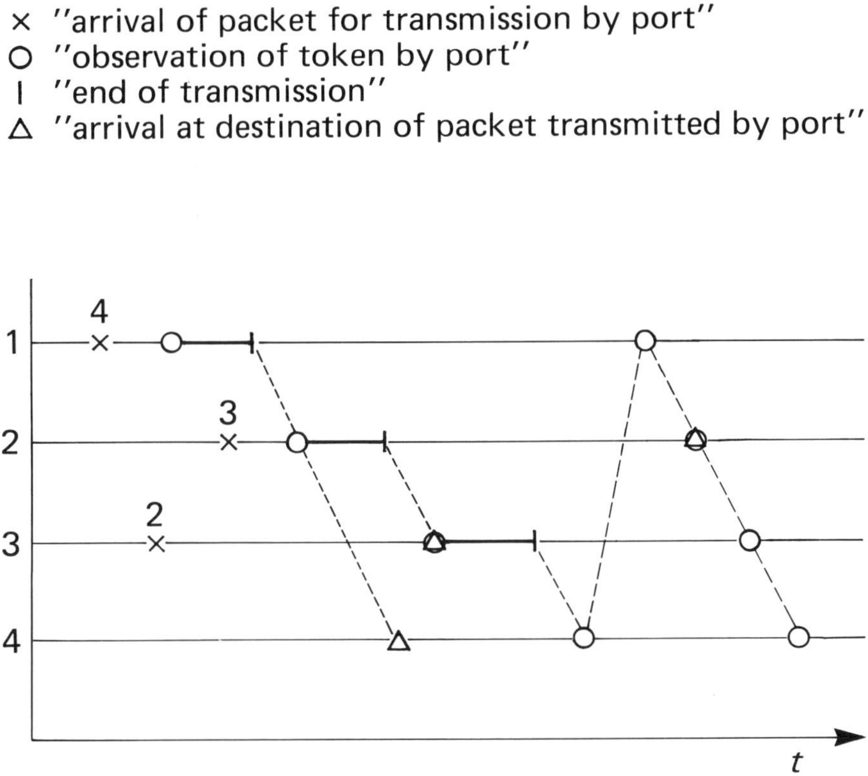


**Figure 6.1.** Time-line diagram for token ring with packet destinations.

port $j$ until the arrival of the next packet for transmission by port $j$ are i.i.d. as a positive random variable $A_j$; with independent probability $p_{jk}$ the destination of the packet is port $k$ ($\neq j$). All interarrival times and transmission times for packets are mutually independent. Observe that there is at most one packet awaiting transmission at any time at any particular port. Figure 6.1 shows a time-line diagram for the token ring with packet destinations. In the figure the number displayed above the symbol × represents the destination of a packet.

Irrespective of transmission times for packets and propagation delays, certain natural events can occur simultaneously. A packet can arrive at its destination (that is, the destination port observes the last bit of the packet) at the same time that the destination port observes the token.

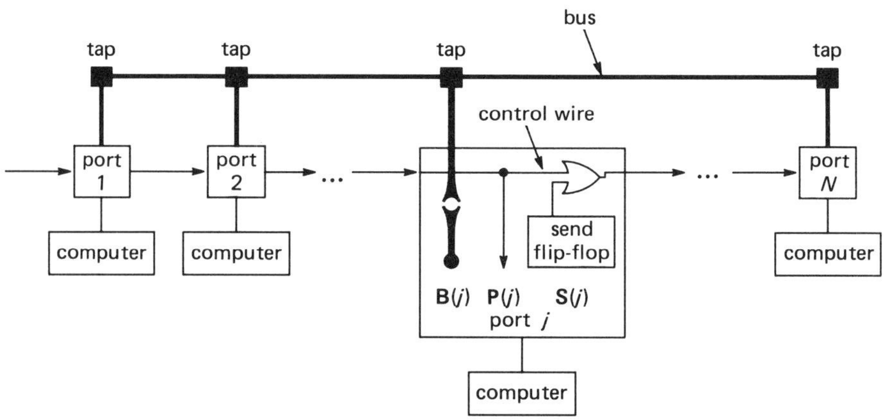


**Figure 6.2.** Collision-free bus network.

**Example 1.2.** (Collision-free bus network) Consider a local area bus network with $N$ ports, numbered $1, 2, \ldots, N$ from left to right; see Figure 6.2. Port $j$ transmits and monitors message packets on the bidirectional bus at tap $\mathbf{B}(j)$. In addition to the bus, a unidirectional (left to right) logic control wire also links the ports. Associated with each port $j$ is a flip-flop $\mathbf{S}(j)$ called the send flip-flop. Port $j$ sets $\mathbf{S}(j)$ to 1 and resets $\mathbf{S}(j)$ to 0. The signal $\mathbf{P}(j)$, called the OR-signal, is tapped at the control wire input to port $j$ and is the inclusive OR of the observed values of the send flip-flops of all ports to the left. Denote by $T$ the propagation delay from end-to-end along the bus plus a small fixed quantity. Let $R(j)$ be the propagation delay along the control wire from port $j$ to port $N$ for $1 \leq j \leq N$; thus, $R(1) > R(2) > \cdots > R(N) = 0$. Assume that signal propagation along the control wire is slower than along the bus in the sense that $R(1) > T$.

Distributed control scheme *A1* is specified in terms of an algorithm for an individual port. When there are no packets awaiting transmission by port $j$ and port $j$ is not transmitting a packet, the arrival of a packet for transmission by port $j$ initiates execution of the algorithm. If there is another packet awaiting transmission by port $j$ when this execution of the algorithm ends, the next execution of the algorithm begins immediately.

**Algorithm A1.**

1. *Set* $\mathbf{S}(j)$ *to 1;*
2. *Wait for a time interval* $R(j) + T$*;*
3. *Wait until the bus is observed at* $\mathbf{B}(j)$ *to be idle and* $\mathbf{P}(j) = 0$*; then start transmission of the packet, simultaneously resetting* $\mathbf{S}(j)$ *to 0.*

Control scheme *A1* is simple and asynchronous and provides collision-free communication among ports; that is, no two ports transmit signals that become electrically superimposed on the bus. Figure 6.3 shows a time-line diagram for the collision-free bus network.

For simplicity, we suppose that there is at most one packet awaiting transmission at any time at any particular port. Assume that the time from end of transmission by port $j$ until the arrival of the next packet for transmission by port $j$ is distributed as a positive random variable $A_j$ with finite mean and absolutely continuous distribution. Also assume that the time for port $j$ to transmit a packet is distributed as a positive random variable $L_j$ with finite mean and transmission times

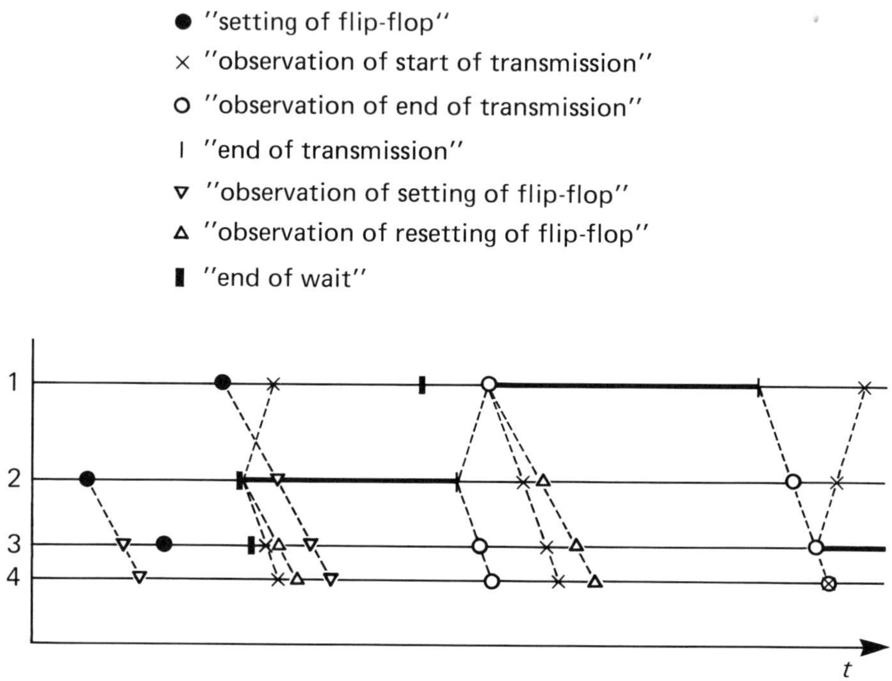


**Figure 6.3.** Time-line diagram for collision-free bus network.

are long in the sense that $P\{L_j > R(1) + T\} = 1$. Successive transmission times at port $j$ are mutually independent as are successive interarrival times for packets. Moreover, transmission times are independent of interarrival times, and random quantities associated with distinct ports are independent.

Denote the propagation delay along the bus between port $i$ and port $j$ by $T(i, j)$. Thus,

$$T(i, j) = T(j, i) < T$$

and

$$T(i, j) + T(j, k) = T(i, k)$$

for $i < j < k$ or $k > j > i$. Denote the propagation delay along the control wire from port $i$ to port $j$ by $R(i,j)$ for $i < j$, so that

$$R(i,j) = R(i) - R(j).$$

Since $R(1) > T$ and signals propagate at uniform rates along the bus and the control wire, necessarily

$$R(i,j)/T(i,j) = R(i,k)/T(i,k)$$

and $R(i,j) > T(i,j)$.

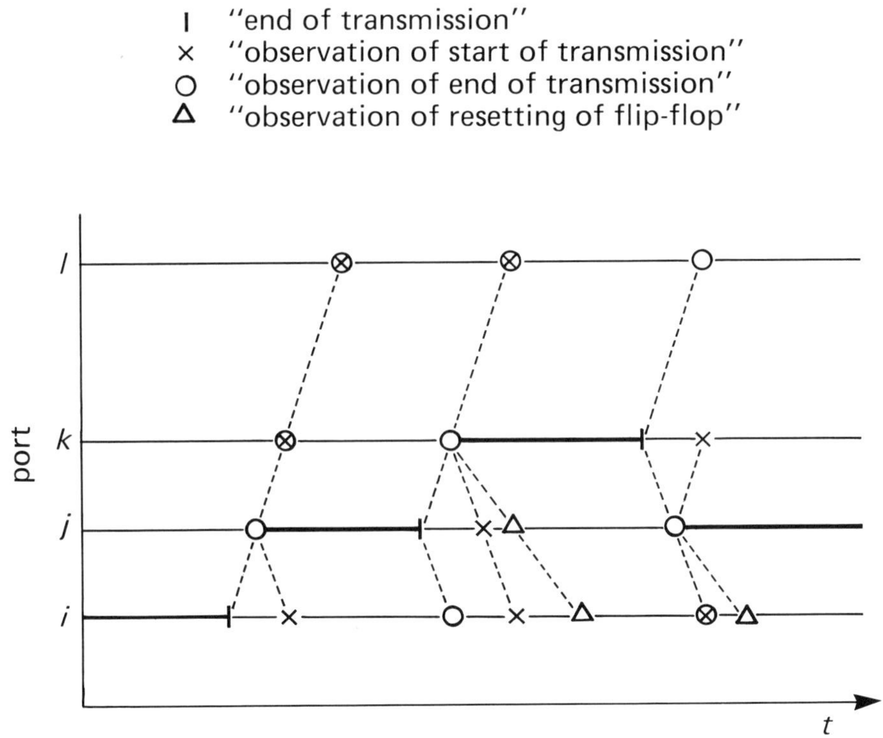


**Figure 6.4.** Simultaneous events in collision-free bus network.

Observe that, for $l < j < k$ and $l > j > k$, the natural events "observation by port $k$ of start of transmission by port $j$" and "observation by port $k$ of end of transmission by port $l$" can occur simultaneously. Moreover, for distinct $l, j, k$ with $T(j, l) = T(j, k)$, events "observation by port $l$ of start of transmission by port $j$" and "observation by port $k$ of start of transmission by port $j$" occur simultaneously as do events "observation by port $l$ of end of transmission by port $j$ and "observation by port $k$ of end of transmission by port $j$." Similarly, for distinct $l, j, k$ with $l < k$ and $T(l, j) = R(l, k)$, events "observation by port $j$ of start of transmission by port $l$" and "observation by port $k$ of resetting (to 0) of flip-flop by port $l$" occur simultaneously. Finally, observe that for $l < j < k$ and $l > j > k$ with $T(k, j) + T(j, k) = T(k, l)$, events "observation by port $l$ of end of transmission by port $k$" and "observation by port $k$ of start of transmission by port $j$" can occur simultaneously. In fact, as many as four natural events can occur simultaneously; see Figure 6.4.

## 6.2. Compound Events

It is sometimes possible to specify the underlying stochastic process of a discrete-event simulation with natural events that can occur simultaneously as a GSMP with unique trigger events. The idea is to use an event set that contains along with the natural events one or more "compound events" that occur when one or more natural events occur simultaneously. When using such compound events, it is often necessary to augment the natural state of the system in order for there to be an $(s \mapsto E(s))$ event-set mapping. In practice, it can be nontrivial to specify the event-set mapping.

**Example 2.1.** (Token ring with packet destinations) Assume as in Example 1.1 that the successive times for port $j$ to transmit a packet are i.i.d. as a positive random variable $L_j$. Also assume that the successive times from end of transmission by port $j$ until the arrival of the next packet for transmission by port $j$ are i.i.d. as a positive random variable $A_j$; with independent probability $p_{jk}$ the destination of the packet is port $k$ ($\neq j$). All interarrival times for packets and transmission times are mutually independent. For ease of exposition, assume that the distribution of each $A_j$ is absolutely continuous and $P\{L_j + R_{jm} = R_{jn}\} = 0$ for all $j$, $m$, and $n$, where

$$R_{jk} = \begin{cases} R_j + R_{j+1} + \cdots + R_{k-1} & \text{if } j \leq k-1 \\ R_j + \cdots + R_N + R_1 + \cdots + R_{k-1} & \text{if } j > k-1. \end{cases}$$

These assumptions ensure that, with probability 1, no two packets arrive simultaneously and no packet can arrive at its destination at the same time that some other port observes the token. Set

$$X(t) = (D_1(t), \ldots, D_N(t), Z_1(t), \ldots, Z_N(t), M(t), N(t)),$$

where

$$D_j(t) = \begin{cases} k & \text{if port } k \text{ is the destination of a packet transmitted by port } j \text{ and the transmitted packet has not yet arrived at port } k \text{ at time } t \\ j & \text{otherwise,} \end{cases}$$

$$Z_j(t) = \begin{cases} 1 & \text{if a packet is queued for transmission at port } j \text{ at time } t \\ 0 & \text{otherwise,} \end{cases}$$

$$M(t) = \begin{cases} j & \text{if port } j \text{ is transmitting a packet at time } t \\ 0 & \text{otherwise,} \end{cases}$$

$N(t) = 1$ if port $N$ is transmitting a packet or the token is propagating to port 1 at time $t$, and $N(t) = j$ if port $j-1$ is transmitting a packet or the token is propagating to port $j$ at time $t$ ($j = 2, 3, \ldots, N$).

Take event $e_{2N+3}$ = "observation of token by port and arrival at destination of transmitted packet," $e_{2N+2}$ = "observation of token by port," $e_{2N+1}$ = "end of transmission," $e_{N+j}$ = "arrival at destination of packet transmitted by port $j$," and $e_j$ = "arrival of packet for transmission by port $j$" ($j = 1, 2, \ldots, N$). Formal specification of the process $\{X(t) \colon t \geq 0\}$ is as a GSMP with (unique trigger events and) a finite state space

$$S = \big\{(d_1, \ldots, d_N, z_1, \ldots, z_N, m, n) \in \{1, 2, \ldots, N\}^N \times \{0, 1\}^N \\ \times \{0, 1, \ldots, N\} \times \{1, 2, \ldots, N\} \colon \quad n = m + 1 \\ \text{when } 0 < m < N; \; n = 1 \text{ when } m = N; \; d_j \neq j \\ \text{when } m = j; \; m = j \text{ only if } z_j = 0 \text{ for } j = 1, 2, \ldots, N\big\}$$

and event set $E = \{e_1, e_2, \ldots, e_{2N+3}\}$. Observe that if event $e_{2N+3}$ occurs in state $s = (d_1, \ldots, d_N, z_1, \ldots, z_N, m, n)$, then $d_j = n$ and $d_j \neq j$ for exactly one $j$. For $s = (d_1, \ldots, d_N, z_1, \ldots, z_N, m, n) \in S$, event $e_{2N+3}$ = "observation of token by port and arrival at destination of transmitted packet" $\in E(s)$ if and only if $m = 0$ and $d_j = n$ for some $j \neq n$. Event $e_{2N+2}$ = "observation of token by port" $\in E(s)$ if and only if $m = 0$ and either $d_j = j$ or $n \neq j$ for all $j$. Event $e_{2N+1}$ = "end of transmission" $\in E(s)$ if and only if $m > 0$. Event $e_{N+j}$ = "arrival at destination of packet transmitted by port $j$" $\in E(s)$ if and only if $d_j \neq j$ and $n \neq d_j$. Event $e_j$ = "arrival of packet for transmission by port $j$" $\in E(s)$ if and only if $z_j = 0$ and $m \neq j$. All speeds $r(s, e)$ for active events are equal to 1.

If $e = e_{2N+3}$ = "observation of token by port and arrival at destination of transmitted packet," then the state transition probability $p(s'; s, e) = p_{nk}$ when

$$s = (d_1, \ldots, d_N, z_1, \ldots, z_{n-1}, 1, z_{n+1}, \ldots, z_N, 0, n) \\ \text{with } d_j = n \;\; (j \neq n, \; n < N) \\ \text{and } s' = (d_1, \ldots, d_{j-1}, j, d_{j+1}, \ldots, d_{n-1}, k, d_{n+1}, \ldots, d_N, \\ z_1, \ldots, z_{n-1}, 0, z_{n+1}, \ldots, z_N, n, n+1),$$

$p(s'; s, e) = p_{Nk}$ when

$$\begin{aligned} s = (&d_1, \ldots, d_{N-1}, N, z_1, \ldots, z_{N-1}, 1, 0, N) \\ &\text{with } d_j = N\,(j \neq N) \\ &\text{and } s' = (d_1, \ldots, d_{j-1}, j, d_{j+1}, \ldots, d_{N-1}, k, \\ &z_1, \ldots, z_{N-1}, 0, N, 1), \end{aligned}$$

and $p(s'; s, e) = 1$ when

$$\begin{aligned} s = (&d_1, \ldots, d_N, z_1, \ldots, z_{n-1}, 0, z_{n+1}, \ldots, z_N, 0, n) \\ &\text{with } d_j = n\,(j \neq n,\ n < N) \\ &\text{and } s' = (d_1, \ldots, d_{j-1}, j, d_{j+1}, \ldots, d_N, z_1, \ldots, z_{n-1}, 0, \\ &z_{n+1}, \ldots, z_N, 0, n+1), \end{aligned}$$

and when

$$\begin{aligned} s &= (d_1, \ldots, d_N, z_1, \ldots, z_{N-1}, 0, 0, N) \text{ with } d_j = N\,(j \neq N) \\ &\text{and } s' = (d_1, \ldots, d_{j-1}, j, d_{j+1}, \ldots, d_N, z_1, \ldots, z_{N-1}, 0, 0, 1). \end{aligned}$$

If $e = e_{2N+2}$ = "observation of token by port," then $p(s'; s, e) = p_{nk}$ when

$$\begin{aligned} s = (&d_1, \ldots, d_{n-1}, n, d_{n+1}, \ldots, d_N, z_1, \ldots, z_{n-1}, 1, \\ &z_{n+1}, \ldots, z_N, 0, n) \text{ with } n < N \\ &\text{and } s' = (d_1, \ldots, d_{n-1}, k, d_{n+1}, \ldots, d_N, \\ &z_1, \ldots, z_{n-1}, 0, z_{n+1}, \ldots, z_N, n, n+1), \end{aligned}$$

$p(s'; s, e) = p_{Nk}$ when

$$\begin{aligned} s &= (d_1, \ldots, d_{N-1}, N, z_1, \ldots, z_{N-1}, 1, 0, N) \\ &\text{and } s' = (d_1, \ldots, d_{N-1}, k, z_1, \ldots, z_{N-1}, 0, N, 1), \end{aligned}$$

and $p(s'; s, e) = 1$ when

$$\begin{aligned} s &= (d_1, \ldots, d_N, z_1, \ldots, z_{n-1}, 0, z_{n+1}, \ldots, z_N, 0, n) \text{ with } n < N \\ &\text{and } s' = (d_1, \ldots, d_N, z_1, \ldots, z_{n-1}, 0, z_{n+1}, \ldots, z_N, 0, n+1), \end{aligned}$$

and when

$$s = (d_1, \ldots, d_N, z_1, \ldots, z_{N-1}, 0, 0, N) \\ \text{and } s' = (d_1, \ldots, d_N, z_1, \ldots, z_{N-1}, 0, 0, 1).$$

If $e = e_{2N+1} =$ "end of transmission," then $p(s'; s, e) = 1$ when

$$s = (d_1, \ldots, d_N, z_1, \ldots, z_N, m, m+1) \text{ with } 0 < m < N \\ \text{and } s' = (d_1, \ldots, d_N, z_1, \ldots, z_N, 0, m+1)$$

and when

$$s = (d_1, \ldots, d_N, z_1, \ldots, z_N, N, 1) \\ \text{and } s' = (d_1, \ldots, d_N, z_1, \ldots, z_N, 0, 1).$$

If $e_{N+j} =$ "arrival at destination of packet transmitted by port $j$," then $p(s'; s, e) = 1$ when

$$s = (d_1, \ldots, d_N, z_1, \ldots, z_N, m, n) \\ \text{and } s' = (d_1, \ldots, d_{j-1}, j, d_{j+1}, \ldots, d_N, z_1, \ldots, z_N, m, n).$$

If $e = e_j =$ "arrival of packet for transmission by port $j$," then $p(s'; s, e) = 1$ when

$$s = (d_1, \ldots, d_N, z_1, \ldots, z_{j-1}, 0, z_{j+1}, \ldots, z_N, m, n) \\ \text{and } s' = (d_1, \ldots, d_N, z_1, \ldots, z_{j-1}, 1, z_{j+1}, \ldots, z_N, m, n).$$

The clock-setting distribution functions are as follows. If the new event $e' = e_{2N+3}$ and $s' = (d_1, \ldots, d_N, z_1, \ldots, z_N, m, n)$, then $F(x; s', e', s, e^*) = 1_{[R_{n-1}, \infty)}(x)$ if $n > 1$ and $F(x; s', e', s, e^*) = 1_{[R_N, \infty)}(x)$ if $n = 1$. If $e' = e_{2N+2}$ and $s' = (d_1, \ldots, d_N, z_1, \ldots, z_N, m, n)$, then $F(x; s', e', s, e^*) = 1_{[R_{n-1}, \infty)}(x)$ if $n > 1$ and equals $1_{[R_N, \infty)}(x)$ if $n = 1$. If $e' = e_{2N+1}$ and $s' = (d_1, \ldots, d_N, z_1, \ldots, z_N, m, n)$, then $F(x; s', e', s, e^*) = \{L_m \leq x\}$. If $e' = e_{N+j}$ and $s' = (d_1, \ldots, d_N, z_1, \ldots, z_N, m, n)$, then $F(x;$

$s', e', s, e^*) = 1_{[R_j+\cdots+R_{k-1},\infty)}(x)$ if $j \leq k-1$ and $F(x; s', e', s, e^*) = 1_{[R_j+\cdots R_N+R_1+\cdots+R_{k-1},\infty)}(x)$ if $j > k-1$. If $e' = e_j$, then $F(x; s', e', s, e^*) = P\{A_j \leq x\}$.

Suppose that at time 0 the token has just arrived at port 1 and all ports have a packet queued for transmission. Then the initial distribution $\mu$ is

$$\mu(A) = 1_{\{s_0\}}(s)P\{L_1 \leq a_{2N+1}\}$$

for all sets

$$A = \{s\} \times \{(c_1, \ldots, c_{2N+3}) \colon 0 \leq c_i \leq a_i \text{ for } 1 \leq i \leq 2N+3\},$$

where $s_0 = (0, \ldots, 0, 0, 1, 1, \ldots, 1, 1, 2)$.

**Example 2.2.** (Collision-free bus network) Assume as in Example 1.2 that the time from end of transmission by port $j$ until the arrival of the next packet for transmission by port $j$ is distributed as a positive random variable $A_j$ with finite mean and absolutely continuous distribution. Also assume that the time for port $j$ to transmit a packet is distributed as a positive random variable $L_j$ with finite mean and transmission times are long in the sense that $P\{L_j > R(1) + T\} = 1$.

Irrespective of propagation delays, a port always can observe an end of transmission and a start of transmission simultaneously, as discussed below. For ease of exposition, assume that $T(l,j) \neq T(l,k)$, $T(l,j) \neq R(l,k)$, $T(k,j) + T(j,k) \neq T(k,l)$, and $T(k,j) + T(j,m) \neq T(k,l)$ for distinct ports $j$, $k$, $l$, and $m$. These assumptions on propagation delays ensure that no two ports ever observe a start (or end) of transmission simultaneously, that no port ever observes a start of transmission simultaneously with the resetting of a flip-flop, and no two ports ever observe a start of transmission and end of transmission simultaneously. The assumption that the distribution of each $A_j$ is absolutely continuous rules out all other such phenomena.

Set

$$X(t) = (W_1(t), \ldots, W_N(t), Z(t), U_1(t), \ldots, U_N(t), \\ V_{2,1}(t), V_{3,1}(t), V_{3,2}(t), V_{4,1}(t), \ldots, V_{N,N-1}(t)),$$

where

$$Z(t) = \begin{cases} k & \text{if some port is transmitting a packet at time } t \\ & \text{and the port started transmission when it} \\ & \text{observed an end of transmission by port } k \\ 0 & \text{otherwise,} \end{cases}$$

$$W_j(t) = \begin{cases} 3 & \text{if port } j \text{ is transmitting a packet at time } t \\ 2 & \text{if port } j \text{ has completed the } R(j)+T \text{ wait} \\ & \text{but has not started transmission at time } t \\ 1 & \text{if port } j \text{ has set its flip-flop but has not yet} \\ & \text{completed the } R(j)+T \text{ wait at time } t \\ 0 & \text{otherwise,} \end{cases}$$

$$U_j(t) = \begin{cases} k & \text{if at time } t \text{ port } j \text{ observes transmission} \\ & \text{of a packet by port } k \\ 0 & \text{otherwise,} \end{cases}$$

and

$$V_{j,k}(t) = \begin{cases} 1 & \text{if } \mathbf{S}(k) = 1 \text{ at time } t - R(k,j) \\ 0 & \text{otherwise} \end{cases}$$

for $k < j$. (Port $j$ observes $\mathbf{P}(j)$ =1 at time $t$ if $V_{j,k}(t) = 1$ for some $k < j$.)

Take event $e_{j,1}$ = "setting (to 1) of flip-flop by port $j$," $e_{j,2}$ = "end of wait for $R(j) + T$," $e_{j,3}$ = "end of transmission by port $j$," $e_{j,l,4}$ = "observation by port $j$ of start of transmission by port $l \neq j$," $e_{j,l,5}$ = "observation by port $j$ of end of transmission by port $l \neq j$," $e_{j,k,6}$ = "observation by port $j$ of the setting (to 1) of flip-flop by port $k$ to the left," $e_{j,k,7}$ = "observation by port $j$ of the resetting (to 0) of flip-flop by port $k$ to the left," and the compound event $e_{j,l,m,8}$ = "observation by port $j$ of end of transmission by port $l$ and start of transmission by port $m$" for $l \neq j$, $k < j$, $l < m < j$ or $j < m < l$, and $j, k, l, m = 1, 2, \ldots, N$. Formal specification of the process $\{X(t): t \geq 0\}$ is as a GSMP with (unique trigger events and) a finite state space $S$ and event set

$$E = \big\{e_{j,1}, e_{j,2}, e_{j,3}, e_{j,l,4}, e_{j,l,5}, e_{j,k,6}, e_{j,k,7}, e_{j,l,m,8}\colon\ l \neq j, k < j,\\ l < m < j \text{ or } j < m < l, \text{ and } j, k, l, m = 1, 2, \ldots, N\big\}.$$

The state space $S$ is the set of all vectors $s = (w_1, \ldots, w_N, z, u_1, \ldots, u_N, v_{2,1}, \ldots, v_{N,N-1}) \in \{0, 1, 2, 3\}^N \times \{0, 1, \ldots, N\}^{N+1} \times \{0, 1\}^{N(N-1)/2}$ satisfying

(i) $w_j = 3$ for at most one $j$;
(ii) $u_j = 3$ if and only if $w_j = 3$;
(iii) $u_j = k$ whenever $w_k = 3$ and $u_l = k$ for some $l$ with $T(k, l) \geq T(k, j)$;
(iv) $u_j = k$ whenever $w_k \neq 3$ and $u_l = k$ for some $l$ with $T(k, j) \geq T(k, l)$;
(v) $u_j = u_k$, $u_j = u_l$, or $u_k = u_l$ whenever $u_j u_k u_l > 0$;
(vi) $v_{j,k} = 1$ for all $j > k$ whenever $w_k = 2$;
(vii) $v_{j,k} = 1$ whenever $w_k = 3$ and $v_{l,k} = 1$ for some $l$ with $l < j$;
(viii) $v_{j,k} = 1$ whenever $w_k \neq 3$ and $v_{l,k} = 1$ for some $l$ with $j < l$;
(ix) $v_{j,k} = 1$ whenever $w_k = 3$ and $u_l \neq k$ for some $l$ with $T(k, l) \leq R(k, j)$;
(x) $v_{j,k} = 0$ whenever $w_k = 3$ and $u_l = k$ for some $l$ with $R(k, j) \leq T(k, l)$; and
(xi) $z > 0$ only if $w_j = 3$ for some $j$.

For $s = (w_1, \ldots, w_N, z, u_1, \ldots, u_N, v_{2,1}, \ldots, v_{N,N-1}) \in S$, event $e_{j,1}$ = "setting (to 1) of flip-flop by port $j$" $\in E(s)$ if and only if $w_j = 0$. Event $e_{j,2}$ = "end of wait for $R(j)+T$" $\in E(s)$ if and only if $w_j = 1$. Event $e_{j,3}$ = "end of transmission by port

$j$"$\in E(s)$ if and only if $w_j = 3$. Event $e_{j,l,4}$ = "observation by port $j$ of start of transmission by port $l \neq j$"$\in E(s)$ if and only if (i) $z = 0$, $w_l = 3$ and $u_j \neq l$; or (ii) $z > 0$, $w_l = 3$, $u_j \neq l$, and either $z < j < l$ or $l < j < z$. Event $e_{j,l,5}$ = "observation by port $j$ of end of transmission by port $l \neq j$"$\in E(s)$ if and only if (i) $z = 0$, $u_j = l$, and $w_l = 0$ or 1; or (ii) $z = l$, $u_j = z$, $w_l = 0$ or 1, $w_m = 3$, and either $m < l < j$ or $j < l < m$. Event $e_{j,k,6}$ = "observation by port $j$ of setting of flip-flop by port $k$ to the left" $\in E(s)$ if and only if $w_k = 1$ and $v_{j,k} = 0$. Event $e_{j,k,7}$ = "observation by port $j$ of resetting of flip-flop by port $k$ to the left" $\in E(s)$ if and only if $w_k = 3$ and $v_{j,k} = 1$. Event $e_{j,l,m,8}$ = "observation by port $j$ of end of transmission by port $l$ and start of transmission by port $m$" $\in E(s)$ if and only if ($l < m < j$ or $j < m < l$ and) $z = l$, $u_j = l$, $w_l = 0$ or 1, and $w_m = 3$. All speeds for active events are equal to 1.

With this event-set mapping, all "observation by port $j$ of setting of flip-flop by port $k$" events become active when port $k$ sets its flip-flop, all "observation by port $j$ of start of transmission by port $k$" events and all "observation by port $j$ of resetting of flip-flop by port $k$" events become active when port $k$ starts transmission of a packet, and all "observations by port $j$ of end of transmission by port $k$" events become active when port $k$ ends transmission of a packet. The event-set mapping also ensures that no "observation by port $j$ of end of transmission by port $l$" and "observation by port $j$ of start of transmission by port $m$" events can occur simultaneously. To see this, suppose that (with $l < m < j$ or $j < m < l$) port $l$ ends transmission of a packet at time $t$ and port $m$ starts transmission of a packet at time $t' = t + T(l,m)$. Then the compound event "observation by port $j$ of end of transmission by port $l$ and start of transmission by port $m$" is scheduled at time $t'$, and the event "observation by port $j$ of end of transmission by port $l$" (that was scheduled at time $t$) is cancelled at time $t'$. The "unnatural" state variable $Z(t)$ is needed for specification of the event-set mapping.

If $e = e_{j,1}$ = "setting (to 1) of flip-flop by port $j$," then the state transition probability $p(s'; s, e) = 1$ when

$$s = (w_1, \ldots, w_{j-1}, 0, w_{j+1}, \ldots, w_N, z, u_1, \ldots, v_{N,N-1})$$
$$\text{and } s' = (w_1, \ldots, w_{j-1}, 1, w_{j+1}, \ldots, w_N, z, u_1, \ldots, v_{N,N-1}).$$

If $e = e_{j,2}$ = "end of wait for $R(j) + T$," then $p(s'; s, e) = 1$ when

$$s = (w_1, \ldots, w_{j-1}, 1, w_{j+1}, \ldots, w_N, z, u_1, \ldots, v_{N,N-1})$$
$$\text{with } u_j > 0 \text{ or } v_{j,1} + \cdots + v_{j,j-1} > 0$$
$$\text{and } s' = (w_1, \ldots, w_{j-1}, 2, w_{j+1}, \ldots, w_N, z, u_1, \ldots, v_{N,N-1})$$

and when

$$s = (w_1, \ldots, w_{j-1}, 1, w_{j+1}, \ldots, w_N, z, u_1, \ldots, u_{j-1}, 0, u_{j+1}, \ldots,$$
$$v_{j-1,j-2}, 0, \ldots, 0, v_{j+1,1}, \ldots, v_{N,N-1})$$
$$\text{and } s' = (w_1, \ldots, w_{j-1}, 3, w_{j+1}, \ldots, w_N, 0,$$
$$u_1, \ldots, u_{j-1}, j, u_{j+1}, \ldots, v_{N,N-1}).$$

If $e = e_{j,3}$ = "end of transmission by port $j$," then $p(s'; s, e) = 1$ when

$$s = (w_1, \ldots, w_{j-1}, 3, w_{j+1}, \ldots, w_N, z, u_1, \ldots, u_{j-1}, j, u_{j+1},$$
$$\ldots, u_N, v_{2,1}, \ldots, v_{N,N-1})$$
$$\text{and } s' = (w_1, \ldots, w_{j-1}, 0, w_{j+1}, \ldots, w_N, 0,$$
$$u_1, \ldots, u_{j-1}, 0, u_{j+1}, \ldots, v_{N,N-1}).$$

If $e = e_{j,l,4}$ = "observation by port $j$ of start of transmission by port $l \neq j$," then $p(s'; s, e) = 1$ when

$$s = (w_1, \ldots, w_N, z, u_1, \ldots, u_{j-1}, 0, u_{j+1}, \ldots, v_{N,N-1})$$
$$\text{with } u_k = l \text{ for all } k \text{ such that } l < k < j \text{ or } j < k < l$$
$$\text{and } s' = (w_1, \ldots, w_N, z, u_1, \ldots, u_{j-1}, l, u_{j+1}, \ldots, v_{N,N-1}).$$

When $s = (w_1, \ldots, w_N, z, u_1, \ldots, u_{j-1}, 0, u_{j+1}, \ldots, v_{N,N-1})$ with $u_k \neq l$ for some $k$ such that $l < k < j$ or $j < k < l$, the probability mass function $p(\,\cdot\,; s, e_{j,l,4})$ may be defined arbitrarily. If $e = e_{j,l,5}$ = "observation by port $j$ of end of transmission by port $l \neq j$," then $p(s'; s, e) = 1$ when

$$
\begin{aligned}
s = (w_1, \ldots, w_N, z, u_1, \ldots, u_{j-1}, l, u_{j+1}, \ldots, v_{N,N-1})& \\
\text{with } u_k = 0 \text{ for all } k \text{ such that } l < k < j \text{ or } j < k < l& \\
\text{and either } w_j \neq 2 \text{ or } v_{j,1} + \cdots + v_{j,j-1} > 0& \\
\text{and } s' = (w_1, \ldots, w_N, z, u_1, \ldots, u_{j-1}, 0, u_{j+1}, \ldots, v_{N,N-1})&
\end{aligned}
$$

and when

$$
\begin{aligned}
s = (w_1, \ldots, w_{j-1}, 2, w_{j+1}, \ldots, w_N, z, u_1, \ldots, v_{j-1,j-2}, 0, \ldots, 0,& \\
v_{j+1,1}, \ldots, v_{N,N-1}) \text{ with } u_k = 0 \text{ for all } k \text{ such that}& \\
l < k < j \text{ or } j < k < l& \\
\text{and } s' = (w_1, \ldots, w_{j-1}, 3, w_{j+1}, \ldots, w_N, l, u_1, \ldots, u_{j-1}, j,& \\
u_{j+1}, \ldots, v_{N,N-1}).&
\end{aligned}
$$

When $s = (w_1, \ldots, w_N, z, u_1, \ldots, u_{j-1}, l, u_{j+1}, \ldots, v_{N,N-1})$ with $u_k \neq l$ for some $k$ such that $l < k < j$ or $j < k < l$ and either $w_j \neq 2$ or $v_{j,1} + \cdots + v_{j,j-1} > 0$ and when $s = (w_1, \ldots, w_{j-1}, 2, w_{j+1}, \ldots, w_N, z, u_1, \ldots, v_{j-1,j-2}, 0, \ldots, 0, v_{j+1,1}, \ldots, v_{N,N-1})$ with $u_k = 0$ for all $k$ such that $l < k < j$ or $j < k < l$, the probability mass function $p(\,\cdot\,; s, e_{j,l,5})$ may be defined arbitrarily. If $e = e_{j,k,6}$ = "observation by port $j$ of the setting (to 1) of flip-flop by port $k$ to the left," then $p(s'; s, e) = 1$ when

$$
\begin{aligned}
s = (w_1, \ldots, w_N, z, u_1, \ldots, v_{j,k-1}, 0, v_{j,k+1}, \ldots, v_{N,N-1})& \\
\text{with } v_{l,k} = 1 \text{ for all } l \text{ such that } k < l < j& \\
\text{and } s' = (w_1, \ldots, w_N, z, u_1, \ldots, u_N, v_{2,1}, \ldots, v_{j,k-1}, 1,& \\
v_{j,k+1}, \ldots, v_{N,N-1}).&
\end{aligned}
$$

When $s = (w_1, \ldots, w_N, z, u_1, \ldots, u_N, v_{2,1}, \ldots, v_{j,k-1}, 0, v_{j,k+1}, \ldots, v_{N,N-1})$ with $v_{l,k} = 0$ for some $l$ such that $k < l < j$, the probability mass function $p(\,\cdot\,; s, e_{j,k,6})$ may be defined arbitrarily. If $e = e_{j,k,7} =$ "observation by port $j$ of the resetting (to 0) of flip-flop by port $k$ to the left," then $p(s'; s, e) = 1$ when

$$
\begin{aligned}
s = (w_1, \ldots, w_N, z, u_1, \ldots, v_{j,k-1}, 1, v_{j,k+1}, \ldots, v_{N,N-1}) \\
\text{with } v_{l,k} = 0 \text{ for all } l \text{ such that } k < l < j \\
\text{and } s' = (w_1, \ldots, v_{j,k-1}, 0, v_{j,k+1}, \ldots, v_{N,N-1}).
\end{aligned}
$$

When $s = (w_1, \ldots, w_N, z, u_1, \ldots, u_N, v_{2,1}, \ldots, v_{j,k-1}, 1, v_{j,k+1}, \ldots, v_{N,N-1})$ with $v_{l,k} = 1$ for some $l$ such that $k < l < j$, the probability mass function $p(\,\cdot\,; s, e_{j,k,7})$ may be defined arbitrarily. If $e = e_{j,l,m,8}$, then $p(s'; s, e) = 1$ when

$$
\begin{aligned}
s = (w_1, \ldots, w_N, z, u_1, \ldots, u_{j-1}, l, u_{j+1}, \ldots, v_{N,N-1}) \\
\text{and } s' = (w_1, \ldots, w_N, z, u_1, \ldots, u_{j-1}, m, u_{j+1}, \ldots, v_{N,N-1}).
\end{aligned}
$$

The clock-setting distribution functions are as follows. If the new event $e' = e_{j,1}$, then $F(x; s', e', s, e^*) = P\{A_j \le x\}$. If $e' = e_{j,2}$, then $F(x; s', e', s, e^*) = 1_{[R(j)+T,\infty)}(x)$. If $e' = e_{j,3}$, then $F(x; s', e', s, e^*) = P\{L_j \le x\}$. If $e' = e_{j,l,4}$ or if $e' = e_{j,l,5}$, then $F(x; s', e', s, e^*) = 1_{[T(l,j),\infty)}(x)$. If $e' = e_{j,k,6}$, then $F(x; s', e', s, e^*) = 1_{[R(k,j),\infty)}(x)$. If $e' = e_{j,k,7}$, then $F(x; s', e', s, e^*) = 1_{[R(k,j),\infty)}(x)$. If $e' = e_{j,l,m,8}$, then $F(x; s', e', s, e^*) = 1_{[T(m,j),\infty)}(x)$.

**Example 2.3.** (Register insertion ring) Consider a unidirectional ring network that contains a fixed number of "idle bits" and has a fixed number of stations, labelled $1, 2, \ldots, N$ in the direction of signal propagation. Packets of fixed length arrive at each station according to a random process and queue externally. When a station has a packet to transmit and is observing

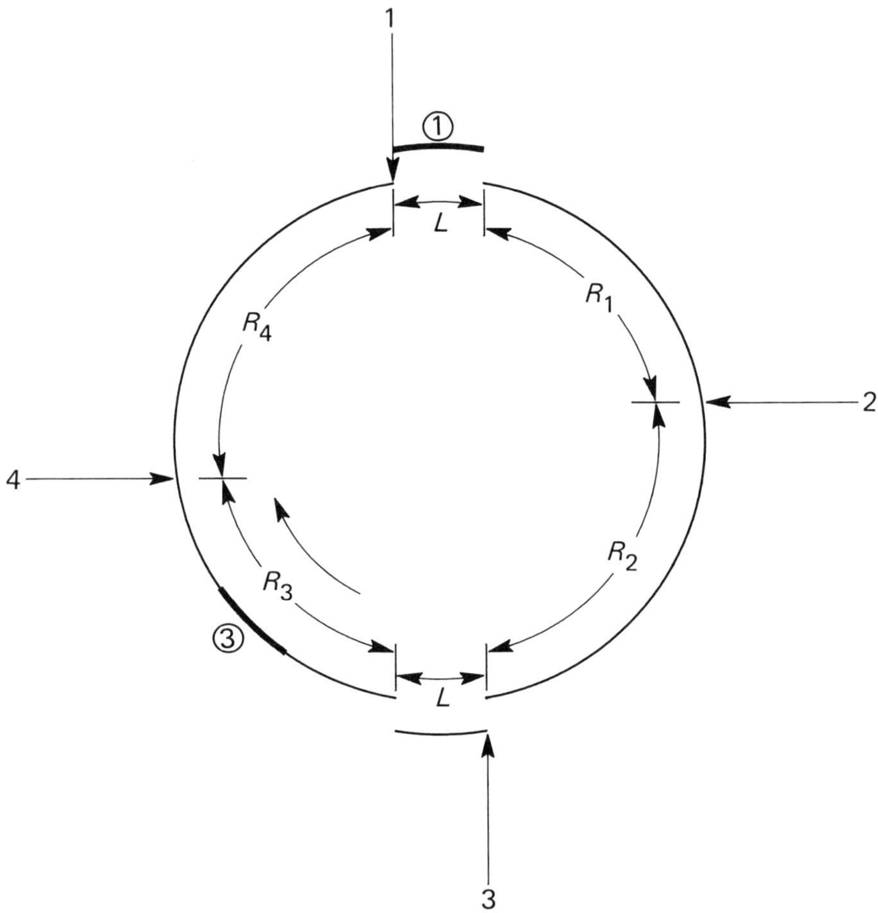


**Figure 6.5.** Register insertion ring.

idle bits (that is, no packet is passing), the station instantaneously inserts a register (buffer) that contains the packet inline on the input side of the delay between the station receiver and station transmitter; see Figure 6.5. If the station is observing packet bits (that is, another packet is passing), the station waits until that packet has passed before inserting a register that contains the new packet. The source station leaves its register inserted until its packet has returned and is contained in the register. The station then instantaneously removes the inserted register from the ring, thus removing the packet.

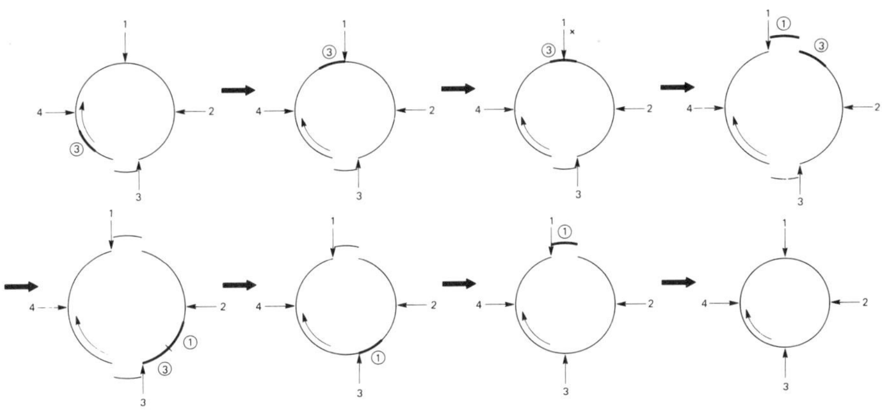

**Figure 6.6.** Propagation of packets in register insertion ring.

Figure 6.6 shows the propagation of packets around a register insertion ring with $N = 4$ stations. A circled number represents the station that has transmitted a packet, and $\times$ represents a packet that is queued for transmission at a station. In the figure, station 3 has a register inserted and its packet is propagating around the ring. Subsequently, station 1 observes the first bit of the packet. While station 1 is observing packet bits, there is an arrival of a packet for transmission by station 1. Station 1 waits until the packet has passed before inserting a register containing the new packet, thereby increasing the size of the ring. The packets transmitted by stations 3 and 1 continue to propagate, and station 3 observes the first bit of its packet. When the last bit of the packet transmitted by station 3 has returned and is contained in its register, station 3 instantaneously removes its register, thereby decreasing the size of the ring. When the last bit of the packet transmitted by station 1 has returned and is contained in its register, station 1 removes its register, and the ring is empty.

Denote the constant propagation delay along the ring from station $j$ to the next station by $R_j$ and the constant propagation delay through an inserted register by $L$. Packets are short in the sense that $L < R_1 + R_2 + \cdots + R_N$. Assume that the time from the removal of the inserted register by station $j$ until the arrival of the next packet for transmission by station $j$ is distributed as a positive random variable $A_j$ with finite mean. With independent probability $p_{jk}$, the destination of a packet that arrives for transmission by station $j$ is station $k$ ($\neq j$). Observe that there can be at most one packet queued for transmission at any time at any particular station.

For ease of exposition, assume that the distribution of each $A_j$ is absolutely continuous and $L \neq R_{jk}$ for all $j$ and $k$, where

$$R_{jk} = \begin{cases} R_j + R_{j+1} + \cdots + R_{k-1} & \text{if } j \leq k-1 \\ R_j + \cdots + R_N + R_1 + \cdots + R_{k-1} & \text{if } j > k-1. \end{cases}$$

These assumptions ensure that, with probability 1, no two packets arrive simultaneously, no packet arrives at the same time that a station observes the first (or last) bit of some other packet, and no station can observe the first bit of a packet at the same time that some other station observes the last bit of the packet.

Set

$$X(t) = (Z_1(t), \ldots, Z_N(t), B_1(t), \ldots, B_N(t), D_1(t), \ldots, D_N(t)),$$

where

$$Z_j(t) = \begin{cases} N+i & \text{if station } j \text{ has a packet queued for} \\ & \text{transmission at time } t \text{ and is observing} \\ & \text{a packet bit transmitted by station } i \\ i & \text{if station } j \text{ has no packet queued for} \\ & \text{transmission at time } t \text{ and is observing} \\ & \text{a packet bit transmitted by station } i \\ 0 & \text{if station } j \text{ is observing an idle bit} \\ & \text{at time } t, \end{cases}$$

$$B_j(t) = \begin{cases} N+i & \text{if station } j \text{ has a packet on the ring} \\ & \text{at time } t \text{ and the first bit of the packet} \\ & \text{transmitted by station } j \text{ follows the last} \\ & \text{bit of a packet transmitted by station } i \\ i & \text{if station } j \text{ has a packet on the ring} \\ & \text{at time } t \text{ and the first bit of the packet} \\ & \text{transmitted by station } j \text{ follows an idle bit} \\ & \text{that is propagating to station } i \\ 0 & \text{if station } j \text{ has no packet on the ring} \\ & \text{at time } t, \end{cases}$$

and

$$D_j(t) = \begin{cases} i & \text{if station } j \text{ has a packet on the ring at time } t \\ & \text{and station } i \text{ is the destination of the packet} \\ 0 & \text{if station } j \text{ has no packet on the ring at time } t. \end{cases}$$

Take event $e_{2N+j}$ = "arrival of packet for transmission by station $j$," $e_{N+j}$ = "observation by station $j$ of last bit of packet transmitted by station," and $e_j$ = "observation by station of first bit of packet transmitted by station $j$" ($j = 1, 2, \ldots, N$). Formal specification of the process $\{X(t)\colon\ t \geq 0\}$ is as a GSMP with (unique trigger events and) a finite state space $S$ and event set

$$E = \{e_1, e_2, \ldots, e_N, e_{N+1}, \ldots, e_{2N}, e_{2N+1}, \ldots, e_{3N}\}.$$

The state space $S$ is the set of all vectors $s = (z_1, \ldots, z_N, \ldots, d_1, \ldots, d_N) \in \left(\prod_{j=1}^{N}\{0, 1, \ldots, N+j-1, N+j+1, \ldots, 2N\}\right)^2 \times \prod_{j=1}^{N}\{0, 1, \ldots, j-1, j+1, \ldots, N\}$ satisfying

(i) $z_j = N + i$ only if $d_j > 0$ and $b_j = 0$;
(ii) $z_j = i$ only if $d_i > 0$;
(iii) $z_j = N + i$ only if $d_i > 0$;
(iv) $z_j = i$ only if $b_i > 0$;
(v) $d_j = 0$ only if $b_j = 0$;
(vi) $b_j = i$ only if $b_i > 0$;
(vii) $b_j = N + i$ only if $b_i > 0$; and
(viii) $b_j = N + i$ and $b_k = N + i$ only if $j = k$.

For $s = (z_1, \ldots, z_N, \ldots, d_1, \ldots, d_N) \in S$, event $e_{2N+j} =$ "arrival of packet for transmission by station $j$" $\in E(s)$ if and only if $d_j = 0$, event $e_{N+j} =$ "observation by station $j$ of last bit of packet transmitted by station" $\in E(s)$ if and only if $z_j = i$ or $z_j = N + i$ for some $0 < i \leq N$, and event $e_j =$ "observation by station of first bit of packet transmitted by station $j$" $\in E(s)$ if and only if $z_j \neq j$ and $b_j = i$ for some $0 < i \leq N$. All speeds for active events are equal to 1.

If $e = e_{2N+j} =$ "arrival of packet for transmission by station $j$," then (with $s' = (z'_1, \ldots, z'_N, \ldots, d'_1, \ldots, d'_N)$) the state transition probability $p(s'; s, e) = p_{jk}$ when

$$s = (z_1, \ldots, z_N, \ldots, d_1, \ldots, d_N) \text{ with } z_j = i$$
$$\text{and } s' = s \text{ except } d'_j = k, z'_j = N + i$$

and when

$$s = (z_1, \ldots, z_N, \ldots, d_1, \ldots, d_N) \text{ with } z_j = 0$$
$$\text{and } s' = s \text{ except } d'_j = k, b'_j = j \pmod{N} + 1.$$

If $e = e_{N+j} =$ "observation by station $j$ of last bit of packet transmitted by station," then $p(s'; s, e) = 1$ when

$$s = (z_1, \ldots, z_N, \ldots, d_1, \ldots, d_N) \text{ with } z_j = i\,(i \neq j, 0 < i \leq N),$$
$$b_k \neq N + i \text{ for all } k$$
$$\text{and } s' = s \text{ except } z'_j = 0,$$

when

$$s = (z_1, \ldots, z_N, \ldots, d_1, \ldots, d_N) \text{ with } z_j = i\,(i \neq j, 0 < i \leq N),$$
$$b_k = N + i \text{ for some } k$$
$$\text{and } s' = s \text{ except } z'_j = k,$$

when

$$s = (z_1, \ldots, z_N, \ldots, d_1, \ldots, d_N) \text{ with } z_j = N + i\,(0 < i \leq N),$$
$$b_k \neq N + i \text{ for all } k$$
$$\text{and } s' = s \text{ except } z'_j = 0, b'_j = N + i,$$

when

$$s = (z_1, \ldots, z_N, \ldots, d_1, \ldots, d_N) \text{ with } z_j = N + i\,(0 < i \leq N),$$
$$b_k = N + i \text{ for some } k$$
$$\text{and } s' = s \text{ except } z'_j = k, b'_j = N + i, b'_k = N + j,$$

when

$$s = (z_1, \ldots, z_N, \ldots, d_1, \ldots, d_N) \text{ with } z_j = j,$$
$$b_k \neq N + j \text{ for all } k$$
$$\text{and } s' = s \text{ except } z'_j = 0, d'_j = 0, b'_j = 0,$$

when

$$s = (z_1, \ldots, z_N, \ldots, d_1, \ldots, d_N) \text{ with } z_j = j,$$
$$b_j = j \pmod{N} + 1, b_k = N + j \text{ for some } k$$
$$\text{and } s' = s \text{ except } z'_j = k, d'_j = 0, b'_j = 0,$$
$$b'_k = j \pmod{N} + 1,$$

and when

$$s = (z_1, \ldots, z_N, \ldots, d_1, \ldots, d_N) \text{ with } z_j = j, b_j = N + i,$$
$$b_k = N + j \text{ for some } k,$$
$$\text{and } s' = s \text{ except } z'_j = k, d'_j = 0, b'_j = 0, b'_k = N + j.$$

If $e = e_j =$ "observation by station of first bit of packet transmitted by station $j$," then $p(s'; s, e) = 1$ when

$$s = (z_1, \ldots, z_N, \ldots, d_1, \ldots, d_N) \text{ with } b_j = i\,(0 < i \leq N), z_i = 0 \text{ and } s' = s \text{ except } z'_i = j, b'_j = i \ (\mathrm{mod}\ N) + 1.$$

The clock-setting distribution functions are as follows. If the new event $e' = e_{2N+j} =$ "arrival of packet for transmission by station $j$," then $F(x; s', e', s, e^*) = P\{A_j \leq x\}$. If $e' = e_{N+j} =$ "observation by station $j$ of last bit of packet transmitted by station," then $F(x; s', e', s, e^*) = 1_{[L,\infty)}(x)$. If $e' = e_j =$ "observation by station of first bit of packet transmitted by station $j$" and $s' = (z'_1, \ldots, z'_N, \ldots, d'_1, \ldots, d'_N)$, then $F(x; s', e', s, e^*) = 1_{[R_k,\infty)}(x)$ when $b'_j = k+1$, $d'_k = 0$, $k \neq 0, j$; $F(x; s', e', s, e^*) = 1_{[L+R_k,\infty)}(x)$ when $b'_j = k + 1$, $d'_k > 0$, $k \neq 0, j$; $F(x; s', e', s, e^*) = 1_{[R_N,\infty)}(x)$ when $b'_j = 1$, $d'_N = 0$, $j \neq N$; $F(x; s', e', s, e^*) = 1_{[L+R_N,\infty)}(x)$ when $b'_j = 1$, $d'_N > 0$, $j \neq N$; $F(x; s', e', s, e^*) = 1_{[R_j,\infty)}(x)$ when $b'_j = j + 1$, $j \neq N$; and $F(x; s', e', s, e^*) = 1_{[R_N,\infty)}(x)$ when $b'_N = 1$.

## 6.3. Formal Specification

Heuristically, a GSMP with simultaneous trigger events is a stochastic process that makes a state transition when one or more events associated with the occupied state occur. Events associated with a state compete to trigger the next state transition, and each set of trigger events has its own distribution for determining the next state. At each state transition of the GSMP, new events may be scheduled. For each of these new events, a clock indicating the time until the event is scheduled to occur is set according to an independent (stochastic) mechanism. If a scheduled event is not in the set of events that triggers a state transition but is associated with the next state, its clock continues to run; if such an event is not associated with the next state, it is cancelled, and the corresponding clock reading is discarded.

Formal definition of a GSMP with simultaneous trigger events is in terms of a GSSMC that describes the process at successive state transition times. Let $S$ be a finite or countably infinite set of *states* and $E = \{e_1, e_2, \ldots, e_M\}$ be a finite set of *events*. (We assume throughout that $S \subseteq \Re^d$ for some $d \geq 1$.) For $s \in S$, let $s \mapsto E(s)$ be a mapping from $S$ to the nonempty subsets of $E$; $E(s)$ denotes the set of all events that are scheduled to occur when the process is in state $s$. When the process is in state $s$, the simultaneous occurrence of one or more events $e \in E(s)$ triggers a state transition to a state $s'$. Denote the probability that the new state is $s'$ given that the events in $E^*$ occur simultaneously in state $s$ by $p(s'; s, E^*)$. For each $s \in S$ and $E^* \subseteq E(s)$, we assume that $p(\cdot\,; s, E^*)$ is a probability mass function. The actual events that trigger a state transition in state $s$ depend on clocks associated with the events in $E(s)$ and the speeds at which these clocks run. Associated with each event is a *clock* with a reading that records the remaining time until the event is scheduled to trigger a state transition. We denote the *speed* (finite, deterministic rate) at which the clock associated with event $e$ runs in state $s$ by $r(s, e)$ ($\geq 0$). By convention, for each $s \in S$, $r(s, e) = 0$ if $e \notin E(s)$, and we assume that $r(s, e) > 0$ for some $e \in E(s)$.

For $s \in S$, define the set $C(s)$ of possible clock-reading vectors in state $s$ as

$$C(s) \stackrel{\text{def}}{=} \{(c_1, \ldots, c_M)\colon\ c_i \geq 0 \text{ and } c_i > 0 \\ \text{if and only if } e_i \in E(s)\,\}. \tag{3.1}$$

The $i$th component of a clock-reading vector $c = (c_1, \ldots, c_M)$ is the clock reading associated with event $e_i$. The conditions in (3.1) allow for the possibility that two or more events can occur simultaneously and trigger a state transition (as defined below). The clock reading $c_i$ and the event $e_i$ are said to be *active* in state $S$ if $e_i \in E(s)$. For $s \in S$ and $c \in C(s)$, set

$$t^* = t^*(s, c) = \min_{\{i\colon\ e_i \in E(s)\}} \{c_i / r(s, i)\}\,, \tag{3.2}$$

where we take $c_i/r(s,e_i)$ to be $+\infty$ when $r(s,i)=0$. Also set

$$c_i^* = c_i^*(s,c) = c_i - t^*(s,c)r(s,e_i) \tag{3.3}$$

for $e_i \in E(s)$ and

$$E^* = E^*(s,c) = \{e_i \in E(s)\colon\ c_i^*(s,c) = 0\}\,. \tag{3.4}$$

Beginning in state $s$ with clock-reading vector $c$, $t^*(s,c)$ is the time to the next state transition and $E^*(s,c)$ is the *trigger event set*.

At a state transition from $s$ to $s'$ triggered by the simultaneous occurrence of the events in the set $E^*$, a clock reading is generated for each new event $e' \in N(s') = N(s';s,E^*) = E(s') - (E(s) - E^*)$. Denote the *clock-setting distribution function* (distribution function of such a new clock reading) by $F(\,\cdot\,;s',e',s,E^*)$, and assume that $F(0;s',e',s,E^*) = 0$. For each old event $e' \in O(s') = O(s';s,E^*) = E(s') \cap (E(s) - E^*)$, the old clock reading is kept after the state transition. For $e' \in (E(s) - E^*) - E(s')$, event $e'$ is cancelled after the state transition, and the corresponding clock reading is discarded. When the set $E^*$ of events that occur simultaneously in state $s$ is $E^* = \{e^*\}$, we write $p(s';s,e^*) = p(s';s,\{e^*\})$, $O(s';s,e^*) = O(s';e,\{e^*\})$, $N(s';s,e^*) = N(s';e,\{e^*\})$, and $F(\,\cdot\,;s',e',s,e^*) = F(\,\cdot\,;s',e',s,\{e^*\})$.

Next consider a GSSMC $\{(S_n,C_n)\colon\ n \geq 0\}$ taking values in the set

$$\Sigma = \bigcup_{s\in S} (\{s\} \times C(s))$$

and representing the state $(S_n)$ and clock-reading vector $(C_n)$ at successive state transition times. The transition kernel of the GSSMC $\{(S_n,C_n)\colon\ n \geq 0\}$ is given by

$$\begin{aligned} &P\big((s,c),A\big) \\ &\quad = p(s';s,E^*) \prod_{e_i\in N(s')} F(a_i;s',e_i,s,E^*) \prod_{e_i\in O(s')} 1_{[0,a_i]}(c_i^*) \end{aligned} \tag{3.5}$$

for all sets

$$A = \{s'\} \times \{(c'_1, \ldots, c'_M) \in C(s')\colon\ 0 \le c'_i \le a_i \text{ for } 1 \le i \le M\},$$

where $N(s') = N(s'; s, E^*)$ and $O(s') = O(s'; s, E^*)$. The right side of (3.5) is the probability beginning with state $s$ and clock-reading vector $c$ that the process hits state $s'$ with the reading $c'_i$ on the clock associated with event $e_i$ set to a value in $[0, a_i]$. Denote the initial distribution of the process by $\mu$; that is, for any (measurable) subset $B \subseteq \Sigma$, $\mu(B)$ represents the probability that $(S_0, C_0) \in B$. For ease of exposition, we restrict attention to initial distributions $\mu$ satisfying

$$\mu(A) = q(s_0) \prod_{e_i \in E(s_0)} F_0\bigl(a_i; e_i, s_0\bigr)$$

for all sets

$$A = \{s_0\} \times \{(c_{0,1}, \ldots, c_{0,M}) \in C(s_0)\colon\ 0 \le c_{0,i} \le a_i \\ \text{for } 1 \le i \le M\},$$

where $q$ is a probability distribution on $S$ and $F_0(\,\cdot\,; e, s)$ is a proper distribution function for $s \in S$ and $e \in E(s)$. We denote by $P_\mu$ the probability distribution for the GSSMC when the initial distribution is $\mu$.

Finally, construct a continuous time process with piecewise constant sample paths from the GSSMC $\{(S_n, C_n)\colon\ n \ge 0\}$ in the following manner. Set $\zeta_0 = 0$ and

$$\zeta_n = \sum_{k=0}^{n-1} t^*(S_k, C_k)$$

for $n \ge 1$, so that $\zeta_n$ is the time of the $n$th state transition. Let $\Delta \notin S$ and set

$$X(t) = \begin{cases} S_{N(t)} & \text{if } N(t) < \infty \\ \Delta & \text{if } N(t) = \infty, \end{cases} \tag{3.6}$$

where

$$N(t) = \max\{n \geq 0\colon\ \zeta_n \leq t\}.$$

The process $\{X(t)\colon t \geq 0\}$ defined by (3.6) is a GSMP with simultaneous trigger events. We assume that

$$P_\mu\Big\{\sup_{n\geq 0} \zeta_n = \infty\Big\} = 1$$

so that, with probability 1, $X(t) \neq \Delta$ for every finite time.

For $E^* \subseteq E$, let $S(E^*) \subseteq S$ be the set of states in which the events in the set $E^*$ are active simultaneously:

$$S(E^*) = \{s \in S\colon\ E^* \subseteq E(s)\}.$$

Also let $S^+(E^*) \subseteq S(E^*)$ be the set of all states $s$ in which there is a positive probability that the events in $E^*$ occur simultaneously:

$$\begin{aligned}&s \in S^+(E^*) \text{ if and only if}\\ &\qquad P_\mu\{\, S_n = s \text{ and } E^*(S_n, C_n) = E^* \text{ for some } n \geq 0\,\} > 0.\end{aligned}$$

Thus, if $s \in S(E^*) - S^+(E^*)$ then, with probability 1, the events in $E^*$ never occur simultaneously in state $s$ even though these events are active simultaneously; in this case the probability mass function $p(\,\cdot\,; s, E^*)$ may be defined arbitrarily.

**Example 3.7.** (Token ring with packet destinations) Assume as in Example 1.1 that the successive times for port $j$ to transmit a packet are i.i.d. as a positive random variable $L_j$. Also assume that the successive times from end of transmission by port $j$ until the arrival of the next packet for transmission by port $j$ are i.i.d. as a positive random variable $A_j$; with independent probability $p_{jk}$ the destination of the packet is port $k$ ($\neq j$). All interarrival times for packets and transmission times are mutually independent. Finally, assume that the distribution of each $A_j$ is absolutely continuous and $P\{L_j + R_{jm} = R_{jn}\} = 0$ for all $j$, $m$, and $n$, where

$$R_{jk} = \begin{cases} R_j + R_{j+1} + \cdots + R_{k-1} & \text{if } j \le k-1 \\ R_j + \cdots + R_N + R_1 + \cdots + R_{k-1} & \text{if } j > k-1. \end{cases}$$

These assumptions ensure that, with probability 1, no two packets arrive simultaneously and no packet can arrive at its destination at the same time that some other port observes the token. Set

$$X(t) = (D_1(t), \ldots, D_N(t), Z_1(t), \ldots, Z_N(t), M(t), N(t)),$$

where

$$D_j(t) = \begin{cases} k & \text{if port } k \text{ is the destination of a packet transmitted by port } j \text{ and the transmitted packet has not yet arrived at port } k \text{ at time } t \\ j & \text{otherwise,} \end{cases}$$

$$Z_j(t) = \begin{cases} 1 & \text{if a packet is queued for transmission at port } j \text{ at time } t \\ 0 & \text{otherwise,} \end{cases}$$

$$M(t) = \begin{cases} j & \text{if port } j \text{ is transmitting a packet at time } t \\ 0 & \text{otherwise,} \end{cases}$$

$N(t) = 1$ if port $N$ is transmitting a packet or the token is propagating to port 1 at time $t$, and $N(t) = j$ if port $j-1$ is transmitting a packet or the token is propagating to port $j$ at time $t$ ($j = 2, 3, \ldots, N$).

Take event $e_{2N+2}$ = "observation of token by port," $e_{2N+1}$ = "end of transmission," $e_{N+j}$ = "arrival at destination of packet transmitted by port $j$," and $e_j$ = "arrival of packet for transmission by port $j$" ($j = 1, 2, \ldots, N$). Formal specification of the process $\{X(t)\colon\ t \geq 0\}$ is as a GSMP with (simultaneous trigger events and) a finite state space

$$\begin{aligned} S = \big\{(d_1, \ldots, d_N, z_1, \ldots, z_N, m, n) \in \{1, 2, \ldots, N\}^N \times \{0, 1\}^N \\ \times \{0, 1, \ldots, N\} \times \{1, 2, \ldots, N\}\colon\ n = m + 1 \\ \text{when } 0 < m < N;\ n = 1 \text{ when } m = N;\ d_j \neq j \\ \text{when } m = j;\ m = j \text{ only if } z_j = 0 \text{ for } j = 1, 2, \ldots, N\big\} \end{aligned}$$

and event set $E = \{e_1, e_2, \ldots, e_{2N+2}\}$.

For $s = (d_1, \ldots, d_N, z_1, \ldots, z_N, m, n) \in S$, event $e_{2N+2} =$ "observation of token by port" $\in E(s)$ if and only if $m = 0$ and $e_{2N+1} =$ "end of transmission" $\in E(s)$ if and only if $m > 0$. Event $e_{N+j} =$ "arrival at destination of packet transmitted by port $j$" $\in E(s)$ if and only if $d_j \neq j$. Event $e_j =$ "arrival of packet for transmission by port $j$" $\in E(s)$ if and only if $z_j = 0$ and $m \neq j$. All speeds $r(s, e)$ for active events are equal to 1.

If $e = e_{2N+2} =$ "observation of token by port," then the state transition probability $p(s'; s, e) = p_{nk}$ when

$$\begin{aligned} s = (d_1, \ldots, d_{n-1}, n, d_{n+1}, \ldots, d_N, z_1, \ldots, z_{n-1}, 1, \\ z_{n+1}, \ldots, z_N, 0, n) \text{ with } n < N \\ \text{and } s' = (d_1, \ldots, d_{n-1}, k, d_{n+1}, \ldots, d_N, z_1, \ldots, z_{n-1}, 0, \\ z_{n+1}, \ldots, z_N, n, n + 1), \end{aligned}$$

$p(s'; s, e) = p_{Nk}$ when

$$\begin{aligned} s = (d_1, \ldots, d_{N-1}, N, z_1, \ldots, z_{N-1}, 1, 0, N) \\ \text{and } s' = (d_1, \ldots, d_{N-1}, k, z_1, \ldots, z_{N-1}, 0, N, 1), \end{aligned}$$

and $p(s'; s, e) = 1$ when

$$\begin{aligned} s = (d_1, \ldots, d_N, z_1, \ldots, z_{n-1}, 0, z_{n+1}, \ldots, z_N, 0, n) \text{ with } n < N \\ \text{and } s' = (d_1, \ldots, d_N, z_1, \ldots, z_{n-1}, 0, z_{n+1}, \ldots, z_N, 0, n + 1), \end{aligned}$$

and when

$$s = (d_1, \ldots, d_N, z_1, \ldots, z_{N-1}, 0, 0, N)$$
$$\text{and } s' = (d_1, \ldots, d_N, z_1, \ldots, z_{N-1}, 0, 0, 1).$$

If $e = e_{2N+1} =$ "end of transmission," then $p(s'; s, e) = 1$ when

$$s = (d_1, \ldots, d_N, z_1, \ldots, z_N, m, m+1) \text{ with } 0 < m < N$$
$$\text{and } s' = (d_1, \ldots, d_N, z_1, \ldots, z_N, 0, m+1)$$

and when

$$s = (d_1, \ldots, d_N, z_1, \ldots, z_N, N, 1)$$
$$\text{and } s' = (d_1, \ldots, d_N, z_1, \ldots, z_N, 0, 1).$$

If $e_{N+j} =$ "arrival at destination of packet transmitted by port $j$," then $p(s'; s, e) = 1$ when

$$s = (d_1, \ldots, d_N, z_1, \ldots, z_N, m, n) \text{ with } d_j \neq j$$
$$\text{and } s' = (d_1, \ldots, d_{j-1}, j, d_{j+1}, \ldots, d_N, z_1, \ldots, z_N, m, n).$$

If $e = e_j =$ "arrival of packet for transmission by port $j$," then $p(s'; s, e) = 1$ when

$$s = (d_1, \ldots, d_N, z_1, \ldots, z_{j-1}, 0, z_{j+1}, \ldots, z_N, m, n)$$
$$\text{and } s' = (d_1, \ldots, d_N, z_1, \ldots, z_{j-1}, 1, z_{j+1}, \ldots, z_N, m, n).$$

If $E^* = \{e_{2N+2}, e_{N+j}\}$, then $p(s'; s, E^*) = p_{nk}$ when

$$s = (d_1, \ldots, d_N, z_1, \ldots, z_{n-1}, 1, z_{n+1}, \ldots, z_N, 0, n)$$
$$\text{with } d_j \neq j \; (j \neq n,\ n < N)$$
$$\text{and } s' = (d_1, \ldots, d_{j-1}, j, d_{j+1}, \ldots, d_{n-1}, k, d_{n+1}, \ldots, d_N,$$
$$z_1, \ldots, z_{n-1}, 0, z_{n+1}, \ldots, z_N, n, n+1),$$

$p(s'; s, E^*) = p_{Nk}$ when

$$s = (d_1, \ldots, d_{N-1}, N, z_1, \ldots, z_{N-1}, 1, 0, N) \text{ with } d_j \neq j\ (j \neq N)$$
$$\text{and } s' = (d_1, \ldots, d_{j-1}, j, d_{j+1}, \ldots, d_{N-1}, k, z_1, \ldots, z_{N-1}, 0, N, 1),$$

and $p(s'; s, E^*) = 1$ when

$$s = (d_1, \ldots, d_N, z_1, \ldots, z_{n-1}, 0, z_{n+1}, \ldots, z_N, 0, n) \text{ with } d_j \neq j\ (j \neq n,\ n < N)$$
$$\text{and } s' = (d_1, \ldots, d_{j-1}, j, d_{j+1}, \ldots, d_N, z_1, \ldots, z_{n-1}, 0, z_{n+1}, \ldots, z_N, 0, n+1),$$

and when

$$s = (d_1, \ldots, d_N, z_1, \ldots, z_{N-1}, 0, 0, N) \text{ with } d_j \neq j\ (j \neq N)$$
$$\text{and } s' = (d_1, \ldots, d_{j-1}, j, d_{j+1}, \ldots, d_N, z_1, \ldots, z_{N-1}, 0, 0, 1).$$

The probability mass function $p(\cdot\,; s, E^*)$ may be defined arbitrarily for all other states $s \in S$ and subsets $E^* \subseteq E(s)$ with $|E^*| \geq 2$.

The clock-setting distribution functions are as follows. If the new event $e' = e_{2N+2}$ and $s' = (d_1, \ldots, d_N, z_1, \ldots, z_N, m, n)$, then $F(x; s', e', s, E^*) = 1_{[R_{n-1}, \infty)}(x)$ if $n > 1$ and equals $1_{[R_N, \infty)}(x)$ if $n = 1$. If $e' = e_{2N+1}$ and $s' = (d_1, \ldots, d_N, z_1, \ldots, z_N, m, n)$, then $F(x; s', e', s, E^*) = P\{L_m \leq x\}$. If $e' = e_{N+j}$ and $s' = (d_1, \ldots, d_N, z_1, \ldots, z_N, m, n)$, then $F(x; s', e', s, E^*) = 1_{[R_j + \cdots + R_{k-1}, \infty)}(x)$ if $j \leq k-1$ and $F(x; s', e', s, E^*) = 1_{[R_j + \cdots R_N + R_1 + \cdots + R_{k-1}, \infty)}(x)$ if $j > k-1$. If $e' = e_j$, then $F(x; s', e', s, E^*) = P\{A_j \leq x\}$.

Suppose that at time 0 the token has just arrived at port 1 and all ports have a packet queued for transmission. Then the initial distribution $\mu$ is

$$\mu(A) = 1_{\{s_0\}}(s) P\{L_1 \leq a_{2N+1}\}$$

for all sets

$$A = \{s\} \times \{(c_1, \ldots, c_{2N+2}) \colon 0 \le c_i \le a_i \text{ for } 1 \le i \le 2N+2\},$$

where $s_0 = (0, \ldots, 0, 0, 1, 1, \ldots, 1, 1, 2)$.

**Example 3.8.** (Collision-free bus network) For the collision-free bus network in Example 2.2, set

$$\begin{aligned} X(t) = (&W_1(t), \ldots, W_N(t), U_1(t), \ldots, U_N(t), \\ &V_{2,1}(t), V_{3,1}(t), V_{3,2}(t), V_{4,1}(t), \ldots, V_{N,N-1}(t)), \end{aligned}$$

where

$$W_j(t) = \begin{cases} 3 & \text{if port } j \text{ is transmitting a packet at time } t \\ 2 & \text{if port } j \text{ has completed the } R(j)+T \text{ wait} \\ & \text{but has not started transmission at time } t \\ 1 & \text{if port } j \text{ has set its flip-flop but has not yet} \\ & \text{completed the } R(j)+T \text{ wait at time } t \\ 0 & \text{otherwise,} \end{cases}$$

$$U_j(t) = \begin{cases} k & \text{if at time } t \text{ port } j \text{ observes transmission} \\ & \text{of a packet by port } k \\ 0 & \text{otherwise,} \end{cases}$$

and

$$V_{j,k}(t) = \begin{cases} 1 & \text{if } \mathbf{S}(k) = 1 \text{ at time } t - R(k,j) \\ 0 & \text{otherwise} \end{cases}$$

for $k < j$. (Port $j$ observes $\mathbf{P}(j)$ =1 at time $t$ if $V_{j,k}(t) = 1$ for some $k < j$.)

Take event $e_{j,1}$ = "setting (to 1) of flip-flop by port $j$," $e_{j,2}$ = "end of wait for $R(j) + T$," $e_{j,3}$ = "end of transmission by port $j$," $e_{j,l,4}$ = "observation by port $j$ of start of transmission by port $l \neq j$," $e_{j,l,5}$ = "observation by port $j$ of end of transmission by port $l \neq j$," $e_{j,k,6}$ = "observation by port $j$ of the setting (to 1) of flip-flop by port $k$ to the left," and $e_{j,k,7}$ = "observation by port $j$ of the resetting (to 0) of flip-flop by port $k$ to the left" for $l \neq j$, $k < j$, and $j, k, l = 1, 2, \ldots, N$. Formal specification of the process $\{X(t)\colon t \geq 0\}$ is as a GSMP with (simultaneous trigger events and) a finite state space $S$ and event set

$$E = \{e_{j,1}, e_{j,2}, e_{j,3}, e_{j,l,4}, e_{j,l,5}, e_{j,k,6}, e_{j,k,7}\colon\ l \neq j,\ k < j,\\ j, k, l = 1, 2, \ldots, N\}.$$

The state space $S$ is the set of all vectors $s = (w_1, \ldots, w_N, u_1, \ldots, u_N, v_{2,1}, \ldots, v_{N,N-1}) \in \{0,1,2,3\}^N \times \{0,1,\ldots,N\}^N \times \{0,1\}^{N(N-1)/2}$ satisfying

(i) $w_j = 3$ for at most one $j$;
(ii) $u_j = 3$ if and only if $w_j = 3$;
(iii) $u_j = k$ whenever $w_k = 3$ and $u_l = k$ for some $l$ with $T(k,l) \geq T(k,j)$;
(iv) $u_j = k$ whenever $w_k \neq 3$ and $u_l = k$ for some $l$ with $T(k,j) \geq T(k,l)$;
(v) $u_j = u_k$, $u_j = u_l$, or $u_k = u_l$ whenever $u_j u_k u_l > 0$;
(vi) $v_{j,k} = 1$ for all $j > k$ whenever $w_k = 2$;
(vii) $v_{j,k} = 1$ whenever $w_k = 3$ and $v_{l,k} = 1$ for some $l$ with $l < j$;
(viii) $v_{j,k} = 1$ whenever $w_k \neq 3$ and $v_{l,k} = 1$ for some $l$ with $j < l$;
(ix) $v_{j,k} = 1$ whenever $w_k = 3$ and $u_l \neq k$ for some $l$ with $T(k,l) \leq R(k,j)$; and
(x) $v_{j,k} = 0$ whenever $w_k = 3$ and $u_l = k$ for some $l$ with $R(k,j) \leq T(k,l)$.

For $s = (w_1, \ldots, w_N, u_1, \ldots, u_N, v_{2,1}, \ldots, v_{N,N-1}) \in S$ the event sets $E(s)$ are as follows. Event $e_{j,1}$ = "setting (to 1) of flip-flop by port $j$"$\in E(s)$ if and only if $w_j = 0$. Event $e_{j,2}$ = "end of wait for $R(j) + T$"$\in E(s)$ if and only if $w_j = 1$. Event $e_{j,3}$ = "end of transmission by port $j$"$\in E(s)$ if and only if $w_j = 3$. Event $e_{j,l,4}$ = "observation by port $j$ of start of transmission by port $l \neq j$" $\in E(s)$ if and only if $w_l = 3$ and $u_j \neq l$. Event $e_{j,l,5}$ = "observation by port $j$ of end of transmission by port $l \neq j$" $\in E(s)$ if and only if $u_j = l$ and $w_l = 1$ or 0. Event $e_{j,k,6}$ = "observation by port $j$ of setting of flip-flop by port $k$ to the left" $\in E(s)$ if and only if $w_k = 1$ and $v_{j,k} = 0$. Event $e_{j,k,7}$ = "observation by port $j$ of resetting of flip-flop by port $k$ to the left" $\in E(s)$ if and only if $w_k = 3$ and $v_{j,k} = 1$. All speeds for active events are equal to 1.

If $e = e_{j,1}$ = "setting (to 1) of flip-flop by port $j$," then the state transition probability $p(s'; s, e) = 1$ when

$$s = (w_1, \ldots, w_{j-1}, 0, w_{j+1}, \ldots, w_N, u_1, \ldots, v_{N,N-1})$$
$$\text{and } s' = (w_1, \ldots, w_{j-1}, 1, w_{j+1}, \ldots, w_N, u_1, \ldots, v_{N,N-1}).$$

If $e = e_{j,2}$ = "end of wait for $R(j) + T$," then $p(s'; s, e) = 1$ when

$$s = (w_1, \ldots, w_{j-1}, 1, w_{j+1}, \ldots, w_N, u_1, \ldots, v_{N,N-1})$$
$$\text{with } u_j > 0 \text{ or } v_{j,1} + \cdots + v_{j,j-1} > 0$$
$$\text{and } s' = (w_1, \ldots, w_{j-1}, 2, w_{j+1}, \ldots, w_N, u_1, \ldots, v_{N,N-1})$$

and when

$$s = (w_1, \ldots, w_{j-1}, 1, w_{j+1}, \ldots, u_{j-1}, 0, u_{j+1}, \ldots,$$
$$v_{j-1,j-2}, 0, \ldots, 0, v_{j+1,1}, \ldots, v_{N,N-1})$$
$$\text{and } s' = (w_1, \ldots, w_{j-1}, 3, w_{j+1}, \ldots, w_N,$$
$$u_1, \ldots, u_{j-1}, j, u_{j+1}, \ldots, v_{N,N-1}).$$

If $e = e_{j,3}$ = "end of transmission by port $j$," then $p(s'; s, e) = 1$ when

$$s = (w_1, \ldots, w_{j-1}, 3, w_{j+1}, \ldots, u_{j-1}, j, u_{j+1}, \ldots, v_{N,N-1})$$
$$\text{and } s' = (w_1, \ldots, w_{j-1}, 0, w_{j+1}, \ldots, w_N, u_1, \ldots,$$
$$u_{j-1}, 0, u_{j+1}, \ldots, u_N, v_{2,1}, \ldots, v_{N,N-1}).$$

If $e = e_{j,l,4}$ = "observation by port $j$ of start of transmission by port $l \neq j$," then $p(s'; s, e) = 1$ when

$$\begin{aligned} s = (&w_1, \ldots, w_N, u_1, \ldots, u_{j-1}, 0, u_{j+1}, \ldots, v_{N,N-1}) \\ &\text{with } u_k = l \text{ for all } k \text{ such that } l < k < j \text{ or } j < k < l \\ &\text{and } s' = (w_1, \ldots, w_N, u_1, \ldots, u_{j-1}, l, u_{j+1}, \ldots, v_{N,N-1}). \end{aligned}$$

When $s = (w_1, \ldots, w_N, u_1, \ldots, u_{j-1}, 0, u_{j+1}, \ldots, v_{N,N-1})$ with $u_k \neq l$ for some $k$ such that $l < k < j$ or $j < k < l$, the probability mass function $p(\,\cdot\,; s, e_{j,l,4})$ may be defined arbitrarily. If $e = e_{j,l,5}$ = "observation by port $j$ of end of transmission by port $l \neq j$," then $p(s'; s, e) = 1$ when

$$\begin{aligned} s = (&w_1, \ldots, w_N, u_1, \ldots, u_{j-1}, l, u_{j+1}, \ldots, v_{N,N-1}) \\ &\text{with } u_k = 0 \text{ for all } k \text{ such that } l < k < j \text{ or } j < k < l \\ &\text{and either } w_j \neq 2 \text{ or } v_{j,1} + \cdots + v_{j,j-1} > 0 \\ &\text{and } s' = (w_1, \ldots, w_N, u_1, \ldots, u_{j-1}, 0, u_{j+1}, \ldots, v_{N,N-1}) \end{aligned}$$

and when

$$\begin{aligned} s = (&w_1, \ldots, w_{j-1}, 2, w_{j+1}, \ldots, w_N, u_1, \ldots, v_{j-1,j-2}, \\ &0, \ldots, 0, v_{j+1,1}, \ldots, v_{N,N-1}) \text{ with } u_k = 0 \\ &\text{for all } k \text{ such that } l < k < j \text{ or } j < k < l \\ &\text{and } s' = (w_1, \ldots, w_{j-1}, 3, w_{j+1}, \ldots, w_N, \\ &u_1, \ldots, u_{j-1}, j, u_{j+1}, \ldots, v_{N,N-1}). \end{aligned}$$

When $s = (w_1, \ldots, w_N, z, u_1, \ldots, u_{j-1}, l, u_{j+1}, \ldots, v_{N,N-1})$ with $u_k \neq l$ for some $k$ such that $l < k < j$ or $j < k < l$ and either $w_j \neq 2$ or $v_{j,1} + \cdots + v_{j,j-1} > 0$ and when $s = (w_1, \ldots, w_{j-1}, 2, w_{j+1}, \ldots, w_N, z, u_1, \ldots, v_{j-1,j-2}, 0, \ldots, 0, v_{j+1,1}, \ldots, v_{N,N-1})$ with $u_k = 0$ for all $k$ such that $l < k < j$ or $j < k < l$, the probability mass function $p(\,\cdot\,; s, e_{j,l,5})$ may be defined arbitrarily. If $e = e_{j,k,6}$ = "observation by port $j$

of the setting (to 1) of flip-flop by port $k$ to the left," then $p(s'; s, e) = 1$ when

$$s = (w_1, \ldots, w_N, u_1, \ldots, v_{j,k-1}, 0, v_{j,k+1}, \ldots, v_{N,N-1})$$
$$\text{with } v_{l,k} = 1 \text{ for all } l \text{ such that } k < l < j$$
$$\text{and } s' = (w_1, \ldots, w_N, u_1, \ldots, u_N, v_{2,1}, \ldots, v_{j,k-1}, 1, v_{j,k+1}, \ldots, v_{N,N-1}).$$

When $s = (w_1, \ldots, w_N, u_1, \ldots, u_N, v_{2,1}, \ldots, v_{j,k-1}, 0, v_{j,k+1}, \ldots, v_{N,N-1})$ with $v_{l,k} = 0$ for some $l$ such that $k < l < j$, the probability mass function $p(\cdot\,; s, e_{j,k,6})$ may be defined arbitrarily. If $e = e_{j,k,7}$ = "observation by port $j$ of the resetting (to 0) of flip-flop by port $k$ to the left," then $p(s'; s, e) = 1$ when

$$s = (w_1, \ldots, w_N, u_1, \ldots, v_{j,k-1}, 1, v_{j,k+1}, \ldots, v_{N,N-1})$$
$$\text{with } v_{l,k} = 0 \text{ for all } l \text{ such that } k < l < j$$
$$\text{and } s' = (w_1, \ldots, v_{j,k-1}, 0, v_{j,k+1}, \ldots, v_{N,N-1}).$$

When $s = (w_1, \ldots, w_N, u_1, \ldots, u_N, v_{2,1}, \ldots, v_{j,k-1}, 1, v_{j,k+1}, \ldots, v_{N,N-1})$ with $v_{l,k} = 1$ for some $l$ such that $k < l < j$, the probability mass function $p(\cdot\,; s, e_{j,k,7})$ may be defined arbitrarily.

If $E^* = \{e_{j,k,4}, e_{j,l,5}\}$ and $j$, $k$, and $l$ are distinct, then $p(s'; s, E^*) = 1$ when

$$s = (w_1, \ldots, w_N, u_1, \ldots, u_{j-1}, l, u_{j+1}, \ldots, v_{N,N-1})$$
$$\text{with } u_k = l \text{ for all } k \text{ such that } l < k < j \text{ or } j < k < l$$
$$\text{and } s' = (w_1, \ldots, u_{j-1}, k, u_{j+1}, \ldots, v_{N,N-1}).$$

The probability mass function $p(\cdot\,; s, E^*)$ may be defined arbitrarily for all other states $s \in S$ and subsets $E^* \subseteq E(s)$ with $|E^*| \geq 2$.

The clock-setting distribution functions are as follows. If the new event $e' = e_{j,1}$, then $F(x; s', e', s, E^*) = P\{A_j \leq x\}$.

If $e' = e_{j,2}$, then $F(x; s', e', s, E^*) = 1_{[R(j)+T,\infty)}(x)$. If $e' = e_{j,3}$, then $F(x; s', e', s, E^*) = P\{L_j \le x\}$. If $e' = e_{j,l,4}$ or if $e' = e_{j,l,5}$, then $F(x; s', e', s, E^*) = 1_{[T(l,j),\infty)}(x)$. If $e' = e_{j,k,6}$, then $F(x; s', e', s, E^*) = 1_{[R(k,j),\infty)}(x)$. If $e' = e_{j,k,7}$, then $F(x; s', e', s, E^*) = 1_{[R(k,j),\infty)}(x)$.

**Example 3.9.** (Register insertion ring) For the register insertion ring in Example 2.3, set

$$X(t) = (Z_1(t), \ldots, Z_N(t), B_1(t), \ldots, B_N(t), D_1(t), \ldots, D_N(t)),$$

where

$$Z_j(t) = \begin{cases} N+i & \text{if station } j \text{ has a packet queued for} \\ & \text{transmission at time } t \text{ and is observing} \\ & \text{a packet bit transmitted by station } i \\ i & \text{if station } j \text{ has no packet queued for} \\ & \text{transmission at time } t \text{ and is observing} \\ & \text{a packet bit transmitted by station } i \\ 0 & \text{if station } j \text{ is observing an idle bit} \\ & \text{at time } t, \end{cases}$$

$$B_j(t) = \begin{cases} i & \text{if station } j \text{ has a packet on the ring at time } t \\ & \text{and the first bit of the packet transmitted} \\ & \text{by station } j \text{ is propagating to station } i \\ 0 & \text{if station } j \text{ has no packet on the ring} \\ & \text{at time } t, \end{cases}$$

and

$$D_j(t) = \begin{cases} i & \text{if station } j \text{ has a packet on the ring at time } t \\ & \text{and station } i \text{ is the destination of the packet} \\ 0 & \text{if station } j \text{ has no packet on the ring at time } t. \end{cases}$$

Take event $e_{2N+j}$ = "arrival of packet for transmission by station $j$," $e_{N+j}$ = "observation by station $j$ of last bit of packet transmitted by station," and $e_j$ = "observation by station of first bit of packet transmitted by station $j$" ($j = 1, 2, \ldots, N$). Formal specification of the process $\{X(t)\colon\ t \geq 0\}$ is as a GSMP with (simultaneous trigger events and) a finite state space $S$ and event set

$$E = \{e_1, e_2, \ldots, e_N, e_{N+1}, \ldots, e_{2N}, e_{2N+1}, \ldots, e_{3N}\}.$$

The state space $S$ is the set of all vectors $s = (z_1, \ldots, z_N, \ldots, d_1, \ldots, d_N) \in \prod_{j=1}^{N} \{0, 1, \ldots, N+j-1, N+j+1, \ldots, 2N\} \times \{0, 1, \ldots, N\}^N \times \prod_{j=1}^{N} \{0, 1, \ldots, j-1, j+1, \ldots, N\}$ satisfying

(i) $z_j = N + i$ only if $d_j > 0$ and $b_j = 0$;
(ii) $z_j = i$ only if $d_i > 0$;
(iii) $z_j = N + i$ only if $d_i > 0$;
(iv) $z_j = i$ only if $b_i > 0$;
(v) $d_j = 0$ only if $b_j = 0$; and
(vi) $b_j = i$ only if $b_i > 0$.

For $s = (z_1, \ldots, z_N, \ldots, d_1, \ldots, d_N) \in S$, event $e_{2N+j}$ = "arrival of packet for transmission by station $j$" $\in E(s)$ if and only if $d_j = 0$, event $e_{N+j}$ = "observation by station $j$ of last bit of packet transmitted by station" $\in E(s)$ if and only if $z_j = i$ or $z_j = N + i$ for some $0 < i \leq N$, and event $e_j$ = "observation by station of first bit of packet transmitted by station $j$" $\in E(s)$ if and only if $z_j \neq j$ and $b_j = i$ for some $0 < i \leq N$. All speeds for active events are equal to 1.

If $e = e_{2N+j}$ = "arrival of packet for transmission by station $j$," then the state transition probability $p(s'; s, e) = p_{jk}$ when

$$s = (z_1, \ldots, z_N, \ldots, d_1, \ldots, d_N) \text{ with } z_j = i$$
$$\text{and } s' = s \text{ except } d'_j = k, z'_j = N + i$$

and when

$$s = (z_1, \ldots, z_N, \ldots, d_1, \ldots, d_N) \text{ with } z_j = 0$$
$$\text{and } s' = s \text{ except } d'_j = k, b'_j = j \pmod{N} + 1.$$

If $e = e_{N+j}$ = "observation by station $j$ of last bit of packet transmitted by station," then $p(s'; s, e) = 1$ when

$$s = (z_1, \ldots, z_N, \ldots, d_1, \ldots, d_N) \text{ with } z_j = i\,(i \neq j, 0 < i \leq N)$$
$$\text{and } s' = s \text{ except } z'_j = 0,$$

when

$$s = (z_1, \ldots, z_N, \ldots, d_1, \ldots, d_N) \text{ with } z_j = N + i\,(0 < i \leq N)$$
$$\text{and } s' = s \text{ except } z'_j = 0, b'_j = j \pmod{N} + 1,$$

and when

$$s = (z_1, \ldots, z_N, \ldots, d_1, \ldots, d_N) \text{ with } z_j = j$$
$$\text{and } s' = s \text{ except } z'_j = 0, d'_j = 0, b'_j = 0.$$

If $e = e_j$ = "observation by station of first bit of packet transmitted by station $j$," then $p(s'; s, e) = 1$ when

$$s = (z_1, \ldots, z_N, \ldots, d_1, \ldots, d_N) \text{ with } b_j = i\,(0 < i < N), z_i = 0$$
$$\text{and } s' = s \text{ except } z'_i = j, b'_j = i \pmod{N} + 1.$$

If $E^* = \{e_{N+j}, e_k\}$, then $p(s'; s, E^*) = 1$ when

$$s = (z_1, \ldots, z_N, \ldots, d_1, \ldots, d_N) \text{ with } z_j = i\,(i \neq j,$$
$$0 < i \leq N), b_k = j, b_j = 0$$
$$\text{and } s' = s \text{ except } z'_j = k, b'_k = j \pmod{N} + 1,$$

when

$$s = (z_1, \ldots, z_N, \ldots, d_1, \ldots, d_N) \text{ with } z_j = N + i\,(i \neq j,$$
$$0 < i \leq N), b_k = j, b_j = 0$$
$$\text{and } s' = s \text{ except } z'_j = k, b'_j = j \pmod{N} + 1,$$
$$b'_k = j \pmod{N} + 1,$$

when

$$\begin{aligned} s = (z_1, \ldots, z_N, \ldots, d_1, \ldots, d_N) \text{ with } z_j = i\,(i \neq j,\\ 0 < i \leq N), b_k = j, b_j > 0\\ \text{and } s' = s \text{ except } z_j' = k, b_j' = j \pmod{N} + 1,\\ b_k' = j \pmod{N} + 1, \end{aligned}$$

and when

$$\begin{aligned} s = (z_1, \ldots, z_N, \ldots, d_1, \ldots, d_N) \text{ with } z_j = j, b_k = j, b_j > 0\\ \text{and } s' = s \text{ except } z_j' = k, b_j' = 0, d_j' = 0,\\ b_k' = j \pmod{N} + 1. \end{aligned}$$

The probability mass function $p(\,\cdot\,; s, E^*)$ may be defined arbitrarily for all other states $s \in S$ and subsets $E^* \subseteq E(s)$ with $|E^*| \geq 2$.

The clock-setting distribution functions are as follows. If the new event $e' = e_{2N+j}$ = "arrival of packet for transmission by station $j$," then $F(x; s', e', s, E^*) = P\{A_j \leq x\}$. If $e' = e_{N+j}$ = "observation by station $j$ of last bit of packet transmitted by station," then $F(x; s', e', s, E^*) = 1_{[L,\infty)}(x)$. If $e' = e_j$ = "observation by station of first bit of packet transmitted by station $j$" and $s' = (z_1', \ldots, z_N', \ldots, d_1', \ldots, d_N')$, then $F(x; s', e', s, E^*) = 1_{[R_k,\infty)}(x)$ when $b_j' = k+1$, $d_k' = 0$, $k \neq 0, j$; $F(x; s', e', s, E^*) = 1_{[L+R_k,\infty)}(x)$ when $b_j' = k+1$, $d_k' > 0$, $k \neq 0, j$; $F(x; s', e', s, E^*) = 1_{[R_N,\infty)}(x)$ when $b_j' = 1$, $d_N' = 0$, $j \neq N$; $F(x; s', e', s, E^*) = 1_{[L+R_N,\infty)}(x)$ when $b_j' = 1$, $d_N' > 0$, $j \neq N$; $F(x; s', e', s, E^*) = 1_{[R_j,\infty)}(x)$ when $b_j' = j+1$, $j \neq N$; and $F(x; s', e', s, E^*) = 1_{[R_N,\infty)}(x)$ when $b_N' = 1$.

Irreducibility for a GSMP with simultaneous trigger events is defined as follows.

**Definition 3.10.** *A generalized semi-Markov process with simultaneous trigger events and state space $S$ and event set $E$ is said to be irreducible if for each pair $s, s' \in S$ there exist states $s = s_0, s_1, \ldots, s_{n-1}, s_n = s' \in S$ and event sets $E_0, E_1, \ldots, E_{n-1} \subseteq E$ such that*

$$p(s_1; s_0, E_0)r(s_0, E_0)p(s_2; s_1, E_1)r(s_1, E_1) \cdots \\ p(s_n; s_{n-1}, E_{n-1})r(s_{n-1}, E_{n-1}) > 0,$$

*where $r(s, E_i) = \min\{r(s, e)\colon\ e \in E_i\}$.*

The direct analogue for GSMP's with simultaneous trigger events of the basic limit theorem (Theorem 4.14 of Chapter 1) for an irreducible GSMP with unique trigger events and a finite state space does not hold. The assumptions that each clock-setting distribution has finite mean and the continuous part of each clock-setting distribution has a density function that is continuous and positive on $(0, +\infty)$ are not sufficient to ensure that an irreducible GSMP with simultaneous trigger events and a finite state space has a limiting distribution.

The form of the transition kernel in (3.5) leads directly to an algorithm for generating sample paths of the underlying GSSMC of a GSMP with simultaneous trigger events.

**Algorithm 3.11.** *Sample path generation*

1. *Generate an initial state $s \in S$ and an initial clock-reading vector $c \in C(s)$ according to the initial distribution $\mu$;*
2. *Compute the time $t^*(s, c)$ to the next state transition according to $t^*(s, c) = \min_{\{i:\ e_i \in E(s)\}} \{c_i / r(s, e_i)\}$. Also determine the set $E^*$ of events that trigger the state transition as $E^* = \{e_i \in E(s)\colon\ c_i / r(s, e_i) = t^*(s, c)\}$;*
3. *Generate the new state $s'$ according to the probability mass function $p(\,\cdot\,; s, E^*)$;*

4. *For each event* $e_i \in N(s'; s, E^*) = E(s') - (E(s) - E^*)$, *generate a new clock reading* $c'_i$ *according to the distribution function* $F(\cdot\,; s', e_i, s, E^*)$;
5. *For each event* $e_i \in O(s'; s, E^*) = E(s') \cap \big(E(s) - E^*\big)$, *set* $c'_i = c_i - t^*(s, c)r(s, e_i)$;
6. *For each event* $e_i \in \big(E(s) - E^*\big) - E(s')$, *set* $c'_i = 0$.
7. *Return to Step 2 and iterate with* $s'$ *playing the role of* $s$ *and* $c'$ *the role of* $c$.

Recall that by construction, the GSMP takes values in the set $S \cup \{\Delta\}$ and has piecewise constant, right-continuous sample paths. A sample path of the GSMP can be represented as a sequence $\{(X(\zeta_n), \zeta_n)\colon n \geq 0\}$, where $\zeta_0 = 0$ and $\zeta_n = \sum_{k=0}^{n-1} t^*(S_k, C_k)$ for $n \geq 1$. The following algorithm produces a sample path of the GSMP.

**Algorithm 3.12.** *Sample path generation*

1. *Set* $n = 0$ *and* $\zeta_n = 0$;
2. *Set* $X(\zeta_n) = S_k$ *and* $\zeta_{n+1} = \zeta_n + t^*(S_k, C_k)$;
3. *Increment* $n$ *by 1 and go to Step 3.*

In principle, Algorithm 3.11 uses all of the probability mass functions that govern state transitions of the GSMP. As the set of probability mass functions can be very large, the algorithm may not be suitable for implementation. However, it is usually the case when events occur simultaneously that the GSMP makes a state transition as if the events had occurred sequentially. There are then compact representations of the set of probability mass functions that govern state transitions. These representations lead to practical algorithms for generating sample paths.

## Exercises

3.1. (Spatial queues) Consider a network of queues with three centers that provides service to a fixed number $N$ $(\geq 2)$ of jobs. Center 1 provides batch service to two jobs; after completion of service, one job walks to the tail of the queue at center 2 and the other job walks to the tail of the queue at center 3. Centers 2 and 3 each provide service to one job at a time; after completion of service, a job walks to the tail of the queue at center 1. The queueing discipline at each center is FCFS. Assume that the successive walking times from center $j$ to center $k$ are i.i.d. as a positive random variable $T_{jk}$, and the successive service times at center $j$ are i.i.d. as a positive random variable $V_j$ with an absolutely continuous distribution. All walking times and service times are mutually independent. For simplicity, suppose that $P\{V_j \leq T_{jk}\} = 0$ and, initially, each of the $N$ jobs is waiting or in service at some center. Thus, at any subsequent time there is at most one job walking from any center to any other. Also suppose that $0 < P\{T_{12} = T_{13}\} < 1$ so that jobs can, but need not, arrive at centers 2 and 3 simultaneously. Set

$$X(t) = (Z_1(t), Z_2(t), Z_3(t), W_{1,2}(t), W_{1,3}(t), W_{2,1}(t), W_{3,1}(t)),$$

where $Z_i(t)$ is the number of jobs waiting or in service at center $i$ at time $t$ and $W_{i,j}(t)$ is the number of jobs walking from center $i$ to center $j$ at time $t$.

The process $\{X(t)\colon\ t \geq 0\}$ has a finite state space $S$ that is the set of all vectors $s = (z_1, z_2, z_3, w_{1,2}, w_{1,3}, w_{2,1}, w_{3,1}) \in \{0, 1, \ldots, N\} \times \{0, 1, \ldots, N-1\}^2 \times \{0, 1\}^4$ satisfying

(i) $0 \leq w_{1,2} + z_2 + w_{2,1} \leq N - 1$;
(ii) $0 \leq w_{1,3} + z_3 + w_{3,1} \leq N - 1$; and
(iii) $z_1 + z_2 + z_3 + w_{1,2} + w_{2,1} + w_{2,1} + w_{3,1} = N$.

Take event $e_i$ = "service completion at center $i$" and $e_{i,j}$ = "arrival at center $j$ of job from center $i$." Specify the process $\{X(t)\colon\ t \geq 0\}$ as a GSMP with simultaneous trigger events and state space $S$ and event set $E = \{e_1, e_2, e_3, e_{1,2}, e_{1,3}, e_{2,1}, e_{3,1}\}$.

## 6.4. Regeneration Theorems

There are analogues for GSMP's with simultaneous trigger events of the regeneration theorems in Chapter 2 for GSMP's with unique trigger events.

**Theorem 4.1.** *Let $\{X(t)\colon t \geq 0\}$ be a generalized semi-Markov process with simultaneous trigger events and finite state space $S$ and event set $E$. Also let $L(t)$ be the last state occupied by $\{X(t)\colon t \geq 0\}$ before jumping to the state occupied at time $t$, and set $V(t) = (L(t), X(t))$. Assume that for $s, s' \in S$, $E^* \subseteq E(s)$, and $e' \in N(s'; s, E^*)$ the clock-setting distribution function $F(\,\cdot\,; s', e', s, E^*)$ has finite mean. Also assume that there exist $s_0, s_0' \in S$ and a set $H^* \subseteq E(s_0)$ such that for $E^* \subseteq E(s_0)$:*

*(i)* $P\{V(\zeta_n) = (s_0, s_0')$ *i.o.*$\} = 1$;
*(ii)* *the set* $O(s_0'; s_0, E^*) = \emptyset$;
*(iii)* *the set* $N(s_0'; s_0, E^*) = N(s_0'; s_0, H^*)$; *and*
*(iv)* *the clock-setting distribution function* $F(\,\cdot\,; s_0', e', s_0, E^*) = F(\,\cdot\,; s_0', e', s_0, H^*)$ *for all* $e' \in N(s_0'; s_0, E^*)$.

*Then $\{X(t)\colon t \geq 0\}$ is a regenerative process in continuous time. Moreover if $E[\beta_n - \beta_{n-1}] < \infty$, where $\beta_n$ is the index of the $n$th state transition from $s_0$ to $s_0'$, then the expected time between regeneration points is finite.*

The conclusion of Theorem 4.1 also holds if Condition (ii) is replaced by

(ii′) $O(s_0'; s_0, E^*) \neq \emptyset$ and for all $e' \in O(s_0'; s_0, E^*)$ each clock-setting distribution function $F(\,\cdot\,; s', e', s, B^*)$ is exponential with rate parameter independent of $s$, $s'$, and $B^*$.

Let $\zeta_n$ be the time of the $n$th state transition, $n \geq 0$. The idea is to show that the successive random times $\{\zeta_{\beta_n}\colon n \geq 0\}$

at which there is a state transition from $s_0$ to $s_0'$ form a sequence of regeneration points for the process $\{X(t)\colon\ t \geq 0\}$. Condition (i) ensures that the GSMP makes a state transition from $s_0$ to $s_0'$ infinitely often with probability 1. Using an argument similar to the one given in the proof of Theorem 4.6 below, Conditions (ii)–(iv) imply that the GSMP probabilistically restarts at each such state transition. The condition $E[\beta_n - \beta_{n-1}] < \infty$ says that the expected value of the number of state transitions in the time interval $[\zeta_{\beta_{n-1}}, \zeta_{\beta_n})$ is finite. Then since $S$ is finite and each clock-setting distribution has finite mean, Wald's identity implies that the expected time between regeneration points is finite ($E[\zeta_{\beta_n} - \zeta_{\beta_{n-1}}] < \infty$).

Now define a finite-valued increasing sequence $\{T_n\colon\ n \geq 0\}$ of state transition times as follows. For (fixed) $H^* \subseteq E$ and $S^* \subseteq S$, set $T_0 = 0$ and

$$T_n = \inf\big\{t > T_{n-1}\colon \text{ at time } t \text{ the events in } H^* \text{ occur simultaneously in some state } s^* \in S^*\big\} \tag{4.2}$$

for $n \geq 1$. Observe that at each random time $T_n$ events in the set $E - H^*$ also can occur. Proposition 4.4 provides conditions under which $\{X(T_n)\colon\ n \geq 0\}$ hits a fixed state $s_0' \in S$ infinitely often with probability 1. Recall that the geometric trials lemma (Lemma 3.8 of Chapter 2) implies that $P\{X(T_n) = s_0' \text{ i.o.}\} = 1$ if

$$P\{X(T_n) = s_0' | X(T_{n-1}), \ldots, X(T_0)\} \geq \delta \text{ a.s.} \tag{4.3}$$

for some $\delta > 0$ and all $n \geq 1$.

To establish (4.3) in the setting of GSMP's with simultaneous trigger events, we postulate the existence of a set $\{e_{i_1}, \ldots, e_{i_m}\}$ of distinguished events along with distinguished state transition times $\big\{\zeta_{\gamma_{n-1}}\colon\ n \geq 0\big\}$ such that $\zeta_{\gamma_{n-1}} \geq T_{n-1}$ a.s.. We impose the following condition on the sample paths: $X(T_n) = s_0'$ when each of the distinguished events $e_{i_k}$ that is active at time $\zeta_{\gamma_{n-1}}$ occurs prior to a specified random deadline. Then under the condition on the sample paths, the geometric trials criterion for recurrence is satisfied if the clock-setting distributions associated with the distinguished events satisfy a positivity condition (Condition (ii) of Proposition 4.4). For ease of exposition, we also assume that the clock-setting distribution functions for the distinguished events satisfy $F(\,\cdot\,; s', e_{i_k}, s, E^*) = F(\,\cdot\,; e_{i_k})$ for $1 \le k \le m$.

We define the partial history $\mathcal{F}_k$ of the underlying GSSMC $\{(S_n, C_n)\colon n \ge 0\}$ up until the $k$th state transition as follows. Set $E_n^* = E^*(S_n, C_n)$ for $n \ge 0$. For $e_i \in E(S_n)$, let $\lambda(n,i)$ be the index (less than or equal to $n$) of the latest state transition at which the clock associated with event $e_i$ is set: $\lambda(0,i) = 0$ and

$$\lambda(n,i) = \max\bigl\{\, l : 1 \le l \le n \text{ and } e_i \in N(S_l; S_{l-1}, E_{l-1}^*)\,\bigr\}$$

for $n \ge 1$. (If $e_i \in E(S_n)$ and the maximum is taken over an empty set, define $\lambda(n,i) = 0$; if $e_i \notin E(S_n)$, set $\lambda(n,i) = n$.) The *partial history* $\mathcal{F}_k$ of the GSSMC $\bigl\{(S_n, C_n)\colon n \ge 0\bigr\}$ up to the $k$th state transition is the $\sigma$-field generated by sets of the form

$$\begin{aligned} A = \bigl\{\, &\lambda(k,i) = l_i \text{ for } e_i \in E,\ S_m = s_m \text{ for } 0 \le m \le k, \\ &E^*(S_m, C_m) = E_m^* \text{ for } 0 \le m \le k,\ C_{m,i} \le x_{m,i} \\ &\text{for } 0 \le m < l_i \text{ and } e_i \in E\,\bigr\}, \end{aligned}$$

where $0 \le l_i \le k$, $0 \le x_{m,i} < \infty$, $s_m \in S$, and $E_m^* \subseteq E(s_m)$. According to this definition, $\mathcal{F}_k$ contains complete information about the underlying GSSMC of the GSMP up to the $k$th state transition, except for the values of the positive clock readings just after the $k$th state transition.

Let $\bigl\{\zeta_{\gamma_{n-1}}\colon n \ge 1\bigr\}$ be a finite-valued increasing sequence of state transition times such that $\zeta_{\gamma_{n-1}} \ge T_{n-1}$ a.s. and each $\gamma_{n-1}$ is a stopping time with respect to $\{\mathcal{F}_n\colon n \ge 0\}$. Denote the state space of the process $\{S_{\gamma_{n-1}}\colon\ n \geq 1\}$ by $S^+$. Let $e_{i_1}, e_{i_2}, \ldots, e_{i_m} \in E$ and set

$$K(s^+) = \{k\colon\ e_{i_k} \in E(s^+)\}$$

for $s^+ \in S^+$. Observe that $C_{\gamma_{n-1},i_k}$ is the clock reading associated with event $e_{i_k}$ just after the $\gamma_{n-1}$st state transition. Set

$$h_{i_k}(x) = \sup_{y \geq 0} \frac{\overline{F}(x+y; e_{i_k})}{\overline{F}(y; e_{i_k})}$$

for $x \geq 0$ and $1 \leq k \leq m$.

**Proposition 4.4.** *For $s^+ \in S^+$, let $\{R_{1,k}(s^+)\colon\ k \in K(s^+)\}$, $\{R_{2,k}(s^+)\colon\ k \in K(s^+)\}, \ldots$ be identically distributed collections of random variables, each independent of $\{(S_l, C_l)\colon\ l \geq 0\}$. Let $s_0' \in S$. Assume that*

*(i) the following condition on the sample paths holds:*

$$\begin{aligned} P\{X(T_n) = s_0' \mid S_{\gamma_{n-1}} &= s^+, \\ X(T_{n-1}) &= x_{n-1}, \ldots, X(T_0) = x_0\} \\ \geq P\Big\{ C_{\gamma_{n-1},i_k} &\leq R_{n,k}(s^+),\ k \in K(s^+) \ \Big|\ S_{\gamma_{n-1}} = s^+, \\ X(T_{n-1}) &= x_{n-1}, \ldots, X(T_0) = x_0 \Big\} \end{aligned}$$

*for $x_0, x_1, \ldots, x_{n-1} \in S$ and $s^+ \in S^+$; and*

*(ii) there exists $\delta > 0$ such that*

$$\delta(s^+) = E\Big[\prod_{k \in K(s^+)} [1 - h_{i_k}(R_{1,k}(s^+))]\Big] \geq \delta$$

*for $s^+ \in S^+$.*

*Then $P\{X(T_n) = s_0' | X(T_{n-1}), \ldots, X(T_0)\} \geq \delta$ a.s. so that $P\{X(T_n) = s_0'$ i.o.$\} = 1$.*

Observe that when $|S^+| < \infty$ Condition (ii) can be replaced by (iia) *there exist real numbers* $x^*_{i_1}, x^*_{i_2}, \ldots, x^*_{i_m} \geq 0$ *such that* $h_{i_k}(x) < 1$ *for* $x > x^*_{i_k}$ *and* $1 \leq k \leq m$; *and*

(iib) $P\{R_{1,k}(s^+) > x^*_{i_k}$ *for* $k \in K(s^+)\} > 0$ *for* $s^+ \in S^+$

For example, if the clock-setting distribution associated with event $e_{i_k}$ is NBU, then $h_{i_k}(x) = \overline{F}(x; e_{i_k})$ and Condition (iia) holds with

$$x^*_{i_k} = \sup\{x\colon\ F(x; e_{i_k}) = 0\}\,.$$

The distribution of an a.s. bounded random variable satisfies Condition (iia). This condition also holds for an absolutely continuous distribution with density function $f$ such that

$$\lim_{x \to \infty} f(x)/be^{-ax} \leq 1$$

for some constants $a, b > 0$.

**Example 4.5.** (Producer/consumer with finite buffers) Consider a system consisting of two producers, two consumers, and two buffers, each numbered 1 and 2. The producers share a channel for transmission, one at a time, of items from buffers to consumers. Producer $i$ creates items for consumer $i$ one at a time. Items created, but not yet transmitted, are placed in buffer $i$ for transmission. An item created by producer $i$ when the system already contains $B_i - 1$ ($\geq 0$) items for consumer $i$ causes the process of creation of items for consumer $i$ to shut down. This process remains shut down until the first subsequent end of transmission to consumer $i$. Items created by producer $i$ are transmitted in order of their creation, and producer/consumer pair 1 has nonpreemptive priority over producer/consumer pair 2 for use of the channel. Assume that the successive times required by producer $i$ to create an item are i.i.d. as a positive random variable $A_i$ with finite mean and a density function that is continuous and positive on $(0, +\infty)$. Also assume that the successive times to transmit an item to consumer $i$ to consumer $i$ are i.i.d. as a positive random variable $L_i$ with finite mean and a density function that is continuous and positive on $(0, +\infty)$. Creation times are independent of transmission times and random quantities associated with producer/consumer pair 1 are independent of random quantities associated with producer/consumer pair 2.

Set

$$X(t) = \big(B_1(t), B_2(t), M(t)\big),$$

where $B_i(t)$ is the number of items in buffer $i$ at time $t$ and

$$M(t) = \begin{cases} 2 & \text{if producer 2 is transmitting at time } t \\ 1 & \text{if producer 1 is transmitting at time } t \\ 0 & \text{if neither producer is transmitting at time } t. \end{cases}$$

Take event $e_{2+i}$ = "end of transmission from buffer $i$" and $e_i$ = "creation of item by producer $i$." Formal specification of the process $\{X(t)\colon\ t \ge 0\}$ is as a GSMP with finite state space

$$S = \big\{(b_1, b_2, m) \in \{0, 1, \ldots, B_1\} \times \{0, 1, \ldots, B_2\} \times \{0, 1, 2\}: \\ m = 0 \text{ only if } b_1 = b_2 = 0\big\}$$

and event set $E = \{e_1, e_2, e_3, e_4\}$. For $s = (b_1, b_2, m) \in S$, event $e_4$ = "end of transmission from buffer 2" $\in E(s)$ if and only if $m = 2$; $e_3$ = "end of transmission from buffer 1" $\in E(s)$ if and only if $m = 1$; event $e_i$ = "creation of item by producer $i$" $\in E(s)$ if and only if $b_i < B_i$. All speeds $r(s, e)$ for active events are equal to 1.

If $e = e_4$ = "end of transmission from buffer 2," then the state transition probability $p(s'; s, e) = 1$ when

$$s = (b_1, b_2, 2) \text{ with } b_1 > 0 \text{ and } s' = (b_1 - 1, b_2 - 1, 1),$$

when

$$s = (0, b_2, 2) \text{ with } b_2 > 1 \text{ and } s' = (0, b_2 - 1, 2),$$

and when

$$s = (0,1,2) \text{ and } s' = (0,0,0).$$

If $e = e_3 =$ "end of transmission from buffer 1," then the state transition probability $p(s'; s, e) = 1$ when

$$s = (b_1, b_2, 1) \text{ with } b_1 > 0 \text{ and } s' = (b_1 - 1, b_2, 1),$$

when

$$s = (1, b_2, 1) \text{ with } b_2 > 0 \text{ and } s' = (0, b_2 - 1, 2),$$

and when

$$s = (1,0,1) \text{ and } s' = (0,0,0).$$

If $e = e_2 =$ "creation of item by producer 2," then the state transition probability $p(s'; s, e) = 1$ when

$$s = (0,0,0) \text{ and } s' = (0,0,2)$$

and when

$$s = (b_1, b_2, m) \text{ with } m > 0 \text{ and } s' = (b_1, b_2 + 1, m).$$

If $e = e_1 =$ "creation of item by producer 1," then the state transition probability $p(s'; s, e) = 1$ when

$$s = (0,0,0) \text{ and } s' = (0,0,1)$$

and when

$$s = (b_1, b_2, m) \text{ with } m > 0 \text{ and } s' = (b_1 + 1, b_2, m).$$

All other state transition probabilities are equal to 0. The clock-setting distribution functions are as follows. If the new event $e' = e_{i+2} =$ "end of transmission from buffer $i$," then $F(x; s', e', s, e^*) = P\{L_i \le x\}$ for $x \ge 0$. If $e' = e_i =$ "creation of item by producer $i$," then $F(x; s', e', s, e^*) = P\{A_i \le x\}$ for $x \ge 0$.

Assume that the distributions of $A_1$, $A_2$, and $L_2$ satisfy Condition (iia). Take $s'_0 = (B_2-1, B_2, 1)$. Also take $H^* = \{e_3\}$ and $S^* = \{(b_1, b_2, m) \in S\colon m = 1\}$, so that $T_n$ in (4.2) is the $n$th time at which event $e_3$ = "end of transmission to consumer 1" occurs. Observe that $X(T_n) = s'_0$ if buffer 2 is full, there are $B_1 - 1$ items in buffer 1, producer 1 starts to create an item, and a transmission from buffer 1 starts at time $T_n$. Set $\nu_n = \min\{k\colon \zeta_k = T_n\}$ for $n \ge 0$, so that $\nu_n$ is the random index of the state transition that corresponds to the occurrence of event $e_3$ at time $T_n$. Also set

$$\alpha_n = \min\{k > \nu_n\colon e_1 \in E^*(S_{k-1}, C_{k-1})\},$$

so that $\alpha_n$ is the random index that corresponds to the first occurrence of event $e_1$ = "creation of item by producer 1" after time $T_n$. Since producer/consumer pair 1 has nonpreemptive priority over producer/consumer pair 2 and each clock-setting distribution is proper, an inductive argument shows that each $\nu_n$, $\alpha_n$, and $T_n$ is a.s. finite.

Let $\zeta_{\gamma_{n-1}}$ be the time of the most recent start of transmission from buffer 1 prior to time $T_n$, so that $\gamma_{n-1}$ is the random index that corresponds to the most recent occurrence of event $e_2$. Suppose that $S_{\gamma_{n-1}} = s^+ = (b_1^+, b_2^+, m^+)$. If $b_1^+ = B_1$ and $b_2^+ = B_2$, then clearly $X(T_n) = s'_0$. Otherwise, $X(T_n) = s'_0$ provided that $B_1 - b_1^+ - 1$ items are created by producer 1 and $B_2 - b_2^+$ items are created by producer 2 during the transmission from buffer 1. Let $\{e_{i_1}, e_{i_2}\} = \{e_1, e_2\}$ be the set of distinguished events. Take $R_{n,1}(s^+)$ to be a random variable distributed as $L_1 - \sum_{j=1}^{B_1-b_1^+-2} A_{1,j}$ and $R_{n,2}(s^+)$ to be a random variable distributed as $L_1 - \sum_{j=1}^{B_2-b_2^+-1} A_{2,j}$, where the $A_{k,j}$ are mutually independent and distributed as $A_k$. Then Condition (i) of Proposition 4.4 is satisfied. It follows that

$$P\{R_{1,k}(s^+) > x^*_{i_k} \text{ for } k \in K(s^+)\} > 0$$

for $s^+ \in S^+$. Proposition 4.4 implies that $P\{X(T_n) = s_0' \text{ i.o.}\} = 1$ provided that

$$P\{x_1^* + A_{1,2} + \cdots + A_{1,B_1-1} < L_1\} > 0$$

and

$$P\{x_2^* + A_{2,2} + \cdots + A_{2,B_2} < L_1\} > 0.$$

Theorem 4.6 provides conditions under which a GSMP with simultaneous trigger events is a regenerative process in continuous time and the expected time between regeneration points is finite. The proof uses a representation for the conditional joint distribution of a set of clock readings, given the partial history of the underlying GSSMC of the GSMP.

**Theorem 4.6.** *Let $\{T_n\colon n \geq 0\}$ be a finite-valued increasing sequence of state transition times as in (4.2). Assume that there exists $s_0' \in S$ such that*

*(i)* $P\{X(T_n) = s_0' | X(T_{n-1}), \ldots, X(T_0)\} \geq \delta$ a.s. *for some* $\delta > 0$ *and all* $n \geq 1$.

*Also assume that for all states $s^* \in S^*$ and sets $E^*$ of events for which $H^* \subseteq E^* \subseteq E(s^*)$ with $p(s_0'; s^*, E^*) > 0$:*

*(ii) $O(s_0'; s^*, E^*) = O^*$, where $O^* \subseteq E$ is a (possibly empty) fixed set of events and for all $e_i \in O^*$ each clock-setting distribution function $F(\,\cdot\,; s', e_i, s, B^*)$ is exponential with rate parameter $\mu_i$ independent of $s'$, $s$, and $B^*$; and*

*(iii) the clock-setting distribution function $F(\,\cdot\,; s', e', s, B^*) = F(\,\cdot\,; e')$ for all $s', s \in S$, $B \subseteq E(s')$, and $e' \in E(s_0') - O^*$.*

*Then $\{X(t)\colon t \geq 0\}$ is a regenerative process in continuous time. Moreover, if $\sup_{n\geq 0} E[T_{n+1} - T_n] < \infty$, then $E[\tau] < \infty$, where $\tau$ denotes the time between regeneration points. If $\sup_{n\geq 0} E[(T_{n+1} - T_n)^{q+\epsilon}] < \infty$ for some $q \geq 0$ and $\epsilon > 0$, then $E[\tau^q] < \infty$.*

The conditions of Theorem 4.6 ensure that the successive random times at which the events in the set $H^*$ occur simultaneously in some state in $S^*$ and trigger a state transition to $s'_0$ form a sequence of regeneration points for the process $\{X(t)\colon\ t \geq 0\}$. We require that, whenever any such state transition occurs, all old events with nonexponential clock-setting distributions become inactive and the set of new events is equal to the fixed set $E(s'_0) - O^*$ (Condition (ii)), and the new clock reading have a fixed distribution function $F(\,\cdot\,; e')$ for each event $e' \in E(s'_0) - O^*$ (Condition (iii)). Conditions (ii) and (iii) thus guarantee that the process $\{X(t)\colon\ t \geq 0\}$ probabilistically restarts whenever $\{X(T_n)\colon\ n \geq 0\}$ hits state $s'_0$. Condition (i) ensures that $\{X(T_n)\colon\ n \geq 0\}$ hits state $s'_0$ infinitely often with probability 1.

The conclusion of Theorem 4.6 also holds if Condition (iii) is replaced by

(iii) the clock-setting distribution function $F(\,\cdot\,; s'_0, e', s^*, E^*) = F(\,\cdot\,; s'_0, e', s, H^*)$ for all $e' \in E(s'_0) - O^*$ and some $s \in S$.

*Proof.* Recall the definition of the partial history $\mathcal{F}_n$ of the underlying GSSMC up to the $n$th state transition and the associated random indices $\lambda(n,i)$. Denote by $Z_{n,i}$ the amount of time that elapses on the clock associated with event $e_i$ between $\zeta_{\lambda(n,i)}$ and $\zeta_n$: $Z_{n,i} = C_{\lambda(n,i),i} - C_{n,i}$ for $e_i \in E$ and $n \geq 0$. Then the conditional joint distribution of clock readings given the partial history of the GSSMC up to the $n$th state transition

$$P_\mu\{\, C_{n,i} > x_i \text{ for } e_i \in B^* \mid \mathcal{F}_n \,\} = \begin{cases} \prod_{e_i \in B^*} \overline{F}\big(x_i + Z_{n,i}; e_i\big)/\overline{F}\big(Z_{n,i}; e_i\big) & \text{if } B^* \subseteq E(S_n); \\ 0 & \text{otherwise} \end{cases} \tag{4.7}$$

with probability 1 for any subset $B^* \subseteq E$, nonnegative real numbers $\{x_i\colon\ e_i \in B^*\}$, and $n \geq 0$.

Also recall that the GSSMC $\{(S_n, C_n)\colon n \geq 0\}$ takes values in the set $\Sigma = \bigcup_{s \in S}(\{s\} \times C(s))$. Define a measure $\phi$ on measurable subsets of $\Sigma$ by

$$\phi(A) = 1_{\{s_0'\}}(s') \prod_{e_i \in N^*} F(a_i; e_i) \prod_{e_i \in O^*} (1 - e^{-\mu_i a_i}),$$

where $N^* = E(s_0') - O^*$ and

$$A = \{s'\} \times \{(c_1', c_2', \ldots, c_M') \in C(s')\colon 0 \leq c_i' \leq a_i \text{ for } 1 \leq i \leq M\}. \tag{4.8}$$

Next define a sequence of random indices $\{\beta_n\colon n \geq 0\}$ as follows. Set $\beta_0 = 0$ if the initial distribution $\mu$ is equal to $\phi$; otherwise, set $\beta_0 = \min\{k > 0\colon X(T_k) = s_0'\}$. Then set

$$\beta_n = \min\{k > \beta_{n-1}\colon X(T_k) = s_0'\}$$

for $n \geq 1$. Condition (i) ensures that each random time $T_{\beta_n}$ is a.s. finite. Also set

$$\gamma_n = \gamma_n(T_{\beta_n}) = \max\{k\colon \zeta_k = T_{\beta_n}\}$$

for $n \geq 0$. Now fix $n \geq 0$, and observe that since each $\gamma_n$ is a stopping time with respect to $\{\mathcal{F}_k\colon k \geq 0\}$, (4.7) and Conditions (ii)–(iv) together imply that

$$\begin{aligned} &P_\mu\{C_{\gamma_n,i} > x_i \text{ for } e_i \in B^* \mid \mathcal{F}_{\gamma_n}\} \\ &\quad = \begin{cases} \prod_{e_i \in B^*} \overline{F}(x_i + Z_{\gamma_n,i}; e_i)/\overline{F}(Z_{\gamma_n,i}; e_i) & \text{if } B^* \subseteq E(S_{\gamma_n}); \\ 0 & \text{otherwise} \end{cases} \end{aligned} \tag{4.9}$$

with probability 1 for any subset $B^* \subseteq E$ and nonnegative real numbers $\{x_i\colon e_i \in B^*\}$. Also observe that $\mathcal{F}_{\gamma_n}$ contains the $\sigma$-field generated by $\{X(t)\colon\ 0 \le t < T_{\beta_n}\}$. It follows from (4.9) that

$$\begin{aligned}
P\{\,(S_{\gamma_n}, C_{\gamma_n}) \in A \mid X(t)\colon\ 0 \le t < T_{\beta_n}\,\} \\
&= E\bigl[P\{\,(S_{\gamma_n}, C_{\gamma_n}) \in A|\mathcal{F}_{\gamma_n}\,\} \mid X(t)\colon\ 0 \le t < T_{\beta_n}\bigr] \\
&= E\bigl[\phi(A) \mid X(t)\colon\ 0 \le t < T_{\beta_n}\bigr] \\
&= \phi(A) \text{ a.s.}
\end{aligned}$$

for all sets $A$ of the form (4.8). Using an inductive argument, it follows that

$$\begin{aligned}
P\{(S_{\gamma_n+n_1}, C_{\gamma_n+n_1}) \in A_1, \ldots, \\
(S_{\gamma_n+n_m}, C_{\gamma_n+n_m}) \in A_m \mid X(t)\colon\ 0 \le t < T_{\beta_n}\} \\
= P_\phi\,\{\,(S_{n_1}, C_{n_1}) \in A_1, \ldots, (S_{n_m}, C_{n_m}) \in A_m\,\} \text{ a.s.}
\end{aligned}$$

for every sequence $0 \le n_1 < n_2 < \cdots < n_m$ $(m \ge 1)$ and sets $A_1, \ldots, A_m$ of the form (4.8). Standard arguments then show that

$$\begin{aligned}
E[f\,(S_{\gamma_n}, C_{\gamma_n}, S_{\gamma_n+1}, C_{\gamma_n+1}, \ldots)| \; X(t)\colon\ 0 \le t < T_{\beta_n}] \\
= E_\phi[f\,(S_0, C_0, S_1, C_1, \ldots)] \text{ a.s.}
\end{aligned} \tag{4.10}$$

for any real-valued measurable function $f$ on $\Sigma^\infty$ provided that the expectations exist. Since $\{(S_k, C_k)\colon\ k \ge \gamma_n\}$ determines the processes $\{X(t)\colon\ t \ge T_{\beta_n}\}$ and $\{T_{\beta_{k+1}} - T_{\beta_k}\colon\ k \ge n\}$, the regenerative property follows from (4.10) and the fact that $n$ was chosen arbitrarily. The usual argument establishes the finiteness of the moments of $\tau$. □

Observe that the regeneration points $\{T_{\beta_n}\colon\ n \ge 0\}$ constructed in the proof of Theorem 4.6 need not be stopping times with respect to the GSMP $\{X(t)\colon\ t \ge 0\}$. However, the random indices $\{\gamma_n\colon\ n \ge 0\}$ are stopping times with respect to the GSSMC $\{(S_n, C_n)\colon\ n \ge 0\}$. Also observe that, for fixed

$S^* \subseteq S$ and $O^*, H^* \subseteq E$, the limiting distribution of the GSMP is independent of the initial distribution if Condition (i) of Theorem 4.6 holds for every initial distribution. This assertion follows from the fact that for any initial distribution, the random times $\{T_{\beta_n}\colon n \geq 0\}$ form a sequence of regeneration points for the process $\{X(t)\colon t \geq 0\}$. Finally, observe that Theorem 4.6 does not guarantee that the distribution of $\tau$ is aperiodic.

**Example 4.11.** (Collision-free bus network) Assume that the distribution of each $A_j$ and each $L_j$ is NBU. Also assume that packet transmission times are long in the sense that $P\{L_j > R(1) + T\} = 1$ for all $j$ and

$$P\{A_j + R(1) + T \leq L_1\} > 0$$

for $j = 2, 3, \ldots, N$. Set

$$\begin{aligned} X(t) = (&W_1(t), \ldots, W_N(t), U_1(t), \ldots, U_N(t), \\ &V_{2,1}(t), V_{3,1}(t), V_{3,2}(t), V_{4,1}(t), \ldots, V_{N,N-1}(t)) \end{aligned}$$

as in Example 3.8. Recall that event $e_{j,1}$ = "setting (to 1) of flip-flop by port $j$," $e_{j,3}$ = "end of transmission by port $j$," and $e_{j,l,5}$ = "observation by port $j$ of end of transmission by port $l \neq j$." Take $H^* = \{e_{1,3}\}$ and

$$\begin{aligned} S^* = \big\{&(w_1, \ldots, w_N, u_1, \ldots, u_N, v_{2,1}, \ldots, v_{N,N-1}) \in S\colon\ w_1 = 3; \\ &u_j = 1 \text{ for } j = 1, 2, \ldots, N; v_{j,1} = 0 \text{ for } j = 2, 3, \ldots, N\big\} \end{aligned}$$

so that $T_n$ defined by (4.2) is the time of the $n$th end of transmission by port 1, $n \geq 0$. Also take $s_0' = (0, 2, \ldots, 2, 0, 1, \ldots, 1, 0, 0, 1, \ldots, 0, 1, \ldots, 1) \in S$. Observe that $X(T_n) = s_0'$ if at time $T_n$ port 1 ends transmission of a packet and each other port $j$ has observed the resetting of flip-flop by port 1, has a packet queued for transmission, and has completed the $R(j)+T$ wait. Also observe that $T_n - T_{n-1}$ can be represented as

$$T_n - T_{n-1} = A_{1,n} + R(1) + T + K_n + L_{1,n},$$

where $L_{1,n}$ is distributed as $L_1$, $A_{1,n}$ is distributed as $A_1$, and $K_n$ is a nonnegative random variable with

$$E[K_n] \leq \sum_{j=2}^{N} E[L_j].$$

Then

$$E[T_n - T_{n-1}] = E[A_{1,n}] + R(1) + T + E[K_n] + E[L_{1,n}] < \infty, \tag{4.12}$$

so that $T_n < \infty$ a.s.

Let $\zeta_{\gamma_{n-1}}$ be the first time after $T_{n-1}$ that port 1 begins transmission of a packet. Then $X(T_n) = s_0'$ if, at least $R(1)+T$ time units before the end of transmission by port 1, there is an arrival of a packet for transmission at each port that does not have a packet queued for transmission at time $\zeta_{\gamma_{n-1}}$. Take $e_{i_j} = e_{j,1}$ and for $s^+ = (w_1^+, \ldots, w_N^+, u_1^+, \ldots, u_N^+, v_{2,1}^+, \ldots, v_{N,N-1}^+) \in S^+$, set

$$K(s^+) = \{k\colon\ w_k^+ = 0\}$$

so that $K^+ = \{2, 3, \ldots, N\}$. Also take $R_{n,k}(s^+) = L_1 - (R(1) + T)$ for all $s^+ \in S^+$ and $k \in K(s^+)$. Then Condition (i) of Proposition 4.4 is satisfied. Since $P\{A_j + R(1) + T \leq L_1\} > 0$ for $j = 2, 3, \ldots, N$,

$$\delta \stackrel{\text{def}}{=} P\{A_j + R(1) + T \leq L_1,\ j = 2, 3, \ldots, N\} > 0.$$

Then

$$\delta(s^+) = P\{A_j + R(1) + T \leq L_1,\ j \in K(s^+)\} > \delta$$

for $s^+ \in S^+$, and $P\{X(T_n) = s_0' \text{ i.o.}\} = 1$ by Proposition 4.4.

The process $\{X(t)\colon\ t \geq 0\}$ hits state $s_0'$ when event $e_{1,3}$ is the trigger event only if the event occurs in state $s_0 =$

$(3,2,\ldots,2,1,\ldots,1,0,1,\ldots,1)$, and in this case there are no old clocks running ($O^* = \emptyset$ and $E(s_0') - O^* = \{e_{1,1}, e_{2,1,5}, \ldots, e_{N,1,5}\}$). Theorem 4.6 implies that the successive random times at which $\{X(T_n)\colon n \geq 0\}$ hits state $s_0'$ form a sequence of regeneration points for the process $\{X(t)\colon t \geq 0\}$. The expected time between regeneration points is finite by (4.12). The time between regeneration points has finite second moment if, in addition, $E[L_k^{2+\epsilon}] < \infty$ for some $\epsilon > 0$ so that

$$\sup_{n \geq 0} E[(T_{n+1} - T_n)^{2+\epsilon}] < \infty.$$

## Exercises

4.1. (Register insertion ring) Consider a unidirectional ring network that contains a fixed number of "idle bits" and has a fixed number of stations, labelled $1, 2, \ldots, N$ in the direction of signal propagation. Packets of fixed length arrive at each station according to a random process and queue externally. When a station has a packet to transmit and is observing idle bits (that is, no packet is passing), the station instantaneously inserts a register (buffer) that contains the packet in-line on the input side of the delay between the station receiver and station transmitter. If the station is observing packet bits (that is, another packet is passing), the station waits until that packet has passed before inserting a register that contains the new packet. The source station leaves its register inserted until its packet has returned and is contained in the register. The station then instantaneously removes the inserted register from the ring, thus removing the packet.

Denote the constant propagation delay along the ring from station $j$ to the next station by $R_j$ and the constant propagation delay through an inserted register by $L$. Packets are short in the sense that $L < R_1 + R_2 + \cdots + R_N$. The time from the removal of the inserted register by station $j$ until the arrival of the next packet for transmission by station $j$ is exponentially distributed as a random variable $A_j$. With independent probability $p_{jk}$, the destination of a packet that arrives for transmission by station $j$ is station $k$ ($\neq j$). Set

$$X(t) = (Z_1(t), \ldots, Z_N(t), B_1(t), \ldots, B_N(t), D_1(t), \ldots, D_N(t)),$$

as in Example 3.9.

Take event $e_{2N+j}$ = "arrival of packet for transmission by station $j$," $e_{N+j}$ = "observation by station $j$ of last bit of packet transmitted by station," and $e_j$ = "observation by station of first bit of packet transmitted by station $j$" ($j = 1, 2, \ldots, N$). Formal specification of the process $\{X(t)\colon\ t \geq 0\}$ is as a GSMP with (simultaneous trigger events and) a finite state space $S$ and event set $E = \big\{e_1, e_2, \ldots, e_N, e_{N+1}, \ldots, e_{2N}, e_{2N+1}, \ldots, e_{3N}\big\}$. Fix $j$ and take $s_0' = (0, 0, \ldots, 0) \in S$. Observe that when $\{X(t)\colon\ t \geq 0\}$ is in state $s_0'$ the system is empty; that is, no station has its register inserted nor has a packet queued for transmission. Take $H^* = \{e_{N+j}\}$ and

$$S^* = \{(z_1, \ldots, z_N, \ldots, d_1, \ldots, d_N) \in S\colon\ z_j = N + j\},$$

so that $T_n$ defined by (4.2) is the time of the $n$th removal of its register by station $j$, $n \geq 0$. Use a geometric trials criterion to show that $P\{X(T_n) = s_0' \text{ i.o.}\} = 1$. Also show that the successive random times at which $\{X(T_n)\colon\ n \geq 0\}$ hits state $s_0'$ form a sequence of regeneration points for the process $\{X(t)\colon\ t \geq 0\}$ and the expected time between regeneration points is finite.

Appendix A

# Limit Theorems for Stochastic Processes

The usual goal in the analysis of simulation output is to obtain point estimates and confidence intervals for characteristics of well defined random variables. Since simulation output generally does not satisfy the assumptions on distributions required for classical parametric statistics, it is necessary to use asymptotic distributions obtained from limit theorems. This Appendix contains the results on probability and stochastic processes needed as background.

## A.1. Probability Spaces and Random Variables

Let $\Omega$ be a set. A collection $\mathcal{F}$ of subsets of $\Omega$ is called a *σ-field* if

(i) $\emptyset \in \mathcal{F}$ and $\Omega \in \mathcal{F}$;

(ii) $A^c \in \mathcal{F}$ whenever $A \in \mathcal{F}$, where $A^c = \{\omega \in \Omega\colon\ \omega \notin A\}$; and

(iii) $\bigcup_{i=1}^{\infty} A_i \in \mathcal{F}$ whenever $A_1, A_2, \ldots, \in \mathcal{F}$.

A *probability measure* $P$ is a nonnegative real-valued function on $\mathcal{F}$ such that $P(\emptyset) = 0$, $P(\Omega) = 1$, and

$$P\Big(\bigcup_n A_n\Big) = \sum_n P(A_n)$$

whenever $A_1, A_2, \ldots$ form a finite or countably infinite collection of disjoint sets in $\mathcal{F}$. A *probability space* is a triple $(\Omega, \mathcal{F}, P)$ where $\Omega$ (the space of elementary outcomes) is a set, $\mathcal{F}$ (the family of events) is a *σ-field* of subsets of $\Omega$, and $P$ is a probability measure on $\mathcal{F}$. A *random variable* on the probability space $(\Omega, \mathcal{F}, P)$ is a real-valued function $X$ that is defined on $\Omega$ and *measurable* with respect to $\mathcal{F}$ in the sense that $\{\omega\colon\ X(\omega) \le x\} \in \mathcal{F}$ for all real $x$. The function $F$ defined by

$$F(x) = P\{X \le x\}$$

for $-\infty < x < \infty$ is called the *distribution function* of the random variable $X$. Formally, the distribution function of $X$ is given by

$$P\{a < X \le b\} = P(\{\omega\colon\ a < X(\omega) \le b\})$$

for all $a < b$. That is, the probability that the random variable $X$ takes a value in $(a, b]$ is calculated as the probability of the set of outcomes $\omega$ for which $a < X(\omega) \le b$.

The σ-field of *Borel sets* is a distinguished σ-field of sets of real numbers and is the smallest σ-field containing every interval of the form $(a, b]$ for $-\infty \le a \le b < +\infty$. A function $h$ from $\Omega_1$ to $\Omega_2$ is said to be *measurable* with respect to σ-fields $\mathcal{F}_1$ of subsets of $\Omega_1$ and $\mathcal{F}_2$ of subsets of $\Omega_2$ if and only if

$$h^{-1}(A) = \{\omega_1\colon\ h(\omega_1) \in A\} \in \mathcal{F}_1$$

for each $A \in \mathcal{F}_2$. If $\mathcal{F}$ is a σ-field of subsets of $\Omega$, $(\Omega, \mathcal{F})$ is called a *measurable space*, and the sets in $\mathcal{F}$ are called *measurable sets*.

Now consider a simulation associated with a sequence of mutually independent random variables with common distribution function $F$. Denote the real line by $\Re$, and set set

$$\Omega = \Re \times \Re \times \Re \times \ldots.$$

Let $\mathcal{F}$ be a suitable family of subsets of $\Omega$ containing subsets of the form $[a_1, b_1) \times [a_2, b_2) \times \ldots \times [a_n, b_n)$. A general elementary outcome $\omega \in \Omega$ is an infinite sequence of real numbers of the form $\omega = (\omega_1, \omega_2, \ldots)$. Define a sequence of random variables $U_i(\omega)$ by

$$U_i(\omega) = \omega_i.$$

Next observe that if $P$ is a probability measure on $\mathcal{F}$ given by

$$\begin{aligned} P\{\omega \in [a_1, b_1) \times \ldots \times [a_n, b_n) \times \Re \times \Re \times \ldots\} \\ = (b_1 - a_1)(b_2 - a_2) \cdots (b_n - a_n) \end{aligned}$$

for $0 \leq a_1, b_1, \ldots, a_n, b_n \leq 1$, then $U_1, U_2, \ldots$ are mutually independent uniform random variables. To obtain the desired sequence of random variables, use the inverse probability integral transformation; that is, set

$$X_i = F^{-1}(U_i),$$

where

$$F^{-1}(u) = \inf\{x\colon\ F(x) \geq u\}$$

for $0 < u < 1$. The sequence $\{X_i\colon\ i \geq 1\}$ of random variables supported by the probability space $(\Omega, \mathcal{F}, P)$ corresponds to the required simulation experiment.

## A.2. Modes of Convergence

Limit theorems relevant to the analysis of simulation output involve three modes of convergence. Unlike real numbers, sequences of random variables can converge in more than one way. The strongest mode of convergence is *convergence with probability 1*, also called *almost sure (a.s) convergence* or *strong convergence*.

**Definition 2.1.** *Let $\{X_n\colon n \geq 1\}$ and $X$ be random variables defined on a common probability space $(\Omega, \mathcal{F}, P)$. Then $X_n$ is said to converge with probability 1 to $X$ if*

$$P\left\{\lim_{n\to\infty} X_n = X\right\} = 1,$$

*and we write $X_n \to X$ a.s. as $n \to \infty$.*

An equivalent characterization of convergence with probability 1 is that for every $\epsilon > 0$

$$\lim_{n\to\infty} P\left\{|X_m - X_n| \leq \epsilon \text{ for all } m \geq n\right\} = 1.$$

Thus, for all $n$ sufficiently large ($n \geq N(\epsilon, \delta)$), $|X_m - X_n| \leq \epsilon$ for all $m \geq n$ with probability greater than $1 - \delta$, where $\delta > 0$ is an arbitrarily small number.

The second mode of convergence is *convergence in probability.*

**Definition 2.2.** *Let $\{X_n\colon n \geq 1\}$ and $X$ be random variables defined on a common probability space $(\Omega, \mathcal{F}, P)$. Then $X_n$ is said to converge in probability to $X$ if for every $\epsilon > 0$*

$$\lim_{n\to\infty} P\left\{|X_n - X| \leq \epsilon\right\} = 1,$$

*and we write $X_n \to X$ in probability as $n \to \infty$.*

Thus for $n$ sufficiently large ($n \geq N(\epsilon, \delta)$), $|\, X_n - X| \leq \epsilon$ with probability greater than $1 - \delta$. This assertion is a statement about the distribution of $|X_n - X|$ for a single value of $n$, whereas convergence with probability 1 is a statement about all $n$ beyond a certain point.

The third mode of convergence is *convergence in distribution* (*weak convergence*). The distribution function $F$ of a random variable $X$ satisfies $F(x) = P\{X \leq x\}$ for $-\infty < x < +\infty$. Since only one distribution at a time is involved in this definition, there is no need to have $\{X_n\colon n \geq 1\}$ and $X$ all defined on the same probability space. In fact, each of the $X_n$ and $X$ can be defined on different probability spaces. Set $F_n(x) = P\{X_n \leq x\}$ and $F(x) = P\{X \leq x\}$.

**Definition 2.3.** *A sequence $\{X_n\colon n \geq 1\}$ of random variables is said to converge in distribution to $X$ if $F_n(x) \to F(x)$ for all $x$ for which $F$ is continuous, and we write $X_n \Rightarrow X$ as $n \to \infty$.*

An equivalent characterization of convergence in distribution of a sequence $\{X_n\colon n \geq 1\}$ to $X$ is that $E[h(X_n)] \to E[h(X)]$ for all bounded and continuous real-valued functions $h$. Let $h$ be a real-valued function, and denote by $D(h)$ the set of discontinuities of $h$. If $X_n \Rightarrow X$ as $n \to \infty$, $h$ is bounded, and $P\{X \in D(h)\} = 0$, then $E[h(X_n)] \to E[h(X)]$.

The notion of convergence in distribution does not require that the $X_n$ "settle down" as does a convergent sequence of ordinary real numbers. Observe that if the $X_n$ and $X$ are random variables defined on a common probability space, convergence of the $X_n$ to $X$ with probability 1 implies convergence of the $X_n$ to $X$ in probability, and convergence of the $X_n$ to $X$ in probability implies convergence of the $X_n$ to $X$ in distribution. Neither converse of these two implications holds. All three modes of convergence can be extended to random vectors.

## A.3. Convergence Tools

The "continuous mapping theorem" and the "converging together lemma" are two powerful tools that can be used to derive new limit theorems from existing ones. Suppose that $\{X_n\colon n \geq 1\}$ is a sequence of $d$-dimensional random vectors (for some $d \geq 1$) and $X$ is a $d$-dimensional limit random vector. Let $h$ be a function that maps $\Re^d$ ($d$-dimensional Euclidean space for some $d \geq 1$) into the real line $\Re = \Re^1$.

**Theorem 3.1.** *(Continuous mapping theorem) Suppose that $X_n \Rightarrow X$ as $n \to \infty$. If $h$ is continuous or if $P\{X \in D(h)\} = 0$, then $h(X_n) \Rightarrow h(X)$ as $n \to \infty$.*

For the converging together lemma, let $\{X_n\colon n \geq 1\}$ and $\{Y_n\colon n \geq 1\}$ be $d$-dimensional random vectors, and let $X$ be a limit random vector. Suppose that for each $n$ the random vectors $X_n$ and $Y_n$ are defined on the same probability space.

**Lemma 3.2.** *(Converging together lemma) If $X_n \Rightarrow X$ and $X_n - Y_n \Rightarrow 0$, then $Y_n \Rightarrow X$ as $n \to \infty$.*

Under certain conditions, convergence in distribution of a two-dimensional random vector follows from convergence in distribution of the two components of the random vector.

**Lemma 3.3.** *Let $\{X_n\colon n \geq 1\}$ and $\{Y_n\colon n \geq 1\}$ be two sequences of random variables defined on the same probability space. If $X_n \Rightarrow X$ and $Y_n \Rightarrow c$, then $(X_n, Y_n) \Rightarrow (X, c)$ as $n \to \infty$.*

Occasionally we deal with convergence in distribution not of a sequence of random variables but of a stochastic proccss indexed by a continuous parameter $t$.

**Lemma 3.4.** *The stochastic process $\{X(t)\colon t \geq 0\}$ converges in distribution to $X$ as $t \to \infty$ if and only if $X_{t_n} \Rightarrow X$ for every subsequence $t_n \to \infty$.*

## A.4. Stochastic Processes

We now review the principal classes of stochastic processes used in the analysis of simulation output. A stochastic process is a family of random variables (or vectors) defined on a common probability space $(\Omega, \mathcal{F}, P)$. Usually the elements of this family are indexed by a time parameter, either discrete or continuous.

The most tractable stochastic process consists of an infinite sequence $\{X_n\colon n \geq 1\}$ of i.i.d. random variables (or vectors). For this type of process all of the limit theorems of probability theory are available; for example, laws of large numbers and central limit theorems. The results of classical statistics are relevant to this type of stochastic process, and the analysis of simulation output is greatly simplified when these results apply.

Markov chains generalize a stochastic process beyond an i.i.d. sequence by providing for a limited amount of dependence between the random variables. Discrete time as well as continuous time Markov chains are of interest.

**Definition 4.1.** *A discrete time Markov chain with finite or countably infinite state space $S$ is a sequence $\{X_n\colon n \geq 0\}$ of random variables that takes values in $S$ and satisfies the Markov property:*

$$P\{X_{n+1} = j \mid X_n, X_{n-1}, \ldots, X_0\} = P\{X_{n+1} = j \mid X_n\} \tag{4.2}$$

*for all $n \geq 0$ and $j \in S$.*

From (4.2) it is clear that the future evolution of a discrete time Markov chain (DTMC) depends on the past only through the present. When, as we assume throughout,

$$P\{X_{n+1} = j \mid X_n = i\} \stackrel{\text{def}}{=} p_{ij}$$

is independent of $n \geq 0$ for all $i, j \in S$, the DTMC is said to be *time-homogeneous.* The $p_{ij}$ are called *transition probabilities.* In addition to the $p_{ij}$, we need to specify the initial state (or distribution) at time 0 ($n = 0$) in order to determine the complete trajectory of the process. Set $p_i = P\{X_0 = i\}$ for $i \in S$. Analytically, the motion of the DTMC depends on the *transition matrix* $\boldsymbol{P} = \{p_{ij}\colon\ i, j \in S\}$ and the *initial probability vector* $\boldsymbol{p} = \{p_i\colon\ i \in S\}$.

A DTMC with state space $S$ is said to be *irreducible* if any state can be reached from any other state in some finite number of state transitions: for all $i, j \in S$ there exists a positive integer $n$ such that $p_{ij}^n > 0$, where $p_{ij}^n$ is the $(i, j)$-entry of the $n$th power $\boldsymbol{P}^n$ of the transition matrix $\boldsymbol{P}$. State $i$ is said to be *periodic* with period $d$ if $p_{ii}^n = 0$ whenever $n$ is not divisible by $d$ and $d$ is the greatest integer with this property. A state with period 1 is said to be *aperiodic.* For $i, j \in S$, set $f_{ij}^0 = 0$ and

$$f_{ij}^n = P\{X_n = j, X_{n-1} \neq j, \ldots, X_1 \neq j | X_0 = i\}$$

so that $f_{ij}^n$ is the probability, starting in state $i$, that the first state transition to state $j$ occurs at time $n$. Then $f_{ij} = \sum_{n=1}^{\infty} f_{ij}^n$ is the probability that the DTMC ever makes a state transition to state $j$, given that the process starts in state $i$. State $i$ is said to be *recurrent* if and only if $f_{ii} = 1$. A recurrent state is *positive recurrent* if the mean time

$$\mu_{ii} = \sum_{n=1}^{\infty} n f_{ii}^n$$

of first return to state $i$ is finite and is *null recurrent* if the mean time of first return is infinite. A state that is not recurrent is said to be *transient*. If a DTMC is irreducible, either all states are transient, or all states are null recurrent, or all states are positive recurrent; either all states are aperiodic or if one state is periodic with period $d$, then all states are periodic with period $d$.

The question of existence and uniqueness of stationary distributions is important in the theory of Markov chains.

**Definition 4.3.** *The stochastic vector $\pi$ is said to be a stationary distribution of a discrete time Markov chain with transition matrix $\boldsymbol{P}$ if $\pi$ is a solution of $\pi \boldsymbol{P} = \pi$.*

**Theorem 4.4.** *Let $\{X_n\colon n \geq 0\}$ be an irreducible discrete time Markov chain with state space $S$. Then there exists a unique stationary distribution $\pi$ given by*

$$\pi_i = \frac{E_j\left[\sum_{n=0}^{\tau(j)-1} 1_{\{j\}}(X_n)\right]}{E_j[\tau(j)]} = \frac{1}{E_i[\tau(i)]},$$

*where $\tau(j) = \inf\{n \geq 1 : X_n = j\}$.*

The limiting behavior of the $p_{ij}^n$ is nontrivial only in the positive recurrent case.

**Proposition 4.5.** *(Foster's criteria) Let $\{X_n\colon n \geq 0\}$ be an irreducible discrete time Markov chain with state space $S$, and let $S_0$ be a finite subset of $S$. Then the chain is positive recurrent if for some nonnegative real-valued function $g$ with domain $S$ and some $\epsilon > 0$*

$$\sum_{j \in S} p_{ij} g(j) < \infty$$

for $i \in S_0$ and

$$\sum_{j \in S} p_{ij} g(j) \leq g(i) - \epsilon$$

for $i \in S - S_0$.

Observe that $\sum_{j \in S} p_{ij} g(j) \leq g(i) - \epsilon$ can be rewritten as

$$E[g(X_{n+1}) | X_n = i] \leq g(i) - \epsilon.$$

**Theorem 4.6.** *(Ergodic theorem for Markov chains) Let $\{X_n\colon n \geq 0\}$ be an irreducible aperiodic positive recurrent discrete time Markov chain with state space $S$, and let $\pi = \{\pi_i\colon i \in S\}$ be the stationary distribution. Then*

$$\lim_{n \to \infty} p_{ij}^n = \pi_j$$

*for all $i, j \in S$.*

A general state space Markov chain (GSSMC) is specified by a transition kernel and an initial distribution (that is, a probability measure) $\mu$ defined on a measurable space $(\Sigma, \mathcal{S})$.

**Definition 4.7.** *The function $P(\cdot, \cdot)$ on $(\Sigma, \mathcal{S})$ is a transition kernel provided that*

*(i) $P(x, \cdot)$ is a probability measure on $\mathcal{S}$ for each $x \in \Sigma$; and*
*(ii) $P(\cdot, A)$ is measurable (with respect to $\mathcal{S}$) for each $A \in \mathcal{S}$.*

**Definition 4.8.** *A general state space Markov chain is a sequence $\{X_n\colon n \geq 0\}$ of random variables that takes values in $\Sigma$ and satisfies the Markov property:*

$$P\bigl\{X_{n+1} \in A \mid X_0 \ldots, X_n\bigr\} = P\{X_{n+1} \in A \mid X_n\}$$

*for all $n \geq 0$ and $A \in \mathcal{S}$.*

Heuristically, $\mu(A)$ is the probability that the initial state of the process is in the set $A$ and for all $x \in \Sigma$ and $A \in \mathcal{S}$,

$$P(x, A) \stackrel{\text{def}}{=} P\{X_{n+1} \in A \mid X_n = x\}$$

is the probability that the next state of the process is in the set $A$ given that the current state is $x$. If $\Sigma$ is a finite or countably infinite set, specification of the transition kernel for a GSSMC is equivalent to specification of the transition matrix for a DTMC.

The sample function of a continuous time Markov chain (CTMC) is a piecewise constant function. The sequence of states visited is governed by a DTMC, and the holding times (level portions of the trajectory) between jumps are exponentially distributed random variables.

**Definition 4.9.** *A continuous time Markov chain with finite or countably infinite state space $S$ is a family $\{X(t)\colon t \geq 0\}$ of random variables that takes values in $S$, has piecewise constant sample paths with probability 1, and satisfies the Markov property:*

$$P\bigl\{X(t+u) = j \mid X(v)\colon 0 \leq v \leq t\bigr\} = P\bigl\{X(t+u) = j \mid X(t)\bigr\}$$

*for $i \in S$ and $t, u \geq 0$.*

When, as we assume throughout,

$$P\bigl\{X(t+u) = j \mid X(t) = i\bigr\} \stackrel{\text{def}}{=} P^{u}(i, j)$$

is independent of $t \geq 0$ for all $i, j \in S$ and $u \geq 0$, the CTMC is said to be *time-homogeneous*. For fixed $i, j \in S$, the function $t \mapsto P^t(i, j)$ is called a *transition function*, and the family of matrices $\boldsymbol{P}^t$ of the transition functions $P^t(i, j)$ is called the transition function of the CTMC $\{X(t)\colon\ t \geq 0\}$. For $t \geq 0$ and $i \in S$, $P^t(i, \cdot)$ is a probability mass function on $S$ and

$$P^{t+u}(i, j) = \sum_{k \in S} P^t(i, k) P^u(k, j)$$

for $t, u \geq 0$ and $i, j \in S$. The quantity $P^t(i, j)$ represents the probability, starting in state $i$, that the chain is in state $j$ exactly $t$ time units later.

The limits

$$q_{ij} = \lim_{h \downarrow 0} \frac{1}{h} \left( P^h(i, j) - P^0(i, j) \right) \tag{4.10}$$

exist for all $i$ and $j$. The matrix $\boldsymbol{Q} = \{q_{ij}\colon\ i, j \in S\}$ describes the infinitesimal behavior of the CTMC and is called the *infinitesimal generator*. When the state space $S$ is a finite set, the matrix $\boldsymbol{Q}$ generates $\boldsymbol{P}^t$ in the sense that

$$\boldsymbol{P}^t = \exp(\boldsymbol{Q}t) = \sum_{n=0}^{\infty} \frac{\boldsymbol{Q}^n t^n}{n!}$$

The converse is more important in the context of simulation. Given a matrix $\boldsymbol{Q}$ and a probability vector $\boldsymbol{p}$, the task is to find a CTMC with infinitesimal generator $\boldsymbol{Q}$ and initial probability vector $\boldsymbol{p}$ and to generate sample paths of the process. The matrix $\boldsymbol{Q}$ must satisfy certain conditions; that is, there are some $\boldsymbol{Q}$ for which there exists no CTMC satisfying (4.10). We impose the following restrictions on $\boldsymbol{Q}$ and $\boldsymbol{p}$:

(i) $q_{ij} \geq 0$ for $i \neq j$;
(ii) $\sum_{j \in S} q_{ij} = 0$ for $i \in S$;

(iii) $0 < q_i < \infty$, where $q_i = -q_{ii}$; and either

(iv) $\sup\{q_i\colon\ i \in S\} < \infty$ or

(iv′) the states $i$ such that $p_i > 0$ are recurrent states for the DTMC with transition matrix $\boldsymbol{R} = \{r_{ij}\colon\ i, j \in S\}$, where

$$r_{ij} = \begin{cases} q_{ij}/q_i & \text{if } j \neq i \\ 0 & \text{if } j = i. \end{cases}$$

The element $r_{ij}$ of the *jump matrix* $\boldsymbol{R}$ represents the probability that the CTMC makes a jump from state $i$ to state $j$, and $q_i$ is the rate parameter of the exponential holding time in state $i$. A CTMC is said to be *irreducible positive recurrent* if its embedded DTMC is irreducible and positive recurrent.

A semi-Markov process (SMP) is a stochastic process that can be viewed as a generalization of a CTMC. The holding time in a state $i$ need not be exponentially distributed and may depend on the next state. The definition of an SMP is in terms of a Markov renewal process (MRP). Let $\boldsymbol{S} = \{S_n\colon\ n \geq 0\}$ be an increasing sequence of random variables ($0 = S_0 \leq S_1 \leq \ldots$) and $\boldsymbol{X} = \{X_n\colon\ n \geq 0\}$ be a stochastic process with finite or countably infinite state space $S$.

**Definition 4.11.** *The process* $(\boldsymbol{X}, \boldsymbol{S}) = \{(X_n, S_n)\colon\ n \geq 0\}$ *is a Markov renewal process with state space* $S$ *if*

$$\begin{aligned} P\big\{X_{n+1} = j, S_{n+1} - S_n \leq t \mid X_n, X_{n-1}, \ldots, X_0;\\ S_n, S_{n-1}, \ldots, S_0\big\} \\ = P\big\{X_{n+1} = j, S_{n+1} - S_n \leq t \mid X_n\big\} \end{aligned}$$

*for all* $n \geq 0,\ j \in S,$ *and* $t \geq 0$.

The given data for a MRP is the *kernel* $\boldsymbol{Q} = \big\{Q(i,j,t)\colon\ i, j \in S, t \geq 0\big\}$ defined by

$$Q(i,j,t) = P\left\{X_{n+1} = j, S_{n+1} - S_n \leq t \mid X_n = i\right\}$$

for $i, j \in S$ and $t \geq 0$, along with an initial probability vector $\boldsymbol{p}$. As with a CTMC, we impose restrictions on $Q(i, j, t)$ and $\boldsymbol{p}$ so as to guarantee reasonable behavior:

(i) $Q(i, \cdot, \cdot)$ is a probability measure on $S \times [0, +\infty)$; and either

(ii) there exist $\alpha, \epsilon > 0$ such that $\sup_{i \in S} Q(i, j, \alpha) < 1 - \epsilon$ or

(ii′) the states $i$ such that $p_i > 0$ are recurrent states for the DTMC with transition matrix $\boldsymbol{R} = \{r_{ij}\colon\ i, j \in S\}$, where $r_{ij} = \lim_{t \to \infty} Q(i, j, t)$.

**Definition 4.12.** *Let $\{(X_n, S_n)\colon\ n \geq 0\}$ be a Markov renewal process with kernel $\boldsymbol{Q}$ and initial probability vector $\mathbf{p}$. The semi-Markov process $\{Y(t)\colon\ t \geq 0\}$ corresponding to $\boldsymbol{Q}$ and $\mathbf{p}$ is given by*

$$Y(t) = X_n \text{ for } t \in [S_n, S_{n+1}).$$

An SMP visits a sequence of states in accordance with the DTMC $\{X_n\colon\ n \geq 0\}$ and stays for a random amount of time in each state. Unlike the embedded jump chain of a CTMC, with positive probability $\{X_n\colon\ n \geq 0\}$ may make a jump from a state to itself. Also as stated above, the holding time in state $i$ has an arbitrary distribution (concentrated on the nonnegative real line) that may depend on the next state. The conditional holding time in state $i$, given that the next state is $j$, has distribution function $F_{ij}$ given by $F_{ij}(t) = Q(i, j, t)/r_{ij}$.

The last type of process discussed here is a renewal process (RP). Consider a system that fails after a random time and then is immediately replaced (renewed). A renewal counting process (RCP) counts the number of replacements in the interval $[0, t]$. The renewal process is the sequence of random times at which the replacements occur. Let $\{X_n\colon\ n \geq 1\}$ be a sequence of positive i.i.d. random variables with common distribution function $F$ defined on a probability space $(\Omega, \mathcal{F}, P)$.

Set $S_0 = 0$ and form the partial sums $S_n = X_1 + \cdots + X_n$ for $n \geq 1$. Also set $\mu = E[X_n]$.

**Definition 4.13.** *The process* $\boldsymbol{S} = \{S_n\colon\ n \geq 0\}$ *is a renewal process.*

**Definition 4.14.** *The process* $\boldsymbol{N} = \{N(t)\colon\ t \geq 0\}$ *defined by*

$$N(t) = \sup\{n \geq 0\colon\ S_n \leq t\}$$

*is a renewal counting process.*

The partial sum $S_n$ is distributed as the $n$-fold convolution $F_n$ of $F$ with itself and

$$P\{N(t) = n\} = F_n(t) - F_{n+1}(t)$$

for all $0 \leq t < \infty$. The *renewal function* $m(t) = E[N(t)]$ satisfies

$$m(t) = \sum_{n=1}^{\infty} F_n(t)$$

and $m(t) < \infty$ for all $0 \leq t < \infty$.

**Theorem 4.15.** *(Elementary renewal theorem)*

$$\lim_{t\to\infty} \frac{m(t)}{t} = \frac{1}{\mu}.$$

A nonnegative random variable (and its distribution function $F$) is said to be *lattice* if there exists $d \geq 0$ such that

$$\sum_{n=0}^{\infty} P\{X = nd\} = 1.$$

**Definition 4.16.** *Let $h$ be a function defined on $[0, +\infty)$. For $\delta > 0$ and $n = 1, 2, \ldots,$ set*

$$\underline{m}_n(\delta) = \inf\{h(t)\colon\ (n-1)\delta \le t \le n\delta\}$$

*and*

$$\overline{m}_n(\delta) = \sup\{h(t)\colon\ (n-1)\delta \le t \le n\delta\}.$$

*The function $h$ is said to be directly Riemann integrable if $\sum_{n=1}^{\infty} \overline{m}_n(\delta)$ and $\sum_{n=1}^{\infty} \underline{m}_n(\delta)$ are finite for all $\delta > 0$ and*

$$\lim_{\delta\to 0} \delta \sum_{n=1}^{\infty} \overline{m}_n(\delta) = \lim_{\delta\to 0} \delta \sum_{n=1}^{\infty} \underline{m}_n(\delta).$$

A function $h$ defined on $[0, +\infty)$ is directly Riemann integrable if (i) $h(t) \ge 0$ for all $t \ge 0$, (ii) $h$ is nonincreasing, and (iii) $\int_0^\infty h(t)dt < \infty$.

**Theorem 4.17.** *(Key renewal theorem) If $F$ is not lattice and $h$ is directly Riemann integrable, then*

$$\lim_{t\to\infty} \int_0^t h(t-u)dm(u) = \frac{1}{\mu} \int_0^t h(t)dm(t).$$

## A.5. Laws of Large Numbers

A law of large numbers in essence asserts that the sample mean of a stochastic process converges either strongly (with probability 1) or weakly (in probability or in distribution) to an appropriate population mean. The usual laws of large numbers deal with sequences of i.i.d. random variables.

**Theorem 5.1.** *(Strong law of large numbers for i.i.d. random variables) Let $\{X_n\colon n \geq 1\}$ be a sequence of independent, identically distributed random variables, and set $S_n = X_1 + X_2 + \cdots + X_n$ for $n \geq 1$. If $E[|X_1|] < \infty$, then*

$$\frac{S_n}{n} \to E[X_1] \text{ a.s.}$$

*as $n \to \infty$. Furthermore, if $E[X_1] = \infty$, then*

$$\frac{S_n}{n} \to \infty \text{ a.s.}$$

*as $n \to \infty$.*

Since this result holds with probability 1, it also holds in probability, and hence in distribution. A similar result holds for RCP's. Let $\{X_n\colon n \geq 1\}$ be a sequence of positive i.i.d. random variables, and let $\boldsymbol{S} = \{S_n\colon n \geq 0\}$ and $\boldsymbol{N} = \{N(t)\colon t \geq 0\}$ be the associated RP and RCP.

**Theorem 5.2.** *(Strong law of large numbers for renewal counting processes) If $E[X_1] < \infty$, then*

$$\frac{N(t)}{t} \to \frac{1}{E[X_1]} \text{ a.s.}$$

*as $t \to \infty$. If $E[X_1] = \infty$, then*

$$\frac{N(t)}{t} \to 0 \text{ a.s.}$$

*as $t \to \infty$.*

Analogous results hold for DTMC's, CTMC's, and SMP's.

**Theorem 5.3.** *(Strong law of large numbers for discrete time Markov chains) Let $\{X_n\colon n \geq 0\}$ be an irreducible positive recurrent discrete time Markov chain with stationary distribution $\pi$. If $f$ is a real-valued function such that*

$$\sum_{i \in S} \pi_i |f(i)| < \infty,$$

*then*

$$\frac{1}{n} \sum_{i=0}^{n-1} f(X_i) \to \sum_{i \in S} \pi_i f(i) \text{ a.s.}$$

*for any initial distribution* $\boldsymbol{p}$.

**Definition 5.4.** *The stochastic vector $\pi$ is said to be a stationary distribution of a continuous time Markov chain with infinitesimal generator $\boldsymbol{Q}$ if $\pi$ is a solution of $\pi \boldsymbol{Q} = 0$.*

**Theorem 5.5.** *(Strong Law of large numbers for continuous time Markov chains) Let $\{X(t)\colon t \geq 0\}$ be an irreducible positive recurrent continuous time Markov chain with stationary distribution $\pi$. If $f$ is a real-valued function such that*

$$\sum_{i \in S} \pi_i |f(i)| < \infty,$$

*then*

$$\frac{1}{t} \int_0^t f(X(u))du \to \sum_{i \in S} \pi_i f(i) \text{ a.s.}$$

*for any initial distribution* $\boldsymbol{p}$.

**Theorem 5.6.** *(Strong law of large numbers for semi-Markov processes) Let $\{Y(t)\colon t \geq 0\}$ be a semi-Markov process for which the matrix $\boldsymbol{R} = \{r_{ij}\colon i,j \in S\}$ is irreducible and positive recurrent with stationary distribution $\gamma$. Suppose that*

$$\sum_{i,j\in S} r_{ij} m_{ij} < \infty,$$

*where $m_{ij}$ is the mean of the distribution function $F_{ij}$. Set $m_i = \sum_{j\in S} r_{ij} m_{ij}$ and $\pi_j = \gamma_j m_j / \sum_{i\in S} \gamma_i m_i$. If $f$ is a real-valued function such that*

$$\sum_{i\in S} \pi_i |f(i)| < \infty,$$

*then*

$$\frac{1}{t}\int_0^t f(Y(u))du \to \sum_{i\in S} \pi_i f(i) \text{ a.s.}$$

*for any initial distribution $\boldsymbol{p}$.*

## A.6. Central Limit Theorems

In a sense central limit theorems (CLT's) are second order results about the random fluctuations (measured in an appropriate scale) of the sample mean about the population mean. However, such theorems always have to do with weak convergence rather than strong convergence. The standard CLT deals with an i.i.d. sequence $\{X_n\colon n \geq 1\}$ of random variables with finite mean $\mu$ and finite variance $\sigma^2$. Again set $S_n = X_1 + X_2 + \cdots + X_n$ for $n \geq 1$, and denote by $N(0,1)$ a standardized (mean 0, variance 1) normal random variable.

**Theorem 6.1.** *(Central limit theorem) Let $\{X_n: n \geq 1\}$ be a sequence of independent, identically distributed random variables. If $\mu = E[X_1] < \infty$ and $\sigma^2 =$ Var $[X_1] < \infty$, then*

$$\frac{(S_n - n\mu)}{\sigma\sqrt{n}} \Rightarrow N(0,1) \tag{6.2}$$

*as $n \to \infty$.*

Theorem 6.3 provides a CLT for a random number of summands. Let $\{\nu_n: n \geq 1\}$ be a sequence of positive, integer-valued random variables and $\{X_n: n \geq 1\}$ be an i.i.d. sequence of random variables with finite mean $\mu$ and finite variance $\sigma^2$. The sequences $\{\nu_n: n \geq 1\}$ and $\{X_n: n \geq 1\}$ need not be independent.

**Theorem 6.3.** *(Central limit theorem for random partial sums) Let $0 < c < \infty$. If $\nu_n/n \Rightarrow c$, then*

$$\frac{(S_{\nu_n} - \nu_n\mu)}{\sigma\sqrt{\nu_n}} \Rightarrow N(0,1) \tag{6.4}$$

*and*

$$\frac{(S_{\nu_n} - \nu_n\mu)}{\sigma\sqrt{cn}} \Rightarrow N(0,1) \tag{6.5}$$

*as $n \to \infty$.*

Observe that (6.4) is obtained simply by replacing $n$ in (6.2) by $\nu_n$. Also observe that (6.5) is obtained from (6.4) by an application of the converging together lemma and the continuous mapping theorem.

Theorem 6.6 is known as the "convergence of types" theorem. Let $\{Z_n: n \geq 0\}$ be a sequence of random variables and $Y$ and $X$ be two nondegenerate random variables.

**Theorem 6.6.** *(Convergence of types) Let $Z_n \Rightarrow X$ and $a_n Z_n + b_n \Rightarrow Y$, where $a_n > 0$ and $b_n$ are constants. Then there exist constants $a$ and $b$ such that $a_n \to a$, $b_n \to b$, and $Y$ has the same distribution as $aX + b$.*

In the context of simulation, the principal application of this result is to show that the variance constants in two CLT's are the same. Suppose that $\{U_n\colon n \geq 1\}$ is a sequence of random variables and we have shown that

$$\frac{U_n}{\sqrt{nc}} \Rightarrow N(0,1)$$

and

$$\frac{U_n}{\sqrt{nc'}} \Rightarrow N(0,1)$$

as $n \to \infty$, where $c, c' > 0$. Take the random variables $X$ and $Y$ to be $N(0,1)$ and set $Z_n = U_n/(\sqrt{nc})$, $a_n = a = c/c' > 0$, and $b_n = 0$. Then by the convergence of types theorem, $Y$ and $(c/c')X$ have the same distribution. Therefore, $c = c'$. If the two CLT's are indexed by a continuous parameter, the same argument goes through.

## A.7. Recurrent Events

Let $(\Omega, \mathcal{F}, P)$ be a probability space, and consider a sequence $\{A_n\colon n \geq 0\}$ of events belonging to $\mathcal{F}$. The Bonferroni inequality asserts that

$$P\{A_1 \cap A_2 \cap \cdots \cap A_K\} \geq 1 - \sum_{i=1}^{K} P\{A_i^c\},$$

where $A_i^c$ denotes the complement of event $A_i$. Set

$$\limsup_n A_n = \bigcap_{n=0}^{\infty} \bigcup_{k=n}^{\infty} A_k.$$

A point $\omega$ lies in $\limsup_n A_n$ if and only if it lies in infinitely many of the $A_n$; that is, the statement that "event $\limsup_n A_n$ has occurred" is equivalent to the statement that "infinitely many of the events $A_n$ have occurred." It is therefore common to write $\limsup_n A_n$ as "$A_n$ i.o.," where i.o. denotes "infinitely often." The Borel-Cantelli lemma can be used to show that a sequence of events $\{A_n : n \geq 0\}$ is *recurrent*; that is, to show that $P\{A_n \text{ i.o.}\} = 1$.

**Theorem 7.1.** *(Borel-Cantelli lemma) If $\{A_n : n \geq 0\}$ is an independent sequence of events and $\sum_{i=1}^{\infty} P\{A_i\}$ diverges, then $P\{A_n \text{ i.o.}\} = 1$.*

## A.8. Wald's Moment Identities

Let $(\Omega, \mathcal{F}, P)$ be a probability space and $\{\mathcal{F}_t : t \geq 0\}$ be an increasing family of sub $\sigma$-fields of $\mathcal{F}$: $\mathcal{F}_u \subseteq \mathcal{F}_t$ for $u \leq t$. A random variable $T$ is a *stopping time with respect to* $\{\mathcal{F}_t : t \geq 0\}$ if $\{T \leq t\} \in \mathcal{F}_t$ for all $t \geq 0$. In this definition the set $\{t \geq 0\}$ can be interpreted as $\{0, 1, \ldots\}$ or as $[0, +\infty)$. Part (i) of Theorem 8.1 is known as Wald's identity and part (ii) as Wald's second moment identity.

**Theorem 8.1.** *Let $\{X_n : n \geq 1\}$ be a sequence of i.i.d. random variables defined on a probability space $(\Omega, \mathcal{F}, P)$ Also let $T$ be an a.s. finite stopping time with respect to an increasing family $\{\mathcal{F}_n : n \geq 0\}$ of sub $\sigma$-fields of $\mathcal{F}$. Suppose that, for any $n$, $X_n$ is measurable with respect to $\mathcal{F}_n$ and $X_{n+1}, X_{n+2}, \ldots$ are independent of $\mathcal{F}_n$. Set $S_n = X_1 + \cdots + X_n$ for $n \geq 1$. Then*

*(i) $E[S_T] = E[X_1]E[T]$ if either $E[|X_1|] < \infty$ and $E[T] < \infty$ or if $X_1 \geq 0$; and*

*(ii) $E[(S_T - T\mu)^2] = \sigma^2 E[T]$ if $\sigma^2 = \text{Var}\,[X_1] < \infty$ and $E[T] < \infty$.*

Appendix B

# Random Number Generation

The basis of most commonly used algorithms for generating uniform random numbers is a recurrence relation in which each number is a deterministic function of the preceding number. True randomness requires independence of successive numbers, but the algorithm produces a deterministic, dependent sequence. If the parameters of the recurrence relation are chosen properly, algorithms of this kind for generating uniform random numbers produce sequences that (statistically) appear to be random. This appearance of randomness is the origin of the term "pseudo-random numbers."

## B.1. Linear Congruential Generators

Most available algorithms for generating uniform random numbers are of a type known as *linear congruential generators*. Such generators employ a recurrence relation of the form

$$X_n = bX_{n-1} + c \pmod{m}. \tag{1.1}$$

All quantities appearing in (1.1) are nonnegative integers. Read "$X_n$ equals $bX_{n-1} + c$ modulo $m$," this equation says that $X_n$ is the remainder when $bX_{n-1}+c$ is divided by $m$. The quantity $b$ is called the *multiplier*, $m$ is the *modulus*, and $c$ is the *increment*. Given a starting value $X_0 \geq 0$ along with values $b \geq 0$,

$c \geq 0$, and $m$ such that $m > X_0$, $m > b$, and $m > c$, successive application of (1.1) produces a sequence of integers $X_1, X_2, \ldots$ taken to be integer uniform random numbers between 0 and $m-1$. Uniform random numbers $U_n$ on the interval (0,1) are obtained by dividing by $m$:

$$U_n = \frac{X_n}{m}$$

for $n = 1, 2, \ldots$. The recurrence relation of (1.1) is called a *mixed congruential* generator, the term "mixed" coming from the fact that the relation involves multiplication by a constant $b$ along with addition of a constant $c$. Many random number generators are *multiplicative* or *pure congruential* in that $c$=0, giving

$$X_n = bX_{n-1} \pmod{m}.$$

The starting value $X_0$ ($> 0$) is called the *seed* of the random number generator.

On first examination it appears that (1.1) produces $m$ distinct numbers. It is characteristic of generators of this type that there is ultimately a cycle of numbers that repeats indefinitely. The length of this repeating cycle of numbers is called the *period* of the generator. As an example, consider the linear congruential generator of (1.1) with $m = 32$. Choose $c$=0, $b$=9, and start with $X_0 = 1$:

| $n$ | $X_{n-1}$ | $bX_{n-1}$ | $bX_{n-1} \pmod{m}$ |
|---|---|---|---|
| 1 | 1 | 9 | 9 |
| 2 | 9 | 81 | 17 |
| 3 | 17 | 153 | 25 |
| 4 | 25 | 225 | 1 |
| 5 | 1 | 9 | 9 |

The modulus is 32, but the recurrence relation realizes a period of only 4. This period is obviously undesirable in a random number generator. If $X_0 = 2$, then $X_1 = 18$, $X_2 = 2$, $X_3 = 18$, $X_4 = 2, \ldots$, and the period is 2. Thus, the period of a linear congruential generator depends on the starting value $X_0$ as well as on the parameters $b$, $c$, and $m$.

It is clear that a congruential sequence used as a source of random numbers should have a large period. Since the period of a congruential generator can never be greater than the modulus $m$, the value of $m$ should be rather large. Even if we only require random 0's and 1's, the choice of $m$=2 is a poor one; at best we obtain a sequence of the form $\ldots, 0, 1, 0, 1, 0, \ldots$. Speed of generation is another consideration in the choice of $m$ in that we want to be able to compute $bX_n + c \pmod{m}$ rapidly.

One criterion for the choice of the multiplier $b$ is to ensure that the generated sequence has a period of maximum length. However, an acceptable appearance of randomness of the generated sequence involves more than just a large period. Very nonrandom sequences can have large periods; for example, the sequence

$$X_n = X_{n-1} + 1 \pmod{m} \tag{1.2}$$

obtained by taking $b = c = 1$ has (maximum) period equal to $m$, but clearly has an unacceptable appearance of randomness. The recurrence relation in (1.2) shows that in a mixed congruential generator with modulus $m$ it is always possible to achieve the maximum period $m$. Observe that the choice of the starting value $X_0$ does not affect the period when the period equals $m$ since every number from 0 through $m-1$ occurs exactly once in the repeating cycle.

Number-theoretic results are available for characterizing the values of $b$ and $c$ that provide a maximum period. For the special case of multiplicative congruential generators ($c = 0$), the principal result concerning the maximum period says that the maximum period ($m$) cannot be achieved. However, it is possible to obtain multiplicative congruential generators with very large periods. Results on the maximum period for this multiplicative case are available, but the number-theoretic considerations are involved.

If the modulus $m$ in a multiplicative congruential generator is prime (that is, has no positive, integer divisors other than 1 and itself), a period equal to $m - 1$ can be achieved. This period, of course, is just 1 less than the maximum possible period for a linear congruential generator. Moreover, if we choose a multiplier $b$ satisfying an appropriate number-theoretic condition with respect to (prime) $m$, then for any starting value $X_0 < m$, a period equal to $m-1$ is achieved. In general, the determination of values for multipliers $b$ satisfying the condition for maximum period in a multiplicative congruential generator involves lengthy calculations.

Although there are an infinite number of positive integers, only a finite number of positive integers are representable in any particular digital computer system, the limitation being the word size of the system. There are multiplicative congruential generators of uniform random numbers that utilize the full word size of IBM computer systems. In such systems, the word size is 32 bits with 1 bit reserved for algebraic sign; an obvious choice for $m$ is thus $2^{31}$.

A multiplicative congruential generator ($c = 0$) with $m = 2^k$ for some positive integer $k$ can have a maximum period equal to $m/4$. Accordingly, with $m = 2^{31}$ the maximum period is $2^{31}/4 = 2^{29}$, and the period may also depend on the starting value. When the modulus $m$ is prime (and $c$=0), the maximum possible period is $m - 1$. It happens (fortuitously) that the largest prime less than or equal to $2^{31}$ is $2^{31} - 1$, and by choosing $m = 2^{31} - 1$, uniform random number generators with a period equal to $m - 1 = 2^{31} - 2$ can be implemented on IBM computer systems.

Observe that the conditions ensuring maximum period do not necessarily guarantee good statistical properties for the generator, although the choice of the particular multiplier $b =$

$7^5$ does satisfy some known conditions regarding the statistical appearance of the generated sequence.

*Learmonth-Lewis Generator.* Let $X_0 > 0$. Then for $n \geq 1$ set

$$X_n = 7^5 X_{n-1} \pmod{2^{31} - 1} = 16807 X_{n-1} \pmod{2^{31} - 1} \tag{1.3}$$

and

$$U_n = \frac{X_n}{2^{31} - 1}.$$

The Learmonth-Lewis uniform random number generator has been tested extensively and the results of the statistical tests indicate that it is very satisfactory.

The sequence of integers generated by (1.3) have an acceptable appearance of randomness. In addition, the integers in the sequence have the empirically derived property that the low-order (right-most) bits are uniformly distributed. This property can be exploited for efficiency in assembler language implementations of algorithms for generating nonuniform random numbers. Observe that only certain congruential generators have uniformly distributed low-order bits.

Multipliers other than 16807 have been proposed and are used in widely available linear congruential generators with modulus $2^{31} - 1$. The multiplier 397204094 has been proposed as an alternative to 16807 when there is concern about the high-dimensional properties of the generator. Generators (with modulus $2^{31} - 1$) that use as multipliers either 16807, 397204094, or 950706376 are also available.

There are other practical considerations in uniform random number generators. A generator needs to be fast and without excessive memory requirements. Generators of uniform random numbers are frequently available as subroutine or function subprograms written in assembler or machine language. A variety of programming tricks (for example, "division simulation" rather than division operation) can be used to speed up routines, but it is necessary to use care in the selection of such tricks. There are examples in the literature of attempts to obtain a fast generator by using a multiplier that has a binary representation with many zeros that yield an unacceptable period.

For most purposes the sequence of numbers produced by the multiplicative congruential generator based on (1.3) has an acceptable appearance of randomness; that is, the output appears to consist of independent numbers, uniformly distributed on (0,1). Of course due to the procedure used to produce the numbers, they are not actually independent, and it is possible that a sequence of such numbers could be "shuffled" or further randomized to yield a better appearance of randomness. From the point of view of statistical tests, some results of shuffling have been encouraging. In particular, the widely available congruential generator RANDU, based on the recurrence relation

$$X_n = (2^{16} + 3)X_{n-1} \pmod{2^{31}} = 65539X_{n-1} \pmod{2^{31}}$$

and known to have some deficiencies, can be improved by shuffling. In the absence of shuffling, all triples $(U_n, U_{n+1}, U_{n+2})$ produced by RANDU and interpreted as coordinates of points in three-dimensional space (with $0 \le x, y, z \le 1$) lie on precisely 15 planes; for a proper appearance of randomness, these points should fill the three-dimensional space. This property of RANDU illustrates the need for proper selection of the multiplier in a congruential generator.

## B.2. Inverse Transformation

Generation of nonuniform random numbers by inverse transformation of a uniform random number rests on properties of the distribution function of a random variable.

**Definition 2.1.** *The distribution function $F$ of a random variable $X$ is defined by $F(x) = P\{X \le x\}$ for $-\infty < x < +\infty$.*

A distribution function $F$ has the following properties:

(i) $\lim_{x \to -\infty} F(x) = 0$ and $\lim_{x \to +\infty} F(x) = 1$;
(ii) $F$ is nondecreasing; and
(iii) $F$ is right-continuous.

In fact, these properties characterize distribution functions in the sense that for any function $F$ satisfying (i), (ii), and (iii) there exists a random variable $X$ such that $F(x) = P\{X \leq x\}$. The proof of this fact provides a means of obtaining a sample of the random variable $X$ from a uniform random number. The method involves the inverse transformation procedure mentioned above.

**Theorem 2.2.** *Let $F$ be a distribution function, and denote by $F^{-1}$ the inverse function corresponding to $F$:*

$$F^{-1}(u) = \inf\{x\colon\ F(x) \geq u\}$$

*for $0 < u < 1$. If $U$ is uniformly distributed on the interval (0,1), then the random variable $X = F^{-1}(U)$ has distribution function $F$.*

*Proof.* First observe that for $0 < u < 1$,

$$F^{-1}(u) \leq x \text{ if and only if } u \leq F(x).$$

Then for any $x$,

$$P\{X \leq x\} = P\left\{F^{-1}(U) \leq x\right\} = P\{U \leq F(x)\}.$$

But the right side equals $F(x)$ since $U$ is uniformly distributed on (0,1). Therefore, $P\left\{F^{-1}(U) \leq x\right\} = F(x)$. □

It follows that to produce samples of a random variable $X$ with distribution function $F$ by inverse transformation of a uniform random number, it is necessary to solve the equation $F(x) = u$ for $0 < u < 1$.

The inverse transformation method can also produce samples from a specified discrete distribution. Let $X$ be a discrete random variable taking values in a countably infinite set $\{x_1, x_2, \ldots\}$, and assume that $x_1 < x_2 < \ldots$. Set

$$p_i = P\{X = x_i\}$$

and

$$F(x) = \sum_{\{i:\ x_i \leq x\}} p_i .$$

**Algorithm 2.3.** *Inverse transformation (discrete random variable)*

1. *Generate $U$ uniformly distributed between 0 and 1;*
2. *Return $X = x_i$, where $i$ is the smallest positive integer such that $U \leq F(x_i)$.*

The crucial point is that the distribution function $F$ must be such that the equation $F(x) = u$ $(0 < u < 1)$ can be solved reasonably easily. In general, finding solutions of $F(x) = u$ is very time-consuming.

The problem of generating random numbers from a specified nonuniform distribution is in principle solved by having a source of uniform random numbers and transforming them by means of the inverse probability integral as discussed above. Since it is not always possible to compute the inverse of a given distribution function (or to compute the inverse efficiently), direct methods for generating nonuniform random numbers are of interest. Desirable properties of such direct methods are that they be exact, very fast, easy to implement, and economical of computer storage. The property of exactness is that any deviation from the specified distribution results from computer round-off error rather than a defect in the method itself. Comparisons are hard to make among particular methods, partly because of machine dependencies. It is often true that with very little cost in complexity it is possible to improve on the inverse probability integral transformation by an order of magnitude in execution time. However, the fastest available algorithms for generating nonuniform random numbers are often quite complex.

## B.3. Rejection-Acceptance Methods

Rejection-acceptance and decomposition (discussed in Section B.4) are systematic methods for generating nonuniform random numbers and allow the generation of most types of nonuniform random numbers simply, if not with full efficiency. Rejection-acceptance techniques are often used in conjunction with decomposition techniques and are applicable to discrete as well as continuous random variables.

Let $X$ be a random variable with a density function $f_X$ that is bounded and concentrated on a finite interval. For definiteness, we take the interval to be (0,1): for some $\alpha < \infty$,

$$f_X(x) \le \alpha$$

for $0 < x < 1$ and $f_X(x) = 0$ otherwise.

**Algorithm 3.1.** *Rejection-acceptance (uniform majorizing density function)*

1. *Generate* $U_1$ *and* $U_2$ *independent and uniformly distributed between 0 and 1;*
2. *Compute* $\alpha^{-1} f_X(U_1)$. *If* $U_2 \le \alpha^{-1} f_X(U_1)$, *return* $X = U_1$. *Otherwise, reject the pair* $(U_1, U_2)$ *and repeat Step 1.*

Observe that not all pairs $(U_1, U_2)$ of uniform random numbers produce a sample of the random variable $X$. Proposition 3.2

shows that Algorithm 3.1 produces a random sample from the density function $f_X$.

**Proposition 3.2.** *The output $X$ of Algorithm 3.1 is a random sample from the density function $f_X$.*

*Proof.* The sample pair $(U_1, U_2)$ is accepted if $U_2 \le \alpha^{-1} f_X(U_1)$ and in this case $X = U_1$. Thus,

$$\begin{aligned} P\{X \le x\} &= P\{U_1 \le x | \text{ Acceptance }\} \\ &= \frac{P\{U_1 \le x, U_2 \le \alpha^{-1} f_X(U_1)\}}{P\{U_2 \le \alpha^{-1} f_X(U_1)\}} \end{aligned}$$

so that

$$\begin{aligned} P\{X \le x\} &= \frac{\int_0^x \int_0^{\alpha^{-1} f_X(u_1)} 1 du_2 du_1}{\int_0^1 \int_0^{\alpha^{-1} f_X(u_1)} 1 du_2 du_1} \\ &= \frac{\int_0^x \alpha^{-1} f_X(u_1) du_1}{\int_0^1 \alpha^{-1} f_X(u_1) du_1} \\ &= \int_0^x f_X(u) du \end{aligned}$$

as required. □

Algorithm 3.3 provides a rejection-acceptance procedure that uses a general majorizing density function to produce samples of a random variable $X$ with a density function $f_X$. Suppose that $f_X$ is concentrated on a finite interval $[a, b]$. Let $g$ be a function such that

$$f_X(x) \le g(x)$$

for $a \le x \le b$ and set

$$\alpha = \int_a^b g(x) dx < \infty.$$

Then $h$ given by

$$h(x) = \frac{g(x)}{\alpha}$$

is a density function concentrated on $[a, b]$.

**Algorithm 3.3.** *Rejection-acceptance (general majorizing density function)*

1. *Generate $Y$ with density function $h$;*
2. *Generate $U$ uniformly distributed between 0 and 1;*
3. *If $U \leq f_X(Y)/g(Y)$, return $X = Y$. Otherwise, reject the pair $(U, Y)$ and repeat Step 1.*

If the density function is concentrated on an interval $(-\infty, b]$, $[a, +\infty)$, or $(-\infty, +\infty)$, Algorithm 3.3 is valid provided that the integral that defines $\alpha$ is finite.

Observe that in Step 3 the probability of acceptance is $\alpha^{-1}$:

$$\begin{aligned} P\{U \leq f_X(Y)/g(Y)\} &= \int_a^b \int_0^{f_X(y)/g(y)} 1 \cdot h(y) du dy \\ &= \int_a^b \frac{f_X(y)}{g(y)} \frac{g(y)}{\alpha} dy \\ &= \alpha^{-1}. \end{aligned}$$

The number of comparisons required for each sample has a geometric distribution (starting at 1) since the number of comparisons arises as a sequence of independent (Bernoulli) trials with probability of success equal to $\alpha^{-1}$. It follows that the expected number of comparisons is $\alpha$. For an efficient algorithm $\alpha$ should be close to 1. We try to accomplish this by choosing $g$ to fit $f_X$ as closely as possible over the interval $[a, b]$. There is of course the basic requirement that it be easy to generate samples of the random variable $Y$ and in particular be easier to generate than samples of $X$ itself. Examples of useful random variables $Y$ are:

(i) uniform random variables;
(ii) *wedges* corresponding to $Y = \max(U_1, U_2)$ and $Y = \min(U_1, U_2)$, where $U_1$ and $U_2$ are independent uniform random variables on the interval (0,1):

$$f_Y(x) = h(x) = \begin{cases} 2x & \text{if } 0 \le x \le 1 \\ 0 & \text{otherwise} \end{cases}$$

and

$$f_Y(x) = h(x) = \begin{cases} 2(1-x) & \text{if } 0 \le x \le 1 \\ 0 & \text{otherwise ;} \end{cases}$$

(iii) exponential random variables; and
(iv) random variables $Y$ of the form $Y = 1/U$, where $U$ is a uniform random variable on the interval (0,1):

$$f_Y(x) = h(x) = \begin{cases} 1/x^2 & \text{if } x \ge 1 \\ 0 & \text{if } x < 1 \, . \end{cases}$$

By way of comparison with the inverse probability integral transformation, rejection-acceptance requires evaluation of the density function $f_X$. This evaluation is almost always easier (for example, a normal distribution) than evaluation of the inverse distribution function.

## B.4. Decomposition Methods

We now consider a very powerful, systematic way of generating nonuniform random numbers that is applicable to discrete as well as continuous random variables. Decomposition ("mixing") techniques rest on viewing samples of the specified random variable $X$ as coming from several populations. The essential idea is to represent the distribution of $X$ as a convex linear mixture of distributions in such a way that most of the time $X$ is realized as a sample of a random variable that is easy to generate. Specific decomposition procedures are usually complex and often incorporate rejection-acceptance techniques.

Let $X$ be a random variable with a density function $f_X$ that can be represented as

$$f_X = p_1 f_1 + \cdots + p_k f_k,$$

where $p_1, p_2, \ldots, p_k \geq 0$, $\sum_{i=1}^{k} p_i = 1$, and $f_1, f_2, \ldots, f_k$ are probability density functions. Also let $Z$ be a discrete random variable such that

$$P\{Z = j\} = p_j$$

for $j = 1, 2, \ldots, k$.

**Algorithm 4.1.** *Decomposition*

1. *Generate* $Z$;
2. *Generate and return* $X$ *with density function* $f_Z$.

The fastest methods available for generating normal and exponential random numbers use decomposition (along with rejection-acceptance). The approach is to divide the density function of the (normal or) exponential random variable into a tail and a large number of rectangles and wedges.

There are many ways of decomposing a specified density function. In selecting a decomposition we employ geometric considerations. Triangular density functions are frequently used in decomposition techniques. Wedges corresponding to $\max(U_1, U_2)$ and $\min(U_1, U_2)$, where $U_1$ and $U_2$ are independent uniform random variables on the interval (0,1), are also useful.

In connection with Algorithm 4.1, observe that if a uniform random number $U$ (on (0,1)) is used to generate $Z$, it may be scaled and re-used to generate a random number with density function $f_Z$; that is, given that

$$P_{j-1} \stackrel{\text{def}}{=} p_1 + \cdots + p_{j-1} < U \leq P_j \stackrel{\text{def}}{=} p_1 + \cdots + p_j,$$

$(U - P_{j-1})/p_j$ may be used as an independent (0,1) uniform random number. This use is permissible because $V$ defined by

$$V = \frac{U - P_{j-1}}{p_j} \text{ for } P_{j-1} < U \leq P_j,$$

(with $P_0 = 0$) is uniformly distributed on (0,1), and $V$ and $Z$ are independent.

Although the Cauchy distribution is not of great importance for simulation of stochastic systems, it serves well to illustrate decomposition. A Cauchy random variable $X$ has density function $f_X$ given by

$$f_X(x) = \frac{1}{\pi(1 + x^2)}$$

for $-\infty < x < +\infty$. Decomposition along with a rejection-acceptance technique can be used to produce a random sample of $|X|$ and a true Cauchy random number is obtained by attaching a random sign. Decompose the area under $f_{|X|}$ into three areas by a vertical line $x = a$ and a horizontal line $f_{|X|}(x) = f_{|X|}(a)$. Call the area between the horizontal and vertical lines region 1, the area above the horizontal line region 2, and the area to the right of $x = a$ region 3. Choose $a = 1$ in order to make region 1 as large as possible and consequently sample from a uniform distribution as often as possible. Take $p_1$, $p_2$, and $p_3$ to be the areas of region 1,2, and 3, respectively, the corresponding uniform function density function $f_1$ to be concentrated on $[0, 1]$, the wedge-shaped density function $f_2$ to be concentrated on $[0, 1]$, and the long-tailed density function $f_3$ to be concentrated on $(1, +\infty)$.

Formally, represent $f_{|X|}$ as a convex linear mixture of density functions:

$$f_{|X|}(x) = p_1 f_1(x) + p_2 f_2(x) + p_3 f_3(x)$$

for $0 \le x < \infty$, where

$$p_1 = \frac{1}{\pi},\ p_2 = \frac{1}{2} - \frac{1}{\pi},\ \text{and } p_3 = \frac{1}{2},$$

$$f_1(x) = \begin{cases} 1 & \text{if } 0 \le x \le 1 \\ 0 & \text{if } x > 1\ , \end{cases}$$

$$f_2(x) = \begin{cases} \left(\frac{2}{\pi(1+x^2)} - \frac{1}{\pi}\right) / \left(\frac{1}{2} - \frac{1}{\pi}\right) & \text{if } 0 \le x \le 1 \\ 0 & \text{if } x > 1\ , \end{cases}$$

and

$$f_3(x) = \begin{cases} \frac{4}{\pi(1+x^2)} & \text{if } x > 1 \\ 0 & \text{if } 0 \le x \le 1\ . \end{cases}$$

To obtain this representation, observe that since $p_1$, $p_2$, and $p_3$ are the areas of regions 1, 2, and 3, respectively,

$$p_1 = \int_0^1 \frac{1}{\pi} dx = \frac{1}{\pi},$$

$$p_2 = \int_0^1 \frac{2}{\pi(1+x^2)} dx - p_1 = \frac{1}{2} - \frac{1}{\pi},$$

and

$$p_3 = \int_1^\infty \frac{2}{\pi(1+x^2)} dx = \frac{1}{2}.$$

Next observe that $f_1(x) = f_2(x) = 0$ for $x > 1$ and $f_3(x) = 0$ for $0 \le x \le 1$. Since

$$\frac{2}{\pi(1+x^2)} = p_1 f_1(x) + p_2 f_2(x) + p_3 f_3(x)$$

for $0 \leq x \leq 1$, we have

$$\frac{2}{\pi(1+x^2)} = p_3 f_3(x)$$

for $x > 1$. This equation determines $f_3$ since the value of $p_3$ is known. Similarly,

$$\frac{2}{\pi(1+x^2)} = p_1 f_1(x) + p_2 f_2(x)$$

and

$$\frac{1}{\pi} = p_1 f_1(x)$$

for $0 \leq x \leq 1$. These two equations determine $f_1$ and $f_2$ since the values of $p_1$ and $p_2$ are known.

Samples from the uniform density function $f_1$ can be obtained directly, and samples from the tail density function $f_3$ can be obtained by a rejection-acceptance technique as follows. Choose $h$ given by $h(x) = 1/x^2$ as a majorizing density function and recall that $h$ is the density function of the reciprocal of a random variable uniformly distributed on the interval (0,1). The remaining wedge-shaped density function $f_2$ is a so-called "nearly linear" density function for which there is an efficient rejection-acceptance technique.

The graph of á nearly linear density function can be concave upward or concave downward; it is reasonably close to a straight line and can be enclosed by two parallel lines.

**Definition 4.2.** *A density function $f$ concentrated on the interval $[s, s+h]$ is a nearly linear density function if there exist constants $a$ and $b$ such that*

$$a - \frac{b(x-s)}{h} \leq f(x) \leq b - \frac{b(x-s)}{h}$$

*for $s \leq x \leq s+h$.*

**Algorithm 4.3.** *Nearly linear density function*

1. *Generate* $U_1$ *and* $U_2$ *uniformly distributed between 0 and 1. If* $U_1 > U_2$, *interchange* $U_1$ *and* $U_2$;
2. *If* $U_2 \leq ab^{-1}$, *go to 4;*
3. *If* $U_2 > U_1 + b^{-1}f(s + hU_1)$, *go to 1;*
4. *Return* $X = s + hU_1$.

Observe that since $f$ is a nearly linear density function,

$$ab^{-1} \leq u + b^{-1}f(s + hu) \leq 1$$

for $0 \leq u \leq 1$, and it is unnecessary to evaluate $f(s + hU_1)$ when $U_2 \leq ab^{-1}$ in Step 2. The sample $(U_1, U_2)$ in Step 1 is accepted if $U_1 \leq U_2$ and $U_2 \leq U_1 + b^{-1}f(s + hU_1)$ (in which case $X = s + hU_1$) or if $U_2 < U_1$ and $U_1 \leq U_2 + b^{-1}f(s + hU_2)$ (in which case $X = s + hU_2$). Therefore, with $V_1 = \min(U_1, U_2)$ and $V_2 = \max(U_1, U_2)$,

$$\begin{aligned} P\{X &\leq s + hu\} \\ &= P\left\{V_1 \leq u | V_1 \leq V_2 \leq V_1 + b^{-1}f(s + hV_1)\right\} \\ &= \frac{2P\left\{U_1 \leq u, U_1 \leq U_2 \leq U_1 + b^{-1}f(s + hU_1)\right\}}{2P\left\{U_1 \leq U_2 \leq U_1 + b^{-1}f(s + hU_1)\right\}} \\ &= \frac{\int_0^u b^{-1}f(s + hv)dv}{\int_0^1 b^{-1}f(s + hv)dv} \\ &= \int_s^{s+hu} f(x)dx \end{aligned}$$

for $0 \leq u \leq 1$. For the density function $f_2$ in the Cauchy decomposition, $s = 0$ and $h = 1$ and we choose

$$a = f_2(0) = \frac{2}{\pi - 2}.$$

The condition that the line $y = b - b(x - s)/h = b - bx$ be tangent to the curve $y = f_2(x)$ determines the constant $b$. It follows from the intersection equation

$$\left(\frac{2}{\pi - 2}\right)\left(\frac{1 - x^2}{1 + x^2}\right) = b - bx$$

that

$$1 + x = \frac{b}{a}(1 + x^2).$$

Then

$$x = \frac{1 \pm \left\{1 - 4\left(\frac{b}{a}\right)\left(\frac{b}{a} - 1\right)\right\}^{1/2}}{\frac{2b}{a}}$$

and the line $y = b - bx$ is tangent to the curve $y = f_2(x)$ if the discriminant is zero:

$$\frac{b}{a} = \frac{1 \pm \sqrt{2}}{2}.$$

Since $b \geq 0$, $b/a = (1 + \sqrt{2})/2$ and therefore

$$b = \frac{1 + \sqrt{2}}{\pi - 2}.$$

It is also possible to use decomposition with continuous mixtures of distributions.

# Bibliography

Ahrens, J. and Dieter, U. (1973). *Uniform Random Numbers.* Institut für Mathematische Statistik. Technische Hochschule in Graz. Graz, Austria.

Ahrens, J. and Dieter, U. (1973). *Nonuniform Random Numbers.* Institut für Mathematische Statistik. Technische Hochschule in Graz. Graz, Austria.

Asmussen, S. (1987). *Applied Probability and Queues.* Wiley. New York.

Asmussen, S. (1989). Validating the heavy traffic performance of regenerative simulation. *Comm. Statist. Stochastic Models* **5**, 617–628.

Billingsley, P. (1979). *Probability and Measure.* Wiley. New York.

Breiman, L. (1968). *Probability.* Addison-Wesley. Reading, Massachusetts.

Chow, Y. S. and Teicher, H. (1978). *Probability Theory: Independence, Interchangeability, Martingales.* Springer-Verlag. New York.

Chung, K. L. (1974). *A Course in Probability Theory.* (2nd edn.) Academic Press. New York.

Çinlar, E. (1975). *Introduction to Stochastic Processes.* Prentice-Hall. Englewood Cliffs, New Jersey.

Crane, M. A. and Iglehart, D. L. (1975). Simulating stable stochastic systems: III, Regenerative processes and discrete event simulation. *Operations Res.* **23**, 33–45.

Crane, M. A. and Lemoine, A. J. (1977). *An Introduction to the Regenerative Method for Simulation Analysis.* Lecture Notes in Control and Information Sciences, Vol. 4. Springer-Verlag. New York.

Devroye, L. (1986). *Non-uniform random variate generation.* Springer-Verlag. New York.

Doob, J. L. (1953). *Stochastic Processes.* Wiley. New York.

Fishman, G. S. and Moore, L. (1982). A statistical evaluation of multiplicative congruential generators with modulus $2^{31} - 1$. *J. Amer. Stat. Assoc.* **77**, 129–136.

Fishman, G. S. and Moore, L. (1985). An exhaustive analysis of multiplicative congruential generators with modulus $2^{31} - 1$. *SIAM J. Scientific Stat. Comp.* **7**, 24–45.

Fossett, L. D. (1979). Simulating generalized semi-Markov processes. Technical Report No. 50. Department of Operations Research. Stanford University.

Fox, B. L. and Glynn, P. W. (1985). Discrete-time conversion for simulating semi-Markov processes. *Operations Res. Lett.* **5**, 191–196.

Glasserman, P. (1991). *Gradient Estimation Via Perturbation Analysis.* Kluwer Academic. Boston.

Glynn, P. W. (1982). Some new results in regenerative process theory. Technical Report No. 16. Department of Operations Research. Stanford University.

Glynn, P. W. (1982). Regenerative simulation of Harris recurrent Markov chains. Technical Report No. 18. Department of Operations Research. Stanford University.

Glynn, P. W. (1983). On the role of generalized semi-Markov processes in simulation output analysis. *Proc. 1983 Winter Simulation Conference*, 39–42.

Glynn, P. W. (1987). Limit theorems for the method of replication. *Comm. Statist. Stochastic Models* **3**, 343–355.

Glynn, P. W. (1989). A GSMP formalism for discrete event systems. *Proc. IEEE* **77**, 14–23.

Glynn, P. W. and Heidelberger, P. (1990). Bias properties of budget constrained simulations *Operations Res.* **38**, 801–814.

Glynn, P. W. and Heidelberger, P. (1991). Analysis of parallel replicated simulations under a completion time constraint. *ACM Trans. Model. Comput. Simul.* **1**, 3–23.

Glynn, P. W. and Iglehart, D. L. (1987). A joint central limit theorem for the sample mean and regenerative variance estimator. *Ann. Operations Res.* **8**, 41–55.

Glynn, P. W. and Iglehart, D. L. (1986). Estimation of steady-state central moments by the regenerative method of simulation. *Operations Res. Lett.* **5**, 271–276.

Glynn, P. W. and D. L. Iglehart (1990). Simulation output analysis using standardized time series. *Math. Oper. Res.* **15**, 1–16.

Haas, P. J. (1985) Recurrence and regeneration in non-Markovian simulations. Ph. D. Dissertation. Department of Operations Research. Stanford University.

Haas, P. J. and Shedler, G. S. (1985). Regenerative simulation methods for local area networks. *IBM J. Res. Develop.* **29**, 194–205.

Haas, P. J. and Shedler, G. S. (1986). Regenerative stochastic Petri nets. *Performance Evaluation* **6**, 189–204.

Haas, P. J. and Shedler, G. S. (1987). Recurrence and regeneration in non-Markovian networks of queues. *Comm. Statist. Stochastic Models* **3**, 29–52.

Haas, P. J. and Shedler, G. S. (1987). Regenerative generalized semi-Markov processes. *Comm. Statist. Stochastic Models* **3**, 409–438.

Haas, P. J. and Shedler, G. S. (1987). Stochastic Petri nets with simultaneous transition firings. *International Workshop on Petri Nets and Performance Models.* IEEE Computer Society Press. Washington, DC.

Haas, P. J. and Shedler, G. S. (1988). Modelling power of stochastic Petri nets for simulation. *Probab. Engrg. Informational Sci.* **2**, 435–459

Haas, P. J. and Shedler, G. S. (1989). Stochastic Petri net representation of discrete event simulations. *IEEE Trans. Software Engrg.* **15**, 381–393.

Haas, P. J. and Shedler, G. S. (1989). Stochastic Petri nets with timed and immediate transitions. *Comm. Statist. Stochastic Models* **5**, 563–600. (Special Issue Devoted to Computer-Experimental Methods in Probability.)

Haas, P. J. and Shedler, G. S. (1991). Stochastic Petri nets: modeling power and limit theorems. *Probab. Engrg. Informational Sci.* **5**, 477–498.

Hall, P. and Heyde, C. C. (1980). *Martingale Limit Theory and its Application.* Academic Press. New York.

Heidelberger, P. (1979). A variance reduction technique that increases regeneration frequency. *Current Issues in Computer Simulation.* N. Adam and A. Dogramaci, eds. Academic Press. New York.

Heidelberger, P. (1980). Variance reduction techniques for the simulation of Markov processes, I: Multiple estimates. *IBM J. Res. Develop.* **24**, 570–581.

Heidelberger, P. and Iglehart, D. L. (1979). Comparing stochastic systems using regenerative simulation with common random numbers. *Adv. in Appl. Probab.* **11**, 804-819.

Heidelberger, P. and Lewis, P. A. W. (1981). Regression-adjusted estimates for regenerative simulations, with graphics. *Comm. Assoc. Comput. Mach.* **24**, 260–273.

Hordijk, A., Iglehart, D. L., and Schassberger, R. (1976). Discrete time methods for simulating continuous time Markov chains. *Adv. in Appl. Probab.* **8**, 772–788.

Iglehart, D. L. (1977). Simulating stable stochastic systems, VII: Selecting the best system. In *Algorithmic Methods in Probability*. M. F. Neuts (ed.), 37–50. North-Holland. Amsterdam.

Iglehart, D. L. (1978). The regenerative method for simulation analysis. In *Current Trends in Programming Methodology Vol. III: Software Engineering*. K. M. Chandy and R. T. Yeh (eds.), 52–71. Prentice-Hall. Englewood Cliffs, New Jersey.

Iglehart, D. L. (1975). Simulating stable stochastic systems, V: comparison of ratio estimators. *Naval Res. Logist. Quart.* **22**, 554–565.

Iglehart, D. L. (1976). Simulating stable stochastic systems, VI: quantile estimation. *J. Assoc. Comput. Mach.* **23**, 347–360.

Iglehart, D. L. and Lewis, P. A. W. (1978). Regenerative simulation with internal controls. *J. Assoc. Comput. Mach.* **26**, 271–282.

Iglehart, D. L. and Shedler, G. S. (1980). *Regenerative Simulation of Passage Times in Networks of Queues*. Lecture

Notes in Control and Information Sciences, Vol. 4. Springer-Verlag. New York.

Iglehart, D. L. and Shedler, G. S. (1983). Simulation of non-Markovian systems. *IBM J. Res. Develop.* **27**, 472–480.

Iglehart, D. L. and Shedler, G. S. (1983). Statistical efficiency of regenerative simulation methods for networks of queues. *Adv. in Appl. Probab.* **15**, 183–197.

Iglehart, D. L. and Shedler, G. S. (1983). Simulation for passage times in non-Markovian networks of queues. *Proc. IFIP Workshop on Stochastic Programming: Algorithms and Applications.* Gargnano, Italy.

Iglehart, D. L. and Shedler, G. S. (1984). Simulation output analysis for local area computer networks. *Acta Inform.* **21**, 321–338.

Iglehart, D. L. and Stone, M. L. (1983). Regenerative simulation for estimating extreme values. *Operations Res.* **6**, 1145–1166.

Jochens, H. W. and Shedler, G. S. (1989). Modelling and simulation of stochastic systems with SPSIM. Technical Report RJ6825. IBM Almaden Research Center. San Jose, California.

Knuth, D. (1981). *The Art of Computer Programming. Vol. 2, Semi-numerical Algorithms.* Second Edition. Ch. 3. Addison Wesley. Reading, Massachusetts.

König, D., Matthes, K., and Nawrotzki, K. (1967). *Verallgemeinerungen der Erlangschen und Engsetschen Formeln.* Akademie-Verlag. Berlin.

König, D., Matthes, K., and Nawrotzki, K. (1971). Unempfindlichkeitseigenschaften von Bedienungsprozessen. Appendix to Gnedenko, B. V. and I. N. Kovalenko, *Introduction to*

*Queueing Theory*. German edition. R. Oldenbourg Verlag. Munich.

Lamperti, J. (1977). *Stochastic Processes: A Survey of the Mathematical Theory*. Springer-Verlag. New York.

Lavenberg, S. and Sauer, C. H. (1977). Sequential stopping rules for the regenerative method of simulation. *IBM J. Res. Develop.* **21**, 545–558.

Lewis, P. A. W. and Orav, E. J. (1989). *Simulation Methodology for Statisticians, Operations Analysts, and Engineers*. Wadsworth and Brooks/Cole. Pacific Grove, California.

Lewis, P. A. W. and Uribe L. C. (1981). The new Naval Postgraduate School random number generator package LLRANDOMII. Naval Postgraduate School Technical Report PS-55-81-005. Monterey, California.

Marsaglia, G. (1985). A current view of random random number generators. In *Computer Science and Statistics: Proc. of the Sixteenth Symposium on the Interface*, L. Billard (ed.), 3–10. North-Holland. Amsterdam.

Marsaglia, G. and Zaman, A. (1991). A new class of random random number generators. *Annals of Appl. Probability* **1**, 462–480.

Matthes, K. (1962). Zur Theorie der Bedienungsprozesse. *Trans. 3rd Prague Conference on Information Theory and Statistical Decision Functions*. Prague.

Meketon, M. S. and Heidelberger, P. (1982). A renewal theoretic approach to bias reduction in regenerative simulations. *Management Sci.* **28**, 173–181.

Miller, D. R. (1972). Existence of limits in regenerative processes. *Ann. Math. Stat.* **43**, 1275–1282.

Miller, R. G. (1974). The jackknife- a review. *Biometrika* **61**, 1–15.

Orey, S. (1971). *Lecture Notes on Limit Theorems for Markov Chain Transition Probabilities.* Van Nostrand Reinhold. London.

Ross, S. M. (1983). *Stochastic Processes.* Wiley. New York.

Ross, S. M. (1990). *A Course in Simulation.* Macmillan. New York.

Seila, A. F. (1982). A batching approach to quantile estimation in regenerative simulations. *Management Sci.* **28**, 573–581.

Shedler, G. S. (1983). Generation methods for discrete event simulation. *Computer Performance Modeling Handbook.* Ch. 5. S. S. Lavenberg (ed.). Academic Press. New York.

Shedler, G. S. (1987). *Regeneration and Networks of Queues.* Springer-Verlag. New York.

Shedler, G. S. and Slutz, D. R. (1981). Irreducibility in closed multiclass networks of queues with priorities: passage times of a marked job. *Performance Evaluation* **1**, 334–343.

Shedler, G. S. and Southard, J. (1982). Simulation for passage times in closed, multiclass networks of queues with unrestricted priorities. *Performance Evaluation* **2**, 257–267.

Shedler, G. S. and Southard, J. (1982). Regenerative simulation of networks of queues with general service times: passage through subnetworks. *IBM J. Res. Develop.* **26**, 625–633.

Sigman, K. (1990). One-dependent regenerative processes and queues in continuous time. *Math. Oper. Res.* **15**, 175–189.

Smith, W. L. (1958). Renewal theory and its ramifications. *J. Roy. Statist. Soc. Ser. B* **20**, 243–302.

Tin, M. (1965). Comparison of some ratio estimates. *J. Amer. Statist. Assoc.* **60**, 249–307.

Whitt, W. (1980). Continuity of generalized semi-Markov processes. *Math. Oper. Res.* **5**, 494–501.

# Index

STATISTICAL MODELING
AND DECISION SCIENCE

Gerald J. Lieberman and Ingram Olkin, editors

Samuel Eilon, *The Art of Reckoning: Analysis of Performance Criteria*

Enrique Castillo, *Extreme Value Theory in Engineering 1,*

Joseph L. Gastwirth, *Statistical Reasoning in Law and Public Policy: Volume Statistical Concepts and Issues of Fairness; Volume 2, Tort Law, Evidence, and Health*

Takeaki Kariya and Bimal K. Sinha, *Robustness of Statistical Tests*

Marvin Gruber, *Regression Estimators: A Comparative Study*

Alexander von Eye, editor, *Statistical Methods in Longitudinal Research: Volume I, Principles and Structuring Change; Volume II, Time Series and Categorical Longitudinal Data*

M. M. Desu and D. Raghavarao, *Sample Size Methodology*

N. Balakrishnan and A. Clifford Cohen, *Order Statistics and Inference: Estimation Methods*

Michel Jambu, *Exploratory and Multivariate Data Analysis*

Elliot R. Lieberman, *Multi-Objective Programming in the USSR*

David M. Eddy, Vic Hasselblad, and Ross Shachter, *Meta-Analysis by the Confidence Profile Method: The Statistical Synthesis of Evidence*

Gerald S. Shedler, *Regenerative Stochastic Simulation*